Cdn

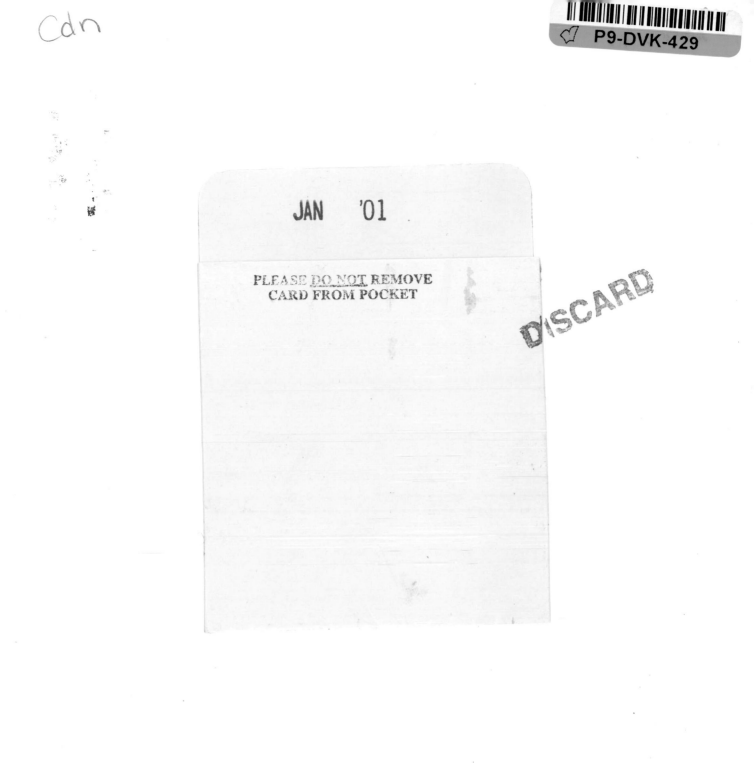

JAN '01

PLEASE DO NOT REMOVE
CARD FROM POCKET

DISCARD

Geography of British Columbia

Geography of British Columbia

People and Landscapes in Transition

Brett McGillivray

UBC Press · Vancouver · Toronto

Printed in Canada on acid-free paper ∞

ISBN 0-7748-0784-9

Canadian Cataloguing in Publication Data

McGillivray, Brett, 1944-
  Geography of British Columbia

  Includes biblographical references and index.
  ISBN 0-7748-0784-9

  1. British Columbia – Geography. I. Title
FC3811.M33 2000   917.11   C00-910266-3
F1087.M33 2000

UBC Press acknowledges the financial support of the Government of Canada through the Book Publishing Industry Development Program (BPIDP) for our publishing activities.
Canadä

We also gratefully acknowledge the support of the Canada Council for the Arts for our publishing program, as well as the support of the British Columbia Arts Council.

UBC Press
University of British Columbia
2029 West Mall, Vancouver, BC V6T 1Z2
(604) 822-5959
Fax: (604) 822-6083
E-mail: info@ubcpress.ubc.ca
www.ubcpress.ubc.ca

Contents

Illustrations, Figures, and Tables

TABLES

Preface

This text is the culmination of years of teaching the geography of British Columbia at Capilano College. It comes at a time when one can question whether all the material gathered here should be offered on-line or in CD-ROM format. I have chosen the "old" textbook style of presentation, in part, because most geography courses at the college level use textbooks. My concern about going on-line stems from an informal survey of my classes. Eighty percent or more of the evening students were connected to the Internet or had CD-ROM capability at home, but only 20 percent or fewer of the day students had the same capability. It would place a huge burden on the library if electronic sources provided the only "text." Future versions may have any number of formats. In its present form, this text includes references to websites where the most up-to-date information can be accessed. It is of course likely that only the government websites and a few others will be maintained in the long run. More important, students are encouraged to use the Internet in conjunction with this text.

Several books have been particularly influential in the creation of this text: Roderick Haig-Brown's *The Living Land* (1961), Mary Barker's *Natural Resources of British Columbia and the Yukon* (1977), and Charles Forward's *British Columbia: Its Resources and People* (1987). Each has done an admirable job in assessing the variety of landscapes and issues in British Columbia. Albert Farley's *Atlas of British Columbia: People, Environment, and Resources* (1979) is another important resource because it not only provides many useful maps but also contains much additional information about the province. I have used *British Columbia: Its Resources and People* as a text for years because of my preference for a geography that examines themes and stresses an historical perspective. References to other works that have influenced this text can be found at the end of each chapter.

Geography of British Columbia: People and Landscapes in Transition is a selective geography. One of its guiding principles is that it is designed for a one-term introductory course, and one can cover only so many topics in thirteen or fourteen weeks. The selection of topics and chapters also shows my bias as a human geographer, and even here there are omissions. Chapters could have been written on transportation, other peoples, and other resources.

Some students have a background in geography and a familiarity with British Columbia; many do not. This has motivated my choice of a topical, or thematic, approach as opposed to a regional one. From my perspective, each theme – such as physical processes, historical geography, and geophysical hazards – has interesting and important stories that help us to understand the processes that have shaped the landscape of this province. The thematic approach is also a means of gaining student interest in geography as a discipline that deals with issues ranging from the local to the global.

The study of British Columbia from a regional perspective is equally valid, and many instructors prefer this approach because it focuses much more on the human and physical features that make individual regions within British Columbia unique. It generally provides a more in-depth view of the many distinctive regions of the province. To address this aspect of BC geography, all chapters in this work, especially the first and last, pay considerable attention to the regional development of the province.

Geography of British Columbia: People and Landscapes in Transition has been created largely by converting my classroom notes to prose. It has not been an easy task, as the spoken word and the written word are two very separate means of communication.

Acknowledgments

Many people are to be thanked, including librarians, my colleagues in the Social Sciences at Capilano College, Gordon Bailey for his insightful comments, and especially Karen Ewing for her assistance with the physical geography chapter. Students were particularly helpful, as the initial version was in draft form and used as a text. Erin Sackney actually returned her "text" to me loaded with editorial comments, which were invaluable. The huge, time-consuming task of editing the many drafts was done by my wife and partner, CarolAnn Glover. I am forever grateful. A special thanks also goes to Susan Harper, a long-time friend, who volunteered to provide editorial comment. Finally, I would also like to thank the readers for UBC Press for their frank and useful comments, which have been taken into account, and Camilla Jenkins for her exceptional, professional editing. Any errors and omissions are the fault of the author.

Geography of British Columbia

British Columbia: A Region of Regions

1

USING GEOGRAPHY TO MAKE SENSE OF THE LANDSCAPE

The focus of geography is the human and physical landscape, and geographical knowledge gives meaning to the changes that are constantly shaping and modifying the landscape. Those who study the human landscape look at where people live, their activities, and how they have modified the landscape. Physical geographers, on the other hand, are interested in the many physical processes that influence the landscape. Of course, landscapes are frequently modified by a combination of physical and human processes, and common to both sides of the discipline is that geography always incorporates a spatial perspective. The two broad divisions in the field – the human and the physical – can be divided into a number of subfields. As Figure 1.1 shows, geography is associated with many other disciplines, but the spatial element keeps it distinct.

British Columbia is a large province, encompassing nearly 950,000 square kilometres, and is also extremely varied from both a human and physical perspective. Many nation states are significantly smaller, and few have such a variety of landscapes. Nevertheless, fewer still have such a small population in relation to area – 4 million in 1999. It is a helpful exercise to compare the size of British Columbia to that of countries around the world and to locate the province in relation to those countries. Next, looking at a map of Canada alone will reveal the size and spatial relationship of British Columbia and other provinces and territories within the country. One theme of this text is to consider how British Columbia is unique within Canada.

One definition of geography is the study of "where things are and why they are where they are" (McCune 1970, 454). "Things" can be physical features, people, places, ideas (or human innovations), or anything in the landscape. "Where" questions concentrate on location as well as recognizing physical and human patterns and the distribution of various activities, people, and features of the landscape. Many of these questions can be answered simply by looking at a map, and students are encouraged to acquire a map of British Columbia as soon as possible. Where is the Peace River? Where is the homeland of the Nisga'a people? Where are the sockeye spawning grounds? Where is Trail? Knowledge of where things are is basic and essential geographical information. A useful beginning to test your knowledge of British Columbia is to draw a map of the province from memory and to place on it the features you consider important. This cognitive mapping exercise reflects individual landscape experiences (which can be shared with others) and demonstrates the importance of location. Using maps to answer where questions is the easiest aspect of geographical study.

Why are things where they are? "Why" questions are far more difficult than where questions and ultimately may verge on metaphysical. Even so, students are encouraged to conduct research about and to analyze the various physical, economic, political, cultural, and historical factors involved in a specific location or locational patterns,

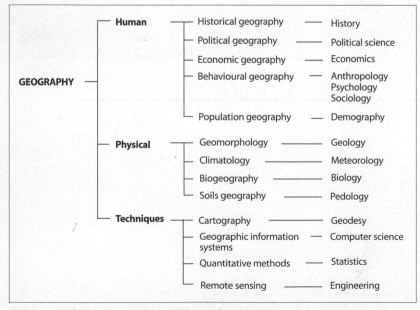

Figure 1.1 Subfields of geography and links to other disciplines

whether it is the location of a type of vegetation, of a community, of a group of people, or of a resource. Why does sage brush dominate the dry southern interior valleys? Why is Vancouver where it is and why has it grown so rapidly? Why did Barkerville become a ghost town? Why did the Doukhobors settle mainly in the Kootenays? Why did the Nechako River get dammed for hydroelectric power? Why is the Peace River region not part of Alberta? Through geography, students can develop critical thinking skills that permit them to recognize relationships and spatial patterns and to unravel their complexity.

The geography of British Columbia is changing constantly. News headlines inform us of these changes and provoke both where and why questions: Earthquake measuring 4.3 hits Vancouver area – Is the "Big One" coming? ... Skeena Cellulose in Prince Rupert requires government assistance to ward off closure ... New aluminum plant planned for British Columbia ... Collapse of Asian economies has had a serious impact on British Columbia's resource exports ... Vancouver-based engineering firm designing a state-of-the-art pulpmill for Indonesia ... British Columbia coho runs face possible extinction because of Alaskan overfishing. Common to such complex issues is the idea that human and physical processes are altering the landscape. The chapters that follow are intended to raise both where and why questions and to help students understand some of the processes influencing change.

One of the themes throughout this text is the importance of history. From a European, colonial perspective, British Columbia has little history of settlement and development compared to eastern Canada or to many other nations in the world. Aboriginal people, however, have approximately 10,000 years of history in British Columbia, and anthropologists and archaeologists are still adding new evidence of their settlement patterns and use of resources. From the viewpoint of physical geography, changes to the landscape are often measured in several hundreds of millions of years, and the BC landscape is no exception.

The combination of physical processes in British Columbia has produced a spectacular variety of mountains, rivers, lakes, islands, fjords, forests, and minerals. Within this physical setting, First Nations people, and later non-Natives, settled, exploited, and altered the landscape,

sometimes irreversibly. These physical-human interactions prompt us to look at the landscape another way, namely, from an environmental perspective.

Meshing with the idea of change over time in such a rugged and physically challenging province is the theme of movement over space. Isolation was a major factor in European settlement and development. As transportation and communication technologies improved, however, accessibility allowed greater opportunity for trade, development of resources, and settlement. The conquering of distance also facilitated the transition to a global economy, especially in trade and investment flowing from the Atlantic to the Pacific. All these changes have affected the regions of British Columbia in vastly different ways and resulted in a much more complex society.

TOPICAL OVERVIEW

This text employs a primarily topical approach. Chapter 1 provides an introduction to geography and an overview of British Columbia's settlement and development. Chapters 2 to 6 introduce the basic physical and human landscapes and some of the processes that change these landscapes. Chapters 7 to 15 take an economic perspective as they concentrate on resources, resource management, and communities that depend on resources.

Most of this text concerns human geography, but the physical landscape is not ignored. Chapter 2 discusses the landscape in terms of the various physical processes that have changed and shaped it. These processes have created distinctive regional variations throughout British Columbia as well as marked contrasts to the rest of Canada.

Human geography themes, in Chapters 3 to 6, include the threat from geophysical hazards to our human use systems, the historical geography of European settlement, the long history from First Nations use of the landscape to its present use, and the story of Asians in British Columbia. Each of these themes takes an historical perspective in examining the factors that influence the many changes to and regional variations in the landscape.

Resources, the theme of Chapters 7 to 15, have played and continue to play a large part in the economic well-being of this province as well as in attracting settlement. Chapter 7 begins with a discussion of resource management issues and the roles played by governments and corporations. Our use of, and dependence on, resources in this province have spatial and regional patterns. Each resource has its own unique history of development and influence on British Columbia. Consequently, forestry, fishing, metal mining, energy, agriculture, water, and tourism are examined in separate chapters. Single-resource communities, the theme of Chapter 15, illuminate the human link between resource development and the people who depend on it for a living.

Chapter 16, the final chapter, summarizes many of the themes and developments by reviewing 200 years of urbanization in this province. An urban view enhances the regional perspective of British Columbia because it shows the growth, and occasionally the decline, of communities throughout the province.

REGIONAL OVERVIEW

Regional geography is a means of assessing an area by categorizing it into smaller geographic regions characterized by distinct physical or human/cultural features. From a regional perspective, British Columbia is a unique province within Canada for a variety of reasons. Physical characteristics set it apart from all other provinces. It has the youngest and highest mountains in the country and is often described as a vertical landscape. It also has the greatest amount of fresh water in Canada, which is an essential resource for the five species of Pacific salmon and provides the potential for hydro-electric power. The highly indented coastline, "punctured by fjords," spans some 41,000 kilometres (Dearden 1987, 259). Weather and climate produce other distinctive patterns. The relatively mild, wet west coast, with the warmest winter

temperatures in Canada, stands in contrast to a considerably colder and drier interior, with desert conditions in the southern river valleys. The interrelationship of climate, soils, and vegetation produces distinctive patterns from west to east in the province and also south to north because of the eleven-degree span of latitude. Vertical change, due to high mountain ranges, produces regional variations similar to latitudinal differences.

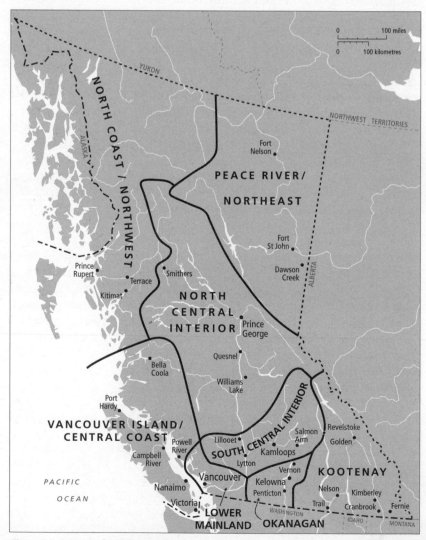

Figure 1.2 Regions of British Columbia

Distinctive physical characteristics and a unique global location have influenced the human characteristics of the province. The population of First Nations, particularly in coastal locations and along salmon-bearing rivers, was greater than anywhere else in Canada (Muckle 1998). Non-Native "discovery" and settlement was also unique in that it occurred from the west rather than from the east. British Columbia went through unique territorial struggles to become a British colony and further political struggles to establish its present boundaries. The connection to the Pacific, and particularly to Asia, increased as transportation systems were developed. No other province has such a long history of immigration by Asians – first Chinese, and later Japanese and Sikhs. Nor did any other province gain the reputation of being so adamantly racist.

British Columbia has an abundant supply of resources, which have been the main attraction for the population and reason for its rapid increase. Yet the physical characteristics of the province initially made resource extraction and export to distant markets difficult, and regionally differentiated patterns of settlement and de-velopment resulted (Robinson 1972). British Columbia is a region of **regions** and can be divided and subdivided, as it is throughout this text, on the basis of both physical and **cultural regions**. Cultural regions are further divided into formal, functional, and vernacular, or perceptual, regions. "A formal cultural region is an area where people have one or more cultural traits in common" (Bone 2000, 4). First Nations groups, such as the Haida of the Queen Charlotte Islands, for example, have the homogeneous characteristics of a common language, religion, and historical experience. Functional regions are often defined by political and/or economic criteria. British Columbia has been divided into areas to administer resources such as fish, forest, and minerals, to police jurisdictions, to provide political representation, and to administer education. The province can also be divided into vernacular, or perceptual, regions. The historical experience of living on the north coast or in the central interior, for example, gives people a sense of belonging to a particular region (Bone 2000, 4).

Figure 1.2 divides the province into eight regions, devised mainly by considering historical development In

Table 1.1

Population by region, 1881-1998

	Vancouver Island/central coast	Lower Mainland	Okanagan	Kootenay	South central interior	North central interior	North coast/ northwest	Peace River/ northeast
1881[a]	18,777	7,949	1,316	863.	4,725	7,550	7,376	923
1891[a]	39,767	23,543	3,360	3,405	6,390	4,889	16,839	n/a
1901[b]	54,629	53,641	7,704	32,733	14,563	5,123	9,270	948
1911[b]	84,786	183,108	21,240	51,993	24,103	9,011	16,595	1,644
1921	119,024	256,579	23,728	53,274	32,232	18,615	18,986	2,144
1931	133,591	379,858	30,919	63,327	37,621	23,236	18,689	7,013
1941	164,751	449,376	40,687	72,949	37,394	26,272	18,051	8,481
1951	233,250	649,238	62,530	93,256	50,363	41,324	20,854	14,395
1961	312,160	907,531	86,230	107,466	57,346	89,085	38,203	31,061
1971	415,254	1,256,425	130,498	125,643	106,993	133,906	63,080	45,155
1981	517,536	1,434,739	196,774	145,412	139,175	186,992	68,376	55,463
1991	611,654	1,829,537	240,291	141,480	142,628	190,141	67,975	58,355
1998	718,407	2,105,215[c]	302,112	154,644[c]	162,809[c]	222,034	73,826	66,109

a   Some approximations for regions as the province was divided into only five electoral areas.
b   Some approximations for regions as the province was divided into only seven electoral areas.
c   1996 data
Sources: Census Canada, various years.

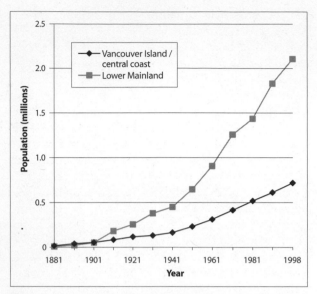

Figure 1.3 Vancouver Island/central coast and Lower Mainland populations, 1881-1998

combination with census subdivisions. Table 1.1 provides the population change figures for each region, spanning nearly 120 years. Maps, graphs, and statistics are some of the key tools of the geographic trade (the "techniques" shown in Figure 1.1). With them, geographers begin to understand the dynamics of where and why questions. One could construct a map of each of the eight regions in Figure 1.2, detailing such features as mountains, rivers, incorporated communities, Aboriginal title boundaries, and transportation systems. Such an exercise engenders familiarity with the vastness of the BC landscape, the physical and human features that distinguish each region, and the factors that integrate these separate regions into the province and with other parts of Canada and the world.

Monitoring regional population change, as Table 1.1 does, is another interesting exercise. Analyzing that change requires research of the various territorial boundaries used by Census Canada for collecting data. Graphing the absolute growth of each region is also useful as it gives a picture of the rate and trend of change and a sense of historical development. A comparison of just two regions – the Lower Mainland and Vancouver Island/central coast – provokes some interesting ques-

tions about rate of growth. Why did the Lower Mainland, a geographically smaller area, outstrip the Vancouver Island/central coast region so rapidly between 1901 and 1911? Bear in mind that the Lower Mainland includes the City of Vancouver, incorporated in 1886. Observe the slope of the graph lines between the years 1921 and 1951 in Figure 1.3, which shows the population increases of the two regions. What accounted for the different rates of population growth during this period? Why has the Lower Mainland population continued to grow at a more rapid rate from the 1960s on?

The answers lie partly in the political decision to locate the Canadian Pacific Railway (CPR) terminus and an international port at Vancouver and in the stimulation this gave to economic growth and thus population. Employment opportunities related to obtaining and processing the primary resources of fish, forests, mines, and agriculture played a role in each region, and the port facility of Vancouver greatly widened the catchment area for exporting resources. The First World War, the Depression of the 1930s, and the Second World War were significant global events that affected each region. Changing technologies altered the way people made a living. Which resources gained importance, how resources were harvested, and the movement of goods, people, and information all had an effect on the populations of these two regions.

A similar comparison could be done for any two regions and would require further analysis of the many factors influencing regional growth. Census figures can also reveal the ethnic composition of each region, and again these figures can form the basis of a host of social, ethnic, political, and economic questions. These lead in turn to geographic analysis of ethnic groups from First Nations and Asians to the British in British Columbia.

The development of communities provides further geographic knowledge about the settlement of British Columbia. Table 1.2 ranks by population the ten largest municipalities from the 1996 census. It is interesting to compare these populations to selected census years in the past (in which the ranking is also given). Many of our present-day communities did not exist before the beginning of the twentieth century; others have changed boundaries; some were much more significant in the past; and others have lost population.

Table 1.2

Population of municipalities for selected years

Municipality		1996		1921		1901		1891		1881
Vancouver (CMA)	(1)	1,891,465	(1)	163,220	(1)	27,010	(2)	13,685		–
Victoria (CMA)	(2)	304,287	(2)	38,727	(2)	20,919	(1)	16,841	(1)	5,925
Abbotsford^a	(3)	105,403		–		–		–		–
Kelowna	(4)	89,442		2,520		261		–		–
Kamloops	(5)	76,394	(7)	4,501	(8)	1,594		–		–
Prince George	(6)	75,150		2,053		–		–		–
Nanaimo	(7)	70,130	(5)	6,304	(4)	6,130	(4)	4,595	(2)	1,645
Chilliwack	(8)	60,186	(9)	3,161		277		–		–
Vernon	(9)	31,817	(8)	3,685		802		–		–
Penticton	(10)	30,987		b		–		–		–
New Westminster		c		c		c	(3)	6,641	(3)	1,500
North Vancouver		c	(3)	7,652		–		–		–
Prince Rupert		16,714	(4)	6,393		–		–		–
Nelson		9,585	(6)	5,230	(5)	5,273		–		–
Rossland		3,802		2,097	(3)	6,156		–		–
Fernie		4,877		2,802	(6)	1,640		–		–
Revelstoke		8,047		2,782	(7)	1,600		–		–
Trail		7,696	(10)	3,020	(9)	1,360		–		–
Greenwood		784		371	(10)	1,359		–		–

a Boundary change and incorporated as a city, 1995.
b Incorporated (1909), but population less than 1,000.
c Included in Vancouver (CMA, Census Metropolitan Area).
Sources: Census Canada, 1881, 1891, 1901, 1921, 1996.

What are the factors responsible for the growth or decline of communities, and how are communities connected to their regions and to other communities?

Comparing the older centre of Victoria to Vancouver illustrates the extremely rapid rise in Vancouver's population since its incorporation in 1886. One sees the powerful influence on Vancouver of its status as a national railway terminus and major port facility, and later, as the focus of other railway, road, and highway systems and site of an international airport. Vancouver has also attracted the major banks, financial institutions, and head offices for many of the resource industries operating throughout the province and the Asia-Pacific region. As the provincial capital, Victoria has experienced the growth of government and services along with infrastructure developments that link it to Vancouver Island's resources and to the mainland, but these economic links are not nearly as extensive as those of Vancouver.

Other communities throughout British Columbia have changed their population ranking over time mainly through the expansion of transportation and resource development and processing. By the 1980s, however, the new urban growth dynamics of the tourism and retirement industries, along with technologies that shrink time and space, had affected some communities more than others. These communities, and their growth (or decline), are intimately tied to the regions in which they are located.

In the sections that follow, the eight regions shown in Figure 1.2 are briefly described in terms of their distinctive physical characteristics and historical development.

Vancouver Island/Central Coast

This region combines Vancouver Island with the central coast, which extends from Powell River north to Bella Coola. Vancouver Island has a rugged spine of mountains,

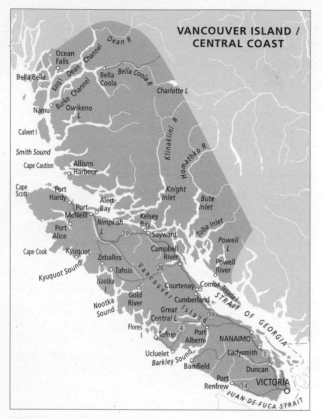

Figure 1.4 Vancouver Island/central coast region

referred to as the Insular Mountains. The central coast is part of the Coast Mountains range, which has peaks reaching considerably higher elevations and includes the highest mountain in British Columbia, Mount Waddington (4,016 metres). These mountains run mainly north-south and influence weather and climate conditions significantly. The prevailing westerlies often result in torrential rains on the west side of Vancouver Island and on the mainland where it is exposed to the open Pacific. The Olympic Mountains of Washington State and the Insular Mountains of Vancouver Island provide a **rain shadow effect** on the east side of Vancouver Island and on the southern end of the central coast. The Pacific Ocean at these latitudes (approximately 48°30' to 52° north) is considerably warmer than the Atlantic Ocean on the east coast of Canada, providing this region with the mildest winter climates in the country.

Precipitation variations due to the rain shadow also influence vegetation and, in particular, the growth of Douglas fir in the drier areas. There are no large rivers on either Vancouver Island or the mainland, but the many small rivers and streams are important for fish habitat.

Historically, the Vancouver Island/central coast region has been home for many First Nations. The peoples from this region experienced the longest exposure to non-Natives, however, and diseases took a huge toll. It was here that the Spanish and British squared off in the 1780s over territorial claims for colonization and the valuable sea otter trade. By the early 1800s, British Columbia was embroiled in an overland fur trade struggle between the aggressive North West Company and the Hudson's Bay Company. In 1821, the merger of these companies resolved the dispute and left the Hudson's Bay Company with a monopoly over the territory.

Non-Native settlement was sparse and temporary until a series of political and economic events in the mid-1800s. The Oregon Treaty of 1846 annexed the British territory and forts south of the forty-ninth parallel, and as a consequence Victoria was established at the south end of Vancouver Island. In 1858, gold was discovered on the lower reaches of the Fraser River, triggering an avalanche of miners seeking their fortune. Victoria became the main port of entry for much of this activity. Further discoveries of gold in the Cariboo region enhanced Victoria's position as the main administrative and service centre. The two separate colonies of Vancouver Island and the mainland were amalgamated in 1866 with Victoria as the capital, a position it has maintained since British Columbia joined Confederation in 1871.

Because of the location of the capital, the southern end of Vancouver Island attracted most of the settlers to the region. The rest of the island was opened up in response to resource discovery, transportation developments, and technological change. Coal was an important resource and pre-dated the discovery of gold. The discovery of other minerals, such as iron ore on Texada Island and copper at the north end of Vancouver Island, created more jobs, but most of this activity did not occur until after the 1960s. Salmon fishing and canneries sprang up along the island and mainland coasts, and farming settlements were established mainly in the southeast of the island. By far the most important industry throughout the region was forestry. Large lumber mills such as the one in Chemainus were in operation by the 1880s, followed by pulpmills in the communities of Ocean Falls, Port Alice, and Powell River in the early 1900s.

Historically, railways were the most important means of land transportation on Vancouver Island. The Esquimalt and Nanaimo Railway (E&N) opened in 1886 and was the most important line as it came with a provincial land grant to over one-quarter of the island. Roads appeared first on the southeast side of Vancouver Island and eventually linked the southern end with the northern as well as sending tentacles across to the few communities on the west side of the island (Wood 1979). Ocean-going transport provides the main link between the mainland coastal communities (e.g., Bella Bella, Bella Coola, Ocean Falls, and Powell River) and Vancouver Island. The highly indented and rugged mainland coastline hindered road development in the early days of settlement. Only Bella Coola was linked by road to the rest of the province.

Today there are in excess of 700,000 people living in this region. The forest industry is still important, especially for those living in the west and north of the island and along the central coast. The southeastern portion of the island has experienced some unique economic dynamics. It has attracted a huge number of retirees because of the mild winters with relatively little snow to shovel and year-round recreational activities. This part of the island also attracts the greatest number of tourists in the region because direct ferry links with the Lower Mainland make it easily accessible. As well, the southern end of Vancouver Island is intimately linked to the Lower Mainland region in the provision of administrative and service functions for western Canada and the Asia-Pacific region, making it part of the core, or heartland, of the province (as opposed to the periphery, or hinterland).

Lower Mainland

The climate of the Lower Mainland is similar to that of the southeast coast of Vancouver Island although it has higher precipitation, including more snow in winter, the higher the elevation. The Fraser River, the largest river

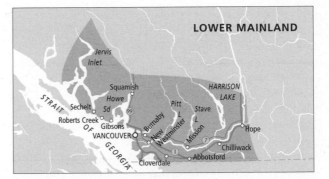

Figure 1.5 Lower Mainland region

system in the province and most significant to the salmon fishing industry, is an important physical feature of this region.

Historically, salmon and other resources of the water and land attracted many First Nations to the region. For non-Natives, gold was the main attraction following its discovery on the Fraser in 1858. Agricultural settlements soon followed, but securing these rich agricultural lands from the threat of floods has not been easy. Major sawmills located on Burrard Inlet exported lumber prior to the existence of Vancouver, and canneries located at the mouth of the Fraser River also pre-dated the city.

Transportation has been a major factor in the growth and development of this region. The completion of the CPR at Port Moody and extension to Vancouver in 1886 was the catalyst for the rapid growth observed in Table 1.1. Vancouver, with its national railway and international port, was the main centre for this relatively small geographic region and was largely responsible for the growth of the adjacent Fraser Valley to the east, the Squamish-Whistler-Pemberton corridor to the north, and the Sunshine Coast to the northwest. The mountains and valleys have been instrumental in framing the transportation links and settlement patterns for this region. The Sunshine Coast has a linear settlement pattern following the Strait of Georgia and is connected to Vancouver via ferry at Horseshoe Bay in West Vancouver. The Pacific Great Eastern Railway (PGE, renamed the British Columbia Railway, or BCR, in 1972) initially ran between Squamish to Quesnel (1921), and then was extended south to North Vancouver (1956) and north beyond Quesnel. This railway line has been an important transportation link to the ports at Squamish and North Vancouver. The Sea-to-Sky Highway is the main transportation system today as it winds its way beside Howe Sound to Squamish and then follows valleys leading past Whistler to Pemberton, Lillooet, and the interior of the province. For the Fraser Valley, the river was the main transportation system at first. The construction of the CPR, and later the Canadian National Railway and British Columbia Electric Railway, all gave accessibility to the region. Road systems were built in the early 1900s, and eventually the Trans-Canada Highway and other highways were constructed, linking Vancouver and the Fraser Valley to the rest of the province and south into the United States.

Favourable climate, superb natural features, highways, railways, port facilities, an international airport, and many commercial links to the rest of Canada, Asia, and the world make Vancouver a world city. Many economic and administrative activities are shared with the southern end of Vancouver Island, making the combined region of the Lower Mainland and southern Vancouver Island the heartland of British Columbia. The Lower Mainland, however, will have the greatest population increases in the future.

Okanagan

The Okanagan Valley lies between the Cascade Mountains to the west and the Monashee Mountains to the east. There are several lakes in this valley, Okanagan Lake being the largest. The region's southern location between

two large mountain chains results in a continental climate with typical temperature extremes of hot summers and relatively cold winters. The vegetation of this arid valley consists mainly of grasses and sage brush and few trees; forests grow on the moister mountain slopes.

The region became the home for the Okanagan First Nations. Some non-Native settlement occurred with the fur trade, but much more took place as the region became recognized for its farming potential as a fruit-growing area. The Okanagan is one of the few places in Canada where apples and "soft fruit" such as peaches and cherries are produced, but irrigation is necessary in this dry belt. A number of communities evolved to serve the growing agricultural settlement. Vernon, Kelowna, and Penticton, all on Okanagan Lake, became the most prominent. Boats on Okanagan Lake served an important transportation function, and later railway lines were built. The Kettle Valley line (1915) provided the link between the Kootenays and the Lower Mainland, giving access to the southern Okanagan, while a branch line of the CPR linking Kamloops to Kelowna (1925) served the northern portion of the region.

Several changes occurred after the 1960s. Tourism, which had mainly been a summer activity, expanded into a year-round endeavour, with golf courses and ski runs. The limited snowfall and relatively low land and housing prices made this a favourable location to retire. These characteristics led to increased population, urban sprawl, and land use conflicts, especially over agricultural land, until the implementation of agricultural land reserves in 1972. The motors of fishing and pleasure boats carried Eurasian milfoil into the region, where it spread through the water systems, converting once sandy beaches to a mass of weeds and jeopardizing the growing tourism industry. Eradication programs in the 1970s made use of herbicides that provided some control but also raised concern about potential carcinogenic effects.

Other industries expanded in the region, including mining and forestry, both of which increased the employment and population base. With the signing of the Free Trade Agreement between Canada and the United States in 1989, agriculture changed rapidly. Fruit crops continued to dominate, and to compete with US producers with access to inexpensive labour, BC producers developed new varieties of apple trees requiring considerably

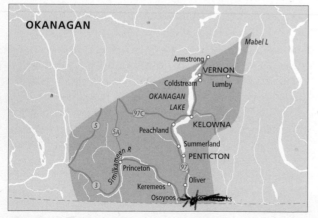

Figure 1.6 Okanagan region

less labour. One of the most significant changes was to the grape and wine industry. Forced to compete globally, it met the challenge by growing new grape varieties. New rules permitting the sale of wine from farms was a large part of the success of the industry, which now attracts many tourists.

The Okanagan has many physical assets and a fairly diversified economy, making it one of the rapid growth regions in the province (Table 1.1). Within the Okanagan, Kelowna has become the most important service, administrative, and manufacturing centre, with a regional airport and a highway link to the coast via the Coquihalla Highway.

Kootenay

Mountains, rivers, and valleys are the main physical features that define the Kootenay region. The rugged mountain chains run north-south: first the Monashees, then to the east of them the Selkirks, farther east the Purcells, and finally the Rockies. All the rivers and lakes in these valleys are part of the Columbia River system, which exits British Columbia at Trail as the Columbia flows into the United States. Climatically, this region is similar to the Okanagan, but slightly colder in winter, not as hot in summer, and with slightly more precipitation.

Several First Nations resided in this region: the Kootenai to the east, the Okanagan to the west, and the Shuswap Nation to the north.

Census Canada has traditionally divided the Kootenays into east and west, but the divisions share a valuable mineral resource base. Gold, silver, coal, copper, lead, and zinc were all discovered and became the lure to settlement and development following the fur trade era. With the Crow's Nest Pass Agreement of 1897, the CPR became the principal land owner, provider of rail transportation, and developer of the resources. The Agreement was a deal struck between the federal and provincial governments and the CPR to run a branch line from Lethbridge, Alberta, through the Crowsnest Pass to the mineral-rich Kootenays, ending initially at Nelson. Through the Agreement and their subsequent purchase of railway grants, the CPR acquired millions of acres of land, coal deposits, metal mines, a major smelter at Trail (Cominco), and West Kootenay Light and Power. The CPR exerted enormous control over this region. Other

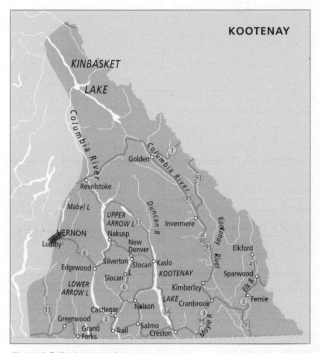

Figure 1.7 Kootenay region

railway companies built lines and acquired land grants, but the CPR purchased most of these over time, consolidating their hold on the Kootenay economy. The changes to population for this region (Table 1.1) hide the boom and bust cycles experienced by individual mines and smelters. The list of **ghost towns** in the Kootenays was sufficient to warrant several books and articles (Barlee 1970; 1978a; 1978b; 1984).

Historically, settlement patterns in the area have been influenced mainly by mineral exploitation, transportation developments, and agricultural opportunities. Other factors also led to settlement. For example, the Kootenays has been home for the Doukhobors, who migrated here in the 1930s in an attempt to escape religious persecution in Europe and political persecution in Saskatchewan. The Kootenays had good agricultural land for their communal lifestyle and appeared to be relatively isolated from government interference. In the early 1940s, many Japanese families, evacuated from the coast, were sent to communities and work camps throughout the Kootenays such as Greenwood, Sandon, New Denver, and Slocan City.

Forestry has been another resource activity in this region. The Kootenays experienced forestry expansion after 1961 with a pulpmill in West Kootenay at Castlegar and in East Kootenay at Skookumchuck. In the 1960s, the provincial government became involved in hydroelectric megaprojects; the Columbia River was a major component of the "two rivers policy." Through this agreement, the Peace River, in the northeast, and the Columbia River, in the southeast, were developed for hydroelectricity simultaneously. British Columbia then used the electrical energy from the Peace and the United States purchased the electricity generated from the dams on the Columbia. The dams also provided flood protection for cities such as Spokane in Washington State. Later other dams, such as the Revelstoke Dam, were constructed to fulfill increased electrical demand by British Columbians.

Currently, the Kootenay region remains dependent on resource development and export. Mining and forestry are vital to the economic well-being of the region. Tourism has offered some diversification, as the region has many hot springs and lakes, and opportunities abound for skiing, hiking, and sightseeing. Nevertheless, at considerable distance from major urban populations in Alberta, British Columbia, and the United States, the region is likely to show slow overall growth in the future.

South Central Interior

The south central interior is largely identified by the Southern Interior Plateau. The region extends west of the Lillooet to Revelstoke on the Columbia River, but the Thompson River system defines this region for the most part. The Thompson River valley is hot in summer and cold in winter, with precipitation occurring mainly on the surrounding mountains. The Shuswap Nation of the Interior Salish have been the traditional users of the land.

Early European interest in this region was over furs and gold. The Thompson River was an important "highway" for the fur trade, and Kamloops was established as a fur trade post in 1812. Small amounts of gold were discovered in the region in the mid-1850s but not in amounts significant enough to create a gold rush. The gold rush to the Cariboo in the 1860s, however, attracted cattle drives and cattle ranching to these interior grasslands.

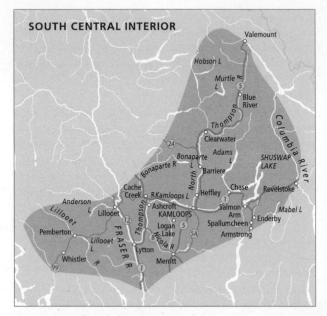

Figure 1.8 South central interior region

When the CPR main line was built, Kamloops became the main supply and service centre for the region. This location, where the North and South Thompson Rivers meet, was enhanced by later transportation developments such as the extension of the Canadian National Railway (CNR) down the North Thompson and through Kamloops on its way to Vancouver. With the development of road systems, Kamloops was on the Trans-Canada Highway, and it is currently at the end of the Coquihalla Highway, which runs to the coast.

Farming and ranching were the main industries in the south central interior until the 1960s, when forestry, mining, tourism, and the retirement industry were added. Forestry has been the most important, with a pulpmill established in Kamloops in the mid-1960s and sawmills and other forest-product manufacturing enterprises located throughout the region. Mining has also played a significant role, with a copper smelter at Kamloops (now closed) and mines through the Highland Valley from Ashcroft to Merritt. Tourism is increasing in importance largely due to the transportation routes that lead to Kamloops and branch north to Jasper and Edmonton and east to Banff and Calgary. Accessibility and many favourable features have also attracted retirees to this region. The south central interior, as shown in Table 1.1, is another region with a relatively rapid increase in population, but it is dominated by one urban centre – Kamloops.

North Central Interior

The north central interior is one of the largest regions in the province and is defined by the Northern Interior Plateau. The northern half of the Fraser River, with its many tributaries, is a large part of this region, although the area extends westward to include Smithers and the Bulkley Valley. Temperature regimes are colder than are southern interior locations in winter and not as hot in summer. The mixing of Pacific and Arctic air masses results in increased precipitation, but this varies through the region in relation to the configuration of the mountain chains.

Historically, the area has been home to many First Nations, such as Sekani, Nat'oot'en, Wet'suwet'en, and Dakelh. The history of non-Native settlement began with the overland fur trade, which relied on the Fraser River for transportation. Forts were erected through the region in the early 1800s, but it was not until the Cariboo gold rush of the early 1860s that more permanent non-Native settlement ensued. A number of routes were used to gain access to the Barkerville area from the Lower Mainland until the Cariboo Road was constructed in the mid-1860s. Gold was also discovered in the northern part of the region, in the Omineca Mountains for example, but these finds were not sustainable. The development of Barkerville and other mining communities also attracted ranching to the southern part of the region, including the Chilcotin Plateau. By the end of the 1860s, however, the gold discoveries were minimal and the miners were disappearing along with the communities. This large region remained relatively uninhabited until the CNR connected it to the port of Prince Rupert to the west in 1914. From the south, the Pacific Great Eastern Railway (PGE) was built in a series of stages: Squamish to Quesnel in 1921, Quesnel to Prince George in 1952, Prince George to the Peace River 1958, and Prince George to Fort St James in 1968. Prince George, like Kamloops, became a transportation hub of rail lines and later highway systems, serving central and northern British Columbia.

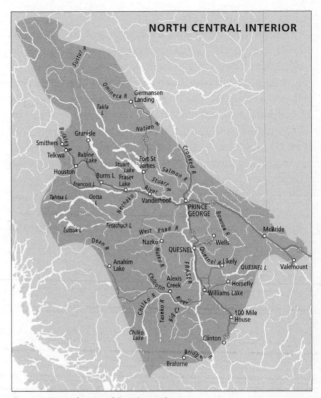

Figure 1.9 North central interior region

Since the 1960s, as the population in Table 1.1 indicates, there has been considerable interest in this region. The mining of copper, molybdenum, and gold brought many workers into the area, but the main industry was forestry. The forests were largely untouched until the 1960s, and then the increase in world demand for forest products saw a massive expansion in the industry throughout the north central interior. The manufacturing of plywood, lumber, and other forest products was integrated with new pulp-and-paper mills built at Quesnel, Prince George, and in the new town of Mackenzie.

Today, some of the mines have closed, and the forest industry is in a downward cycle, reminding us that much of the region is extremely resource dependent. Prince George has emerged as the largest centre. With its accessibility, its new university, and its many services, it has become an important service centre to central and northern British Columbia, performing a role similar to the one that Edmonton performs for Alberta.

North Coast/Northwest

The north coast/northwest is another large and, for much of its area, isolated region because of its rugged, mountainous landscape. The Coast Mountains, which extend the length of this region, are among the highest in British Columbia, and the coastline is highly irregular. The heavily forested Queen Charlotte Islands are also part of the region. Climatically, much of the area is exposed to the Pacific and to the westerly flows of wind (that is, wind flowing from the west). Its northern location brings considerable exposure to the Aleutian low pressure system, which brings much rain in the summer and snow in the winter. The Skeena, Nass, and Stikine are the largest of the very significant river systems flowing to the Pacific.

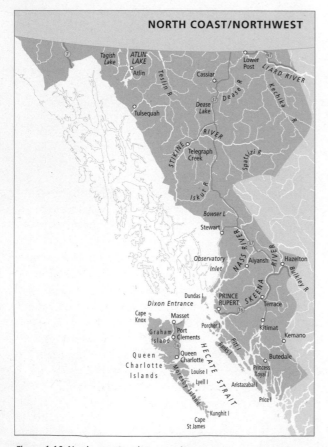

Figure 1.10 North coast/northwest region

The coast has been home to high density populations of First Nations, whereas First Nations in the northwest have had much lower populations. A number of factors led to early European development of north coast/northwest region. In the late 1700s, the sea otter trade aroused considerable interest. The Russians, who erected fur trade posts across Alaska and the Panhandle – the strip of Alaska that extends south along the coast – were the first to exploit these valuable furs. They laid territorial claim to Alaska, thus cutting off the northwest from the sea. Subsequently, Russia sold this territory to the Americans in 1867. Without access to the Pacific, the northwest portion of this region was left largely to overland fur trade interests.

Other resources encouraged temporary settlement. Salmon was valuable, for example, and canneries became evident by the late 1800s and early 1900s, especially at the mouth of the Skeena and the Nass and along the coast. Except for those at the mouth of the Skeena, most disappeared in the 1950s with improvements in fishing technology. Coal was discovered north of Prince Rupert and was mined for several years until the richer and more accessible coal mines of Nanaimo were developed in the 1850s. Small amounts of gold were discovered on the Queen Charlottes in 1850 and on the Skeena River in 1863, but no gold rush occurred in either case. In 1898, the Yukon gold rush opened up the northernmost portion of the region. The famous Chilkoot Pass, and the building of the White Pass and Yukon Railway in 1902, meant that most gold seekers passed through the northern tip of British Columbia before entering the Yukon. There was plenty of interest in this region in these early days but little permanent settlement.

The completion of the CNR to Prince Rupert in 1914 was another important event, and it increased permanent settlement. Prince Rupert was also the site of an early pulpmill, which encouraged growth as well. A copper smelter was built at Anyox, north of Prince Rupert, but only lasted until the Depression. The Second World War prompted construction of the Alaska Highway, giving some accessibility to the remote northwest portion of the region. One of the first areas outside the Lower Mainland to expand with the post-Second World War boom was Kitimat. This planned town was built to house the workers for a new aluminum plant in the early 1950s.

As they did in other regions, the mining and forest industries in the north coast/northwest region underwent major expansion beginning in the 1960s. The Stewart-Cassiar road linked the Alaska Highway through the northwest to Prince Rupert. A major copper mine near Stewart, a large open pit mine for asbestos at Cassiar, and a number of small gold mines brought employment to these rather isolated locations. The development of Quintette and Bullmoose mines and the new town of Tumbler Ridge in the Peace River region – a mining region known as northeast coal – in the early 1980s resulted in the rail line being double tracked to Prince Rupert and a large coal port being built at nearby Ridley Island. A coal port is a "super port" with facilities for unloading coal from trains and loading it onto large ocean-going vessels. A planned expansion of the aluminum smelter at Kitimat was cancelled in 1995 for environmental reasons, but aluminum production remains important to the region. The forest industry expanded in a number of directions. Much of the valuable timber from the Queen Charlottes was harvested and barged south for the mills of the Lower Mainland. A new pulpmill was opened at Kitimat. The pulpmill at Prince Rupert was upgraded, and logging and sawmilling expanded throughout the region. Prince Rupert has emerged as the largest centre for the region, but the threat in 1998 that its pulpmill might close put into perspective the importance of the forest industry to this city and to the region as a whole. If the closure had gone ahead, 450 direct jobs and many more service-related jobs would have been lost to Prince Rupert, affecting over 1,000 people in this community of 17,000.

Northeast of Prince Rupert, the Nisga'a of the Nass River Valley were the first to enter into a modern peace treaty in British Columbia, setting a precedent for future treaties throughout the province. This treaty will result in a cash settlement, a Native-owned forestry company, and a substantial share of the Nass River commercial salmon fishing.

There are few roads or rail lines through this area even today, and its growth is tied closely to resource development. It is a slow-growing region.

Peace River/Northeast

Most of the Peace River/northeast region does not fit the broad physical description of British Columbia as a mountainous, vertical landscape. This flat, sedimentary region east of the Rockies is physiographically similar to the prairies. The two major rivers, the Peace River in the south and Liard River in the north, are part of the Mackenzie River system, which drains into the Arctic Ocean. The region contains areas of permafrost, and much of the vegetation is scrub, boreal forest. Temperatures are cold in winter and surprisingly warm in summer, when the days are long, inducing **convection precipitation.**

North West Company fur traders were the first non-Natives to enter the region, and it is here that the earliest

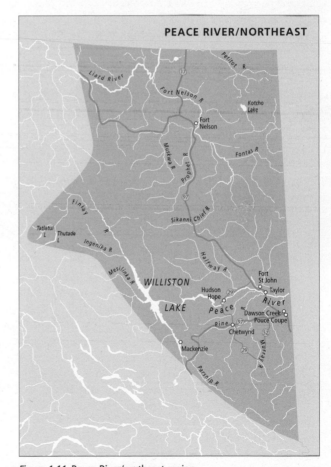

Figure 1.11 Peace River/northeast region

fur trade forts in British Columbia were erected. Few agricultural settlers ventured this far north until the homesteads of the south and central Prairies were no longer available. The development of hardy, early maturing wheat facilitated agricultural homesteads by the 1920s and '30s. Problems of accessibility were improved when the rail line from Alberta was extended to Dawson Creek in 1930. The Alaska Highway was constructed during the war and Dawson Creek became Mile 0, helping to open up the region. The federal government made more farmland available at the end of the Second World War for returning service men. The PGE was finally extended to Fort Nelson by 1971. Wheat farming on the excellent soil and cattle rearing have been the main agricultural activities of the Peace River area.

The discovery and development of oil and natural gas in the 1950s encouraged investment, the building of pipelines, and considerably more people to reside in this region. By the 1960s, the Peace River had become the target of the massive hydroelectric plan referred to as the two rivers policy. The plan involved constructing the W.A.C. Bennett Dam, which in turn created the largest reservoir, or artificial lake, in the province, Williston Lake. As well, transmission lines were built to connect the hydroelectric power source to southwestern British Columbia. All this attracted even more people to the Peace River. The energy crisis, beginning in the early 1970s, sparked another round of oil and gas exploration and development.

The sedimentary basin also contained coal, and in the early 1980s, northeast coal began being developed in the area. The new town of Tumbler Ridge housed the miners, an electric rail line of the British Columbia Railway (formerly the PGE) was constructed, and millions of dollars were spent in upgrading the CNR line from Prince George to Prince Rupert and, as discussed already, building a coal port at Ridley Island. The costs borne by both provincial and federal governments to export the coal were massive. Unfortunately, the world market demand and price for coal declined, and by early 2000 the two coal mines in this area had significantly reduced production. Many workers and their families have had to move out of the community.

Today Fort St John and Dawson Creek are the largest centres in the Peace River/northeast region. The forest industry employs many people in its new pulpmills and other wood-manufacturing plants. Agriculture and the oil and gas industry are also important. Tourism, particularly with the paving of the Alaska Highway in the late 1980s, is another expanding industry. The economy and growth of the Peace River/northeast region is tied to its diverse number of resources. It is a large region, but the population is small and growth is slow.

SUMMARY

Geography is a very practical discipline, as it examines the processes that change and shape our human and physical landscape. The field has a broad scope, but the focus here is on one geographic region, British Columbia. *Geography of British Columbia: People and Landscapes in Transition* uses primarily a topical, or thematic, approach to address the many influences on this diverse landscape. Although most of the topics fall within the realm of human geography, physical processes are not ignored. The physical characteristics of this rugged territory are always present and have influenced the many patterns of settlement and the development of the province. The book also uses a regional approach to analyze a province known as a region of regions. Various themes stress the regional patterns of development and how they change over time. Geographical jargon is kept at a minimum to allow a clear understanding of the many human and physical characteristics that distinguish the various regions throughout the province and make British Columbia unique within Canada.

REFERENCES

Barlee, N.L. 1970. *Gold Creeks and Ghost Towns of Southern British Columbia.* Summerland, BC: Self-published.

–. 1978a. *The Best of Canada West.* Langley, BC: Stagecoach.

–. 1978b. *Similkameen: The Pictograph Country.* Summerland, BC: Self-published.

–. 1984. *West Kootenays, the Ghost Town Country.* Surrey, BC: Canada West Publishers.

Bone, R.M. 2000. *The Regional Geography of Canada.* Don Mills, ON: Oxford.

Dearden, P. 1987. "Marine-Based Recreation." In *British Columbia: Its Resources and People,* ed. C.N. Forward, 259-80.

Western Geographical Series vol. 22. Victoria: University of Victoria.

McCune, S. 1970. "Geography: Where? Why? So What?" *Journal of Geography* 7 (69): 454-7.

Muckle, R.J. 1998. *The First Nations of British Columbia.* Vancouver: UBC Press.

Robinson, J.L. 1972. "Areal Patterns and Regional Character." In *Studies in Canadian Geography: British Columbia,* 1-8. Toronto: University of Toronto Press.

Wood, C.J.B. 1979. "Settlement and Population." In *Vancouver Island: Land of Contrasts,* ed. C.N. Forward, 3-32. Western Geographical Series vol. 17. Victoria: University of Victoria.

INTERNET
For on-line statistics see the following websites:
www.statcan.ca
www.bcstats.gov.bc.ca

Physical Processes and Human Implications

2

Much of the attraction, and beauty, of British Columbia comes from its great variety of physical features. Its many scenic landscapes of rugged mountains, large and turbulent rivers, diverse vegetation, variety of fauna, and often isolated settings – all changing from season to season – are both great assets and great challenges. The province has a rich resource base of minerals, energy, forests, fish, agriculture, and tourism. The physical setting has also been a powerful influence on settlement, the urban system, and transportation corridors, as well as posing a major risk to the human population.

This chapter begins the discussion of physical processes by drawing attention to the geologic time scale and the distinctive rock structures that make up the crust of the earth. The ability to assess the age of and distinguish between different rock types – even those billions of years old – gives a sense of the generalized geological makeup of British Columbia. The study of processes that create these landforms and transform the surface of the earth is known as **geomorphology.**

Another major focus in this chapter is the set of physical processes related to weather and climate. The spatial and seasonal patterns of temperature and precipitation vary greatly throughout the province, with broad distinctions between the coast and the interior and between southern and northern locations.

Soils and vegetation influence each other and are a product of many other physical processes, including variations in climate and geomorphology. The vegetation patterns of British Columbia are very distinct and obvious, ranging from huge coastal coniferous forests to southern interior valleys of cactus and sage brush. These physical landscapes have major implications for human habitation of the province.

From a human perspective, significant changes occur within periods of ten to twenty years in terms of dress, popular music, and employment opportunities. From a physical perspective, it is necessary to think in terms of tens, or even hundreds of millions of years when considering changes to the landscape (Table 2.1). This province has undergone profound transformations over the last 200 million years. Climatic changes have caused ocean levels to rise and fall and glaciers to scrape and modify the landscape; tectonic processes have been responsible for mountain building, **volcanic activity,** and

Table 2.1

Geologic time scale

Era	Period	Years before present
Cenozoic	*Quaternary*	
	Holocene	Present
	Pleistocene	1,500,000
	Tertiary	
	Pliocene	13,000,000
	Miocene	25,000,000
	Oligocene	36,000,000
	Eocene	58,000,000
	Paleocene	66,000,000
Mesozoic	Cretaceous	135,000,000
	Jurassic	180,000,000
	Triassic	230,000,000
Paleozoic	Permian	310,000,000
	Carboniferous	345,000,000
	Devonian	405,000,000
	Silurian	425,000,000
	Ordovician	500,000,000
	Cambrian	600,000,000
Precambrian		Up to 4 billion

the addition of islands to British Columbia. The landscape is by no means static.

THE ROCK CYCLE:
FORMATION AND DESTRUCTION OF ROCKS

Igneous rocks are created when molten **magma** cools and hardens into a solid state. As the temperature drops, crystals form in the liquid, similar to ice crystals forming in water. These crystals are called minerals and have different compositions. Some typical minerals found in igneous rocks are feldspar, quartz, hornblende, and olivine. Because hornblende and olivine contain iron, they are darker than feldspar and quartz. Igneous rocks containing mostly feldspar and quartz minerals are usually light in colour, an example being granite. Igneous rocks that have mostly iron-rich minerals are often dark grey or black; basalt is an example. Basalt is denser than granite since it contains larger proportions of iron and other heavy minerals. Granite is a typical continental rock, while the majority of oceanic rocks are basalt.

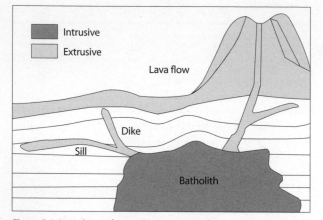

Figure 2.1 Intrusive and extrusive igneous rock

The size of the minerals in an igneous rock reflects the rate of cooling of the magma. If the magma cools relatively fast (over days, years, or even decades), the crystals are so small that a microscope is needed to see them. These quickly cooled rocks are found near or at the earth's surface, primarily as a result of volcanic activity, and are known as extrusive or volcanic igneous rocks. Basalt is an extrusive igneous rock and is commonly found in lava flows, dikes, and sills. Dikes and sills form as magma pushes into older rocks; sills form parallel to surrounding rock layers, while dikes cut across the rock structure (Figure 2.1). Sometimes the cooling process produces an hexagonal pattern of cracks in extrusive rocks, creating what is known as columnar jointing. Extrusive rocks can be seen along many of British Columbia's highways, an indication of the province's extensive volcanic history.

Magma that cools tens of kilometres below the earth's surface may require millions of years to solidify. This time allows the crystals to grow to a size where they can be seen by the human eye. These intrusive or plutonic rocks, of which granite is an example, are often found in huge plutonic masses known as batholiths (Figure 2.1); much of the Coast Mountains range of western British Columbia is batholithic. Although formed deep below the surface, granitic batholiths may be pushed upward during large-scale crustal movements. As surface weathering and erosion strip off the overlying rock formations, the batholiths are exposed.

Rocks at the surface are exposed to weathering processes caused by water, air, and vegetation growth and decay. **Weathering** may cause a rock to fragment into small, sedimentary pieces called **clastics.** Weathering by surface water and groundwater may also cause the minerals within rock to dissolve, making water "hard" and oceans salty. After weathering, sediments and dissolved minerals may be moved by streams, waves, wind, glaciers, and gravity to a new location, a process known as **erosion.** The sediments are eventually deposited, often in layers. The layering effect happens when different sizes and shapes of sediments are deposited, such as small, rounded particles on top of large, angular ones. This usually occurs when the agent of erosion changes, such as from wind to water. As sedimentary layers accumulate, the bottom ones are pushed deeper into the crust, causing compaction and heating. As well, groundwater can infiltrate the sediments and precipitate (or solidify) some of the dissolved minerals.

The precipitated minerals act as a cement to bond the sediments together. Compaction, heating, and cementing change the sediments into rock. Not surprisingly, rocks formed from unconsolidated materials bonded together are known as **sedimentary rocks.** Some common examples of clastic sedimentary rocks are sandstone (made from sand-sized sediments) and shale (made from clay and silt-sized sediments). Precipitated minerals may also form a second type of sedimentary rock called evaporatives. Rock salt, for example, forms in the shallow waters of warm oceans; as the water evaporates, the salts are precipitated. A third sedimentary rock type is formed through organic processes. As plant and animal products accumulate, their remains may eventually turn into rock. Coal is an example. As well, tiny organisms utilize minerals dissolved in lakes and oceans to form their shells and skeletons (similar to the way in which humans extract calcium from milk to make bones). When the organisms die, their hard parts sink to the bottom of the water body and accumulate to form limestone.

Sedimentary rocks are located throughout British Columbia. Some are found in their original horizontal positions, but many are folded, tilted, and fractured. These deformations occur during times of mountain building, a prime example being the Rocky Mountains on the east side of British Columbia.

The third category of rocks, after igneous and sedimentary, is called **metamorphic.** When rocks are exposed to extreme pressures and temperatures, or to chemical infusions from nearby magma bodies, the minerals within the rocks change in shape and chemistry. The original rock does not melt; it metamorphoses into a new rock. Limestone, for example, will metamorphose into marble. Metamorphic rocks are mostly associated with mountain building and igneous intrusions. In British Columbia, metamorphic rocks can be found in the Coast Mountains range in conjunction with their huge batholithic cores.

The **rock cycle** reminds us that rocks are not static in geologic time, and that all can be related (Figure 2.2). Beginning with the molten state, igneous rocks are "born." Like all other rocks, however, they are subject to the processes of weathering and erosion. As particles are deposited and/or precipitated, igneous rocks can become sedimentary rocks. Both sedimentary and igneous rocks can be subject to intense heating, pressure, and chemical action, becoming metamorphic rock in the process. All three rock groups may go back to the molten form.

From this basic information about rocks, the generalized geology of British Columbia can be examined, recognizing that rocks and landforms were created during different eras, as Figure 2.3 demonstrates. Much of the Coast Mountains range and south central British Columbia is made up of intrusive igneous and is approximately 100 million years old. Consequently this region should be characterized by granitic types of rock. Of course, the actual geology of any region is much more complex, and those familiar with the Coast Mountains just north of Vancouver around Squamish and Whistler will recognize Mount Garibaldi, the Black Tusk, and obvious basalt columns as extrusive igneous formations. Mount Fissal, behind Blackcomb Mountain, has sedimentary rocks containing fossils of sea shells over 2,000 metres in elevation. Many mountain-building processes have been at work in the various regions of British Columbia. The generalized geological map gives an overview of the dominant rock type only.

The interior of the province and the north end of the Queen Charlottes is a mix of flat-lying lava and some sedimentary rock. Much of the interior of the province was under water for considerable periods, and sedimentary rock should be found in these locations. About 40 to 50 million years ago, however, significant volcanic activity produced huge lava flows that covered the sedimentary levels. In areas like the spectacular Helmcken Falls on the Murtle River in Wells Grey Park, the much harder basalt rock, twenty to thirty metres thick, can be seen quite clearly over the top of the much softer sedimentary levels.

The third category, flat-lying or gently dipping sedimentary rocks formed some 65 million years ago, forms a plains-like landscape such as the Peace River/northeast region, which is structurally part of the Great Interior Plains. The mineral resources found in this region include oil, natural gas, and coal.

The Rockies are made up of folded sedimentary rock. Some 250 million years ago there was an enormous amount of pressure on flat-lying sedimentary rock that folded it up to elevations in excess of 4,000 metres. Subsequent mountain-building episodes occurred, the last one approximately 65 million years ago. A drive through the Rockies not only gives the impression of a vertical landscape but also shows the folding of the sedimentary layers.

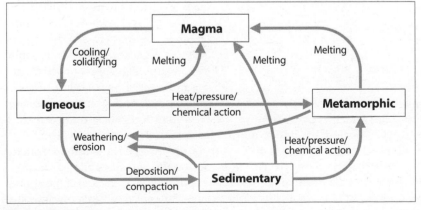

Figure 2.2 Rock cycle

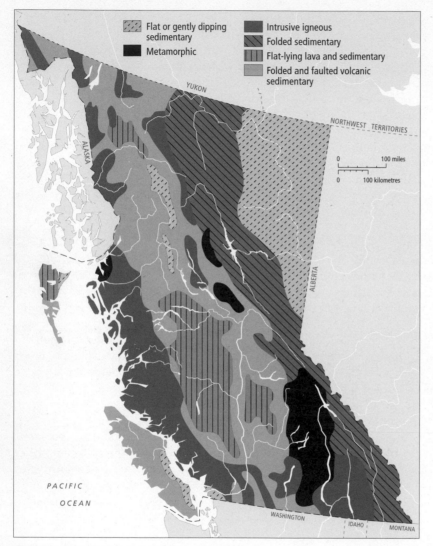

Figure 2.3 Generalized geology of British Columbia
Source: Modified from Rider (1978), 13.

The final category, found mainly in southeast British Columbia in the Kootenays, is metamorphic rock. These very hard rocks are the oldest rock structures in the province, dating back one billion or more years.

Recognizing the processes that formed the three basic rock structures, having some idea of the age of landforms, and being able to generalize about where these rocks and landforms can be found is the beginning in understanding the geomorphology of British Columbia. In the context of millions of years, mountains have been built up and worn down, and it is important to examine the factors that shape and change the physical landscape in more detail.

DEFORMATION OF THE EARTH'S SURFACE

Deformation is a broad term describing the rather complex geologic processes of mountain building and changes to the surface of the earth. The forces that fold, tilt, and fracture rocks are produced deep below the earth's surface and are explained by the theory of **plate tectonics.** Figure 2.4 shows a cross section of the earth, which is made of several components. At the centre of the earth is a solid inner core ringed by a liquid outer core. This core is ringed by the mantle, which is solid but behaves like a plastic substance because of the extreme pressures and temperature. The mantle is also composed of a number of layers, each with its own characteristics.

Close to the top of the mantle is the asthenosphere, which has temperatures and pressures enough to melt the mantle rocks partially. Because the asthenosphere is in a semi-molten state, convection currents develop. The moving plumes of molten material put pressure on the

Most of Vancouver Island, the south end of the Queen Charlottes, and much of the interior is formed of folded and faulted volcanic and sedimentary rock. These areas were covered by water in the last 500 million years and are characterized by sedimentary rock. Approximately 150 million years ago, volcanic activity and pressures were exerted to push these sedimentary and volcanic structures into very rugged landscapes.

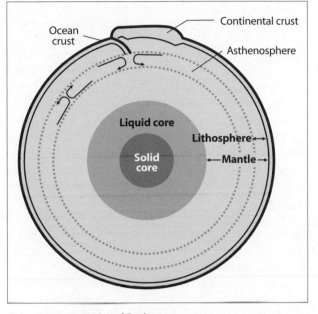

Figure 2.4 Cross section of Earth

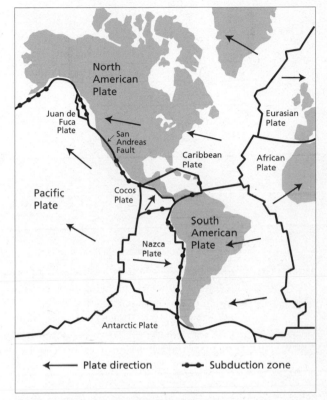

Figure 2.5 Major oceanic and continental plates

solid rocks above, forcing them to break and shift position. The brittle rocks above the asthenosphere belong to a layer known as the lithosphere. The bottom part consists of rocks from the upper mantle, while the top part of the lithosphere contains ocean crust and continental crust. Each of these regions – core, mantle, and crust – has a different mineral composition. The core is composed mainly of iron; the mantle has large amounts of a rock called peridotite; the ocean crust is primarily basalt; and the continental crust contains mainly granite. Sedimentary and metamorphic rocks are also found throughout the crust. The lithosphere, which contains both mantle and crustal materials, is thinnest beneath the oceanic crust, where rising convection plumes from the asthenosphere can force the lithosphere to fracture and move. The broken parts of the lithosphere are known as plates, and most **volcanic** and **earthquake activity** occurs at their edges.

The theory of plate tectonics suggests that the earth's lithosphere consists of seven large plates, as well as many small ones. Several of the large plates are capped primarily by oceanic crust; others contain both ocean and continental crust. The main plates making up North and South America are shown in Figure 2.5.

Where magma finds its way to the surface of the earth and splits the lithosphere apart, it creates a **rift zone** (Figure 2.6). The cooling magma forms new igneous rock in the rift zone. This rock, in turn, may be broken apart by new magma upwelling from below. Rocks formed and fractured in the rift zone make up what are known as the trailing edges of plates. Roughly half of the fractured rock is attached to the trailing edge of one plate and the other half to the second plate. The new rocks, although solid, are still very hot; they may have temperatures of several hundred degrees Celsius. The high temperatures cause the rocks to expand. As the rocks are forced away from the rift zone, they begin to cool, contracting and shrinking in volume. A look at the cross section of a rift zone shows that the ocean becomes deeper on either side of the zone due to the cooling crust as it contracts and thus compresses downward (Figure 2.6). The oceanic crust also becomes smoother the farther it is away

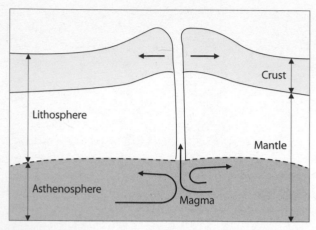

Figure 2.6 Rift zone

from a rift zone because sediments from the continents are deposited in the ocean basins, level the irregularities of the igneous crust, and form a continental shelf.

Rift zones separate the plates, but in doing so, they create collisions in other parts of the globe. When two plates are forced together, the edges may fracture and buckle, forming mountain ranges. This process is known as **orogeny.** Prime examples of orogeny occur when the edge of a plate contains continental crust. The Andes on the west coast of South America were formed in this manner. The Pacific northwest, comprising British Columbia, Washington, Oregon, and northern California, is another region where orogenic processes are at work. Besides colliding and sliding past each other, sometimes one of the plates, usually a denser, oceanic one, is forced downward and underneath the other. (Recall that oceanic rocks such as basalt are much denser, or heavier, than continental rocks such as granite.) This process is called **subduction,** and the major subduction zones for North and South America are shown in Figure 2.5. Continental rocks rarely subduct, which is why they are some of the oldest in the world – up to 3.8 billion years old. The subducting plate forms a trench in the ocean floor as much as five kilometres deep. As the plate subducts deeper, it warms. Eventually the temperature is high enough to allow some of the plate to melt. If enough magma is produced, it will flow upward to create plutonic batholiths within the crust and volcanic eruptions at the surface.

The new rocks created at a rift zone (the trailing edge of a plate) are transported to a subduction zone (the leading edge of a plate), where they are eventually destroyed (returned to the magma state). On large plates, it may take an average of 250 million years to form, transport, and finally destroy the rocks. In this way, the plate acts as a giant conveyor belt. This analogy applies only to some parts of the world, however, as in other regions the plates collide and slip past each other forming **transform faults.**

The leading edge of the Juan de Fuca Plate, as can be observed in Figure 2.7, is in close proximity to Vancouver Island, where it is then subducted under the North American Plate. For the coast of British Columbia this means that there is a relatively narrow continental shelf, and therefore little build-up of sediments. By comparison, the trailing edge of the North American Plate, shown in Figure 2.5, is in the middle of the Atlantic Ocean. Consequently, there is a very wide continental shelf, with a build-up of sediments off the Maritime provinces. Sedimentary build-up also provides potential for oil and natural gas deposits, and the wider the continental shelf the greater the potential.

Rift zones, subduction zones, and transform faults all occur along the west coast of North America. California

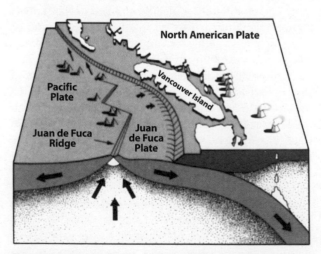

Figure 2.7 Subduction process off the southern BC coast
Source: Energy, Mines and Resources Canada (n.d.), n.p., courtesy of Pacific Geoscience Centre, Geological Survey of Canada.

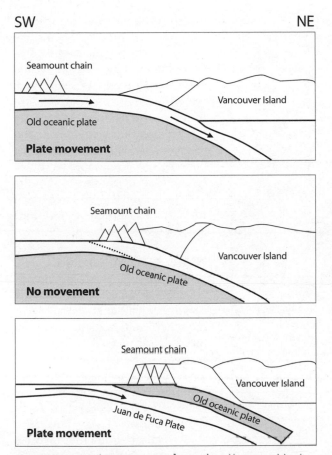

SW   NE

Seamount chain

Vancouver Island

Old oceanic plate

Plate movement

Seamount chain

Vancouver Island

Old oceanic plate

No movement

Seamount chain

Vancouver Island

Old oceanic plate

Juan de Fuca Plate

Plate movement

Figure 2.8 Accreted terrane process for southern Vancouver Island
Source: Modified from Monger (1990), 9.

is marked by the famous San Andreas fault, where the Pacific Plate moves in a northerly direction past the North American Plate. Each movement produces an **earthquake,** and one of the characteristics of the San Andreas fault is how it appears to "lock" in specific locations, such as San Francisco. Here the pressure builds up until a major shift takes place. In 1906, this had catastrophic consequences. From northern California to the end of Vancouver Island, a much smaller plate can be observed. This is the Juan de Fuca Plate, whose jagged rift zone is only 200 to 300 kilometres west of Vancouver Island (Figure 2.7). The subduction process also produces earthquakes, and sometimes large ones, as the oceanic plate slides under the continental plate. Deep oceanic troughs,

orogeny, batholiths, and evidence of volcanic activity (some of it quite recent, such as Mount St Helens) are characteristic of the northwest Pacific region.

North of Vancouver Island, the Pacific Plate continues its northerly migration past the North American Plate with a transform fault in immediate proximity to the Queen Charlotte Islands. This movement, with its history of major earthquakes, continues all the way to Alaska, where another very active subduction zone occurs. The evidence of the highest mountains in Canada – the Mount St Elias range in southwestern Yukon – attests to the mountain-building pressures of this collision. This region is also one of the most seismically active areas of Canada and the United States.

Subducting plates sometimes carry fragments of oceanic and continental crusts from other regions. These fragments, of differing compositions and ages, are known as **terranes.** Instead of being subducted along with the plate, the terranes are pushed up, or accreted, against the edge of the second plate. Figure 2.8 shows an example of a small terrane, the Seamount chain of mountains, being accreted on to the south end of Vancouver Island. The diagram shows not only the dynamics of mountain building and easterly migration of the Juan de Fuca Plate, but also that the southern end of Vancouver Island is composed of geologic structures formed in other regions at other geologic times.

Terranes are a big part of British Columbia. Geologic evidence suggests that the whole Cordilleran region, from the Rockies to the Insular Mountains, is a series of accreted terranes (Figure 2.9). These dynamic tectonic forces have resulted in a highly complex geology for the province and a very rugged landscape rich in minerals.

Deformation of the earth's surface may occur as a result not only of tectonic factors but also of **isostasy,** a process of loading and unloading the surface of the earth with sediments and ice. As sediments erode from higher elevations such as the Rockies (which, keep in mind, are made up of sedimentary rock), these majestic mountains are gradually losing their lofty elevations. Erosion over the past 250 million years might lead one to expect that the "soft" sedimentary rocks would have been reduced to plains. The Rockies should be thought of instead as a floating raft piled high with cargo: as the cargo, or weight, is removed, the raft rises up. For every three

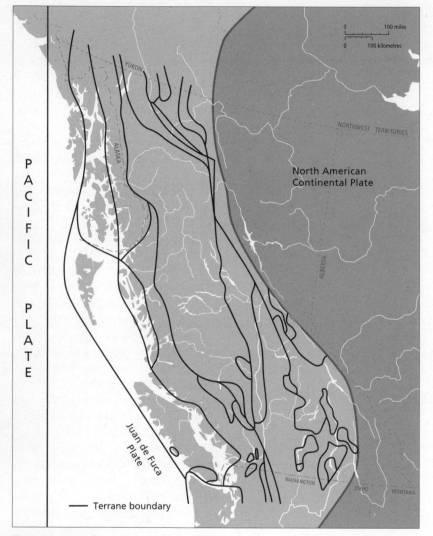

Figure 2.9 Accreted terranes in British Columbia
Sources: Modified from Internet, UBC Earth and Ocean Sciences; Canning and Canning (1999), 8; Tarbuck and Lutgens (1992), 531.

duce this volume of ice, but receded only tens of metres. As the ice melted – which happened in British Columbia only in the last 10,000 years – the ocean levels rose and, eventually, so did the land. This uplifting of the land, or isostatic rebound, takes thousands of years and still continues today. The lag in time between a relatively rapid rise in sea level and relatively slow rise in the level of the land has produced beach formations at different elevations around the world.

WEATHERING AND EROSION

Weathering describes a number of processes responsible for breaking down rock over time. **Erosion,** on the other hand, refers to the movement of rock materials via a number of agents – gravity, water, wind, and ice – working singularly or in combination. Erosion continues the weathering process and helps to form sediments.

Mechanical weathering involves the physical destruction of rocks into smaller and smaller components. Freezing and thawing form one common and powerful type of mechanical weathering in British Columbia. Rain or melting snow runs into cracks and fissures of rocks, where it is subject to freezing. The resulting ice expands and often fragments the rock particles.

metres of erosion to the Rockies, there is approximately two metres of uplift, or isostasy.

The last glacial age is the most recent and obvious example of isostasy. The glaciers built up to a depth of 1,500 to 2,000 metres, an enormous weight pushing the surface of the earth down perhaps several hundred metres. The level of the oceans of the world fell also to pro-

Chemical weathering, as the name describes, refers to a chemical action that breaks rock down. Since rock is made up of various minerals, the common substances of water, oxygen, and carbon dioxide, or some combination of these agents can dissolve and chemically react with minerals. Sands, silts, and clays are often the product of weathering, as is the red "oxidized" colour of soil.

The agents of erosion often work in combination. Gravity, for example, pulls broken rock and other weathered material downward, but when these materials are lubricated with water they are often carried away much more readily. Gravity is a powerful force and responsible for a host of erosion activities, referred to as mass wasting. These can include major landslides, such as one near Hope in 1965, rockfalls, debris torrents, slumping, and soil creep. In all cases, the landscape is altered by particles moving down slope, sometimes rapidly and sometimes extremely slowly.

Flowing water, in the fluvial process, is an effective agent of erosion in British Columbia because of the many turbulent streams and rivers. Water carries particles ranging from boulders to clay particles, carving up the landscape and depositing the materials along stream beds where the stream slows, and eventually into the deltas of lakes or into the ocean. In the southern half of the province the Fraser and Columbia River systems dominate, and in the north the Skeena, Nass, and Stikine are the major rivers flowing to the Pacific. The northeastern portion of British Columbia is drained toward the Arctic by the Peace and Liard River systems (Figure 2.10). Where these rivers and streams flow through mountainous terrain they cut the land into V-shaped valleys. Where there is less relief, the river systems tend to meander and form broad river valleys.

The greatly indented coastline of British Columbia, along with its many islands, is subject to continual erosion from the actions of the ocean. Waves, in particular, are relentless in pounding and modifying the coastline.

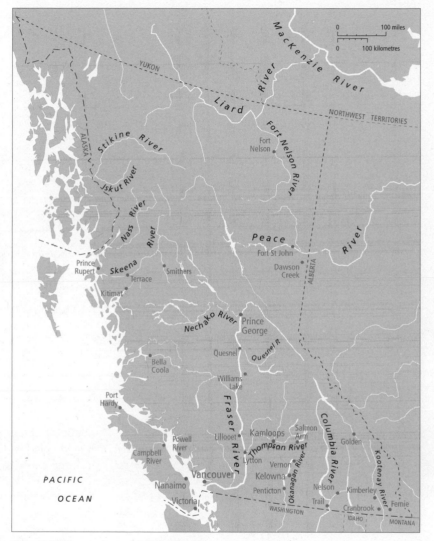

Figure 2.10 River systems of British Columbia

Currents and tides also assist in the erosion process, wearing down and carrying away materials. In other regions of the coast, where these materials are deposited, islands and coastlines are built up.

Wind plays a somewhat minor role in the erosion process. It picks up and disperses fine silts. In the dry interior of British Columbia dust storms are not uncommon. Wind-blown material may also act as an abrasive agent as it "sand blasts" landform structures.

The final factor, and a highly noticeable one because of its recent activity, is glaciation. Most of British Columbia has been subject to glacial erosion. With a cooling climate at the beginning of the glacial age, heavy snowfall in the high alpine areas of the mainland gradually compacted and transformed into icy glaciers. These moved down through stream and river valleys, all the way out to the Strait of Georgia to coalesce with others flowing from the mountains of Vancouver Island. These huge accumulations of ice reached 1,500 to 2,000 metres in height, carrying a great deal of material with them and carving the landscape with their movement. Glacial erosion acted like a bulldozer. Glacial movement down through existing river valleys scoured out the V-shaped valleys and turned them into U-shaped valleys. The process did not end at the coastline but continued along the ocean floor. Subsequent global warming melted the glaciers, raising sea levels, and despite isostatic rebound, left a coastal landscape of "drowned" glacial valleys called fjords – a significant asset for deep harbour ports and coastal navigation. In the Coast Mountains and other ranges, peaks under 2,000 metres have been rounded off through the movement of glaciers.

Moraines are another common landform in British Columbia. These are the large, linear piles of boulders, gravel, and sand that accumulated around the edges of glaciers. Erratics, large boulders that have been transported by glaciers, appear most obviously on flat landscapes such as the Fraser Valley.

Today, glaciers at high elevation in alpine areas are the remnants of what once covered the province.

WEATHER AND CLIMATE OF BRITISH COLUMBIA

Weather refers to the day-to-day atmospheric conditions that influence plans for any number of outdoor activi-

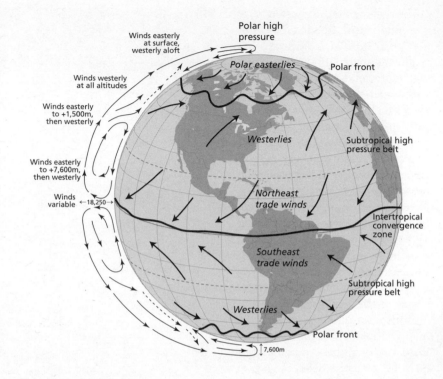

Figure 2.11 Global climate
Source: Modified from map by W. Heibert in Welsted, Everitt, and Stadel (1996), 32, with permission.

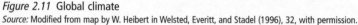

ties or travel. Climate is the longer term effect of weather. It is revealed by the collection of weather statistics over months and years that enable climatologists to assess the averages, and extremes, of atmospheric conditions for any location or region. For example, the climate for the southwestern corner of British Columbia, in comparison to the rest of the province or of Canada as a whole, makes this region one of the most attractive in the country. Climatic conditions have been observed to have cycles, some of which are much longer than others. El Niño conditions, for example, occur approximately every eight to ten years, resulting in much greater precipitation for coastal British Columbia. An ice age represents a much longer climatic cycle.

Figure 2.11 illustrates the global movement of air currents. The sun is the engine, or driving force, of weather and climate variations throughout the world. At the equator, a huge volume of air rises because of the intense surface heating. As the air rises thousands of metres, a low pressure zone is produced at the surface. This is typically characterized by precipitation, as the warm moist air rises, expands, and cools, causing condensation. High above the surface, the rising air diverges to the north and south poles. At approximately thirty degrees north and south of the equator these air masses descend, producing high pressure zones as they descend and compress. These form subtropical high pressure belts, the driest regions of the world. In the northern hemisphere some of the descending air moves back along the surface in a southerly direction, attracted to the low pressure zone. Other portions of the air masses continue in a polar direction. The surface air mass collides with an air mass driven toward the equator by polar high pressure zones. Where these collisions take place a low pressure zone is produced, referred to as the polar front. Thousands of metres above the polar front are high velocity, easterly moving winds known as the jet stream.

The prevailing wind patterns shown in Figure 2.11 are influenced by flows of air between high and low pressure zones, by the rotation of the earth, and by the seasons of the year. During summer in the northern hemisphere the days are longer, the sun is higher in the sky, and the wind patterns shift northward. During the winter there is less sun and the wind patterns shift southward. British Columbia is largely influenced by winds known as the westerlies and by two pressure zones: the Aleutian low (producing cloudy and wet weather) to the north, and the Pacific high (producing clear and dry weather) to the south. The jet stream demarcates these two pressure zones and, as Figure 2.12 shows, moves in seasonal patterns.

There are other influences on the weather and climate of British Columbia. The relatively warm Pacific Ocean directly affects the coastal regions. Farther inland the land mass and mountains have the greatest influence. Land heats up more rapidly than water and to higher temperatures, and cools down more rapidly and to lower temperatures. As Figure 2.13 illustrates, the province has three climate regimes, or climate regions: Pacific, Cordilleran, and Boreal (Hare 1974). Pacific Canada is the coastal region. Unique in Canada for its mild winters, it is often referred to as having a modified Mediterranean climate. The Cordilleran is defined by the many mountain chains between the Coast Mountains and the Rockies. The Boreal regime

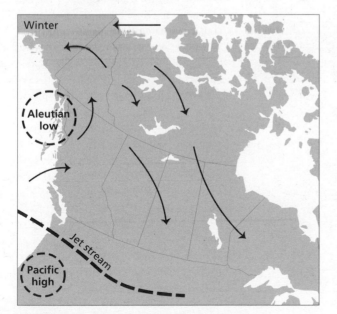

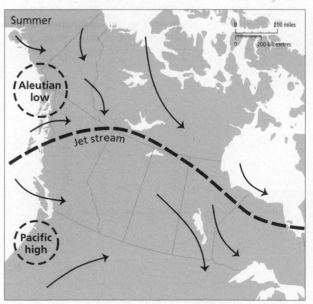

Figure 2.12 Jet stream influences in winter and summer

occurs in the plain-like northeast region of the province. Boreal also defines a vegetation of scrub forest typical of northern climates. Both Cordilleran and Boreal climate regimes are less subject to maritime influences than to continental air masses, which give them far greater extremes.

Prince Rupert, Vancouver, Fort Nelson, and Penticton, can be found in Figure 2.13 and represent north and south coastal locations and north and south interior locations. Table 2.2 gives their mean monthly temperature and precipitation. Graphing the data would quickly show the similarities and differences between these communities and the regions they represent.

Temperature differences between the coastal and interior locations are the greatest in the winter months, when Vancouver and Prince Rupert have considerably milder temperatures than Penticton and Fort Nelson. Winter temperatures in the coastal region are mainly influenced by a relatively warm, large body of water – the Pacific

Ocean – and the prevailing westerlies. In the summer, the ocean is slow to heat up and therefore provides a modifying influence. One of the most noticeable and dramatic influences on coastal temperatures occurs in the winter months. Frigid polar air, being dense and heavy, moves as a high pressure air mass from the Arctic down through the interior of British Columbia and funnels through the mountain passes to the coast, bringing freezing temperatures to Vancouver and Victoria.

As one travels inland, the moderating effect of the Pacific Ocean becomes less and less influential and the speed and intensity with which land can heat and cool becomes more and more influential. Consequently, temperatures in Fort Nelson and Penticton have greater variation than those in Prince Rupert and Vancouver.

There are considerable differences in winter temperatures between the two interior communities of Penticton and Fort Nelson, mainly due to latitude. Fort Nelson, just south of the sixtieth parallel, has only a few daylight hours in winter and, even then, the sun is at an extremely low angle. Penticton, just north of the forty-ninth parallel, gets a good deal more incoming solar radiation because of the increased number of sunlight hours in the winter. Fort Nelson is also much closer to the freezing Arctic high pressure air masses, which are not hindered by any physical barriers, while Penticton is protected by a number of mountain ranges that act as barriers to these Arctic air masses. Extremely cold winter temperatures (below -25°C) are relatively rare in Penticton.

In a vertical landscape such as British Columbia, elevation also influences temperature. Those participating in alpine activities will experience considerably cooler temperatures than elsewhere year round. At the very high elevations (over 3,000 metres) extremely cold temperatures can be experienced even in summer.

Considerable differences in precipitation can be seen throughout

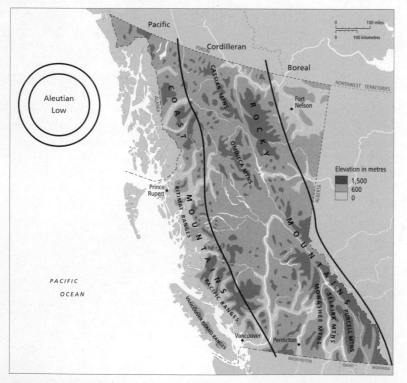

Figure 2.13 Climate regimes of British Columbia
Source: Modified from Hare (1974), 13.

Table 2.2

Climate data for selected communities

	Vancouver 49°11'N 123°10'W Elevation: 5 m		Prince Rupert 54°17'N 130°23'W Elevation: 52 m		Penticton 49°28'N 119°30'W Elevation: 342 m		Fort Nelson 59°50'N 122°W Elevation: 364 m	
	°C	mm	°C	mm	°C	mm	°C	mm
January	2.4	147	1.8	214	-2.9	32	-23.2	26
February	4.4	117	2.7	209	0.3	21	-17.1	24
March	5.8	94	3.7	180	3.7	16	-9.2	25
April	8.9	61	6.2	184	8.7	23	1.2	22
May	12.4	48	9.6	123	13.4	28	9.7	38
June	15.3	45	11.8	107	17.1	36	14.5	64
July	17.4	29	13.6	121	20.1	25	16.7	75
August	17.1	37	13.9	147	19.2	22	14.8	56
September	14.2	61	12.0	242	14.7	18	8.8	39
October	10.1	122	8.7	359	8.7	20	1.3	26
November	6.1	141	5.1	269	3.1	26	-12.3	27
December	3.8	165	2.9	259	-0.4	31	-20.7	26

Source: Internet, Climate Normals 1961-90.

the year between Vancouver and Prince Rupert even though both have coastal locations. A number of factors are at play here. For the Vancouver area, the rain shadow effect has a significant role (Figure 2.14). The relatively warm prevailing westerlies move over the Pacific, absorbing moisture as it evaporates. This air mass is forced to rise (a phenomenon known as the **orographic effect**) over either the Olympic Mountains of Washington State or the Insular Mountains of Vancouver Island, where the mass cools and contracts. This causes condensation, resulting in a great deal of precipitation on the western slopes – some of the highest precipitation in Canada. As the air mass descends the eastern slopes, it expands and warms and has the ability to absorb more moisture as it crosses the Strait of Georgia and Vancouver. The North Shore Mountains and other ranges of the Coast Mountains then force the air mass to repeat the orographic effect. The rain shadow region receives considerably less precipitation. It is worth noting that within Greater Vancouver significant differences in precipitation exist between south Delta and North Vancouver, just thirty-five kilometres away (Figure 2.15).

Even though Prince Rupert has the Queen Charlotte Islands to intercept the westerlies, there is little rain shadow effect. The Queen Charlottes do not have high mountains like Vancouver Island, and the distance from the Queen Charlottes over the Hecate Strait to the mainland is considerably farther than across the Strait of Georgia. Consequently the air mass is still laden with moisture as it is forced to rise up the rugged Coast Mountains, which form the backdrop to Prince Rupert.

The Aleutian low brings rainy and turbulent weather, but its position is changeable, influenced by seasonal changes in the jet stream (see Figure 2.12). In the summer, Prince Rupert is still to the north of the jet stream wave and very much under the influence of the Aleutian low, whereas Vancouver is influenced by the Pacific high.

Precipitation patterns in the interior are very different from those of the coast, in both quantity and season. The greatest influence in the summer is from incoming solar radiation, which heats up the land rapidly, leading moisture to evaporate, rise, cool, and then condense into dark thunder showers. This convection process is particularly common on the prairie-like landscape of Fort

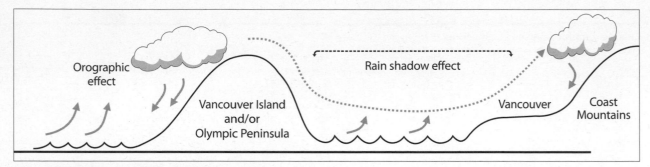

Figure 2.14 The rain shadow effect

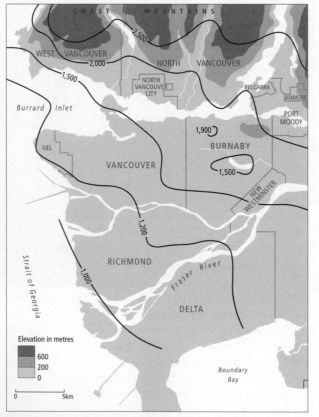

Figure 2.15 Vancouver area annual precipitation in millimetres
Source: Modified from Wynn and Oke (1992), 25.

peaks and relatively little reaches the dry valley bottom. What moisture does fall in the summer months is mainly due to convection.

During the winter, precipitation in the interior is predominantly in the form of snow. In northern, flat locations such as Fort Nelson, the surface has high reflectivity. Incoming solar radiation is therefore reflected back rather than absorbed and the atmosphere has little chance to build up moisture. It is difficult for Pacific flows of moisture to penetrate this far inland. Penticton, on the other hand, does have a marginal increase in precipitation during the winter months. The more southerly swing of the Aleutian low at this time of the year can, and does, bring more moisture. The Monashee Mountains to the east of Penticton accumulate the precipitation in the form of snow, which serves the skiing industry.

SOILS AND VEGETATION

Soils are the most fundamental element for growing plants, but fertile soils are in short supply in British Columbia: "Approximately two-thirds of the area of British Columbia consists of mountain slopes, rocky land, and water areas. Over the remaining third of the province, the parent materials of the soils have been largely formed by glacial drifts" (Dalichow 1972, 9).

Soils consist of mineral and rock fragments that have undergone varying degrees of physical and chemical weathering along with decaying organic matter, resulting in many combinations of minerals, water, and air. Soils also have a variety of textures and are continually changing over time. A cross section, or profile, of soils reveals the characteristics in various layers, or horizons (Figure 2.16). The top layer, or O-Horizon, is made up of undecayed organic matter and gives way to the

Nelson. Penticton is nearly a desert. Its location in the southern Okanagan Valley is well inland, and the many mountains between the Pacific and Penticton wring out most of the moisture. This rugged topography ensures that most of the precipitation falls on the mountain

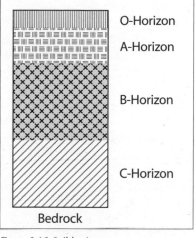

Figure 2.16 Soil horizons

A-Horizon, where the organic matter decays and is added to weathered rock. These uppermost layers are the most subject to weathering and erosion. Below the A-Horizon is the B-Horizon, which accumulates minerals that have been filtered, or leached, out of the A-Horizon. The C-Horizon is the most stable layer and is often called the parent material (Lavkulich and Valentine 1978, 56). Owing to glaciation, the parent material in much of British Columbia has been transported and mixed so that it rarely bears much resemblance to underlying bedrock.

Vegetation is tied closely to soils and climate and, as both vary greatly throughout the province, it will come as no surprise that there are also many vegetation zones. British Columbia is known for its forests and, in particular, the coniferous forests that dominate many regions. There are considerable differences between the types of forests on the coast and in the interior. Forests also differ from north to south with latitude change, often mirroring the changes to forest species wrought by altitude in this mountainous province (Figure 2.17).

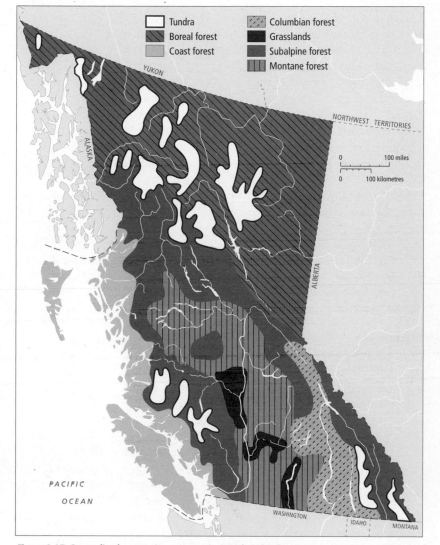

Figure 2.17 Generalized vegetation patterns in British Columbia
Sources: Modified from Barker (1977), 79; Dalichow (1972), 12; and Jones and Annas (1978), 37.

The coast forest is dominated by hemlock, cedar, and in the drier southern portions, Douglas fir. The trees here grow to an enormous size. Subalpine forest refers to vegetation affected by mountain elevation; species such as alpine fir, hemlock, and yellow cedar grow on coastal mountains, while lodgepole pine and spruces occupy interior mountain elevations. Even higher elevations, with extreme climate and scarce soil, give way to tundra vegetation of moss and lichens. The montane forest covers much of the south and central interior. Because it is considerably drier than the coast and has much greater extremes in temperatures, the region is more prone to forest fires. To the south, forests of Ponderosa pine dominate, while varieties of spruce are found in the more northerly regions. Grasslands are also part of this landscape and occupy the Interior Plateau along with southern arid valleys. Here, bunch grass was the dominant vegetation until massive overgrazing occurred from the 1860s Cariboo gold rush on. Sagebrush, which cattle do not eat, has become the most common vegetation today. Columbian forest is found in the mountainous Kootenay region, which has considerably more moisture than the montane forest region. Forests of hemlock, cedar, and Douglas fir again appear but not at nearly the size of those on the coast. Finally, to the north is the boreal forest, characterized primarily by stunted spruce and aspen because of latitude and permafrost.

SUMMARY

The physical processes of British Columbia have had, and continue to have, a powerful influence on shaping the landscape of the province. These processes cannot be ignored in examining the human development of the landscape. The historical use of the land by First Nations, and later by the first non-Natives to the region, was greatly influenced by the physical landscape. Over time, with the creation of new technologies, the physical environment has played less of a role. Yet our modern economy is still very much tied to the resources of the land, as our activities are to the daily weather conditions. And whenever there is a flood, earthquake, avalanche, or wind storm, the power of nature is very apparent.

REFERENCES

Barker, M.L. 1977. *Natural Resources of British Columbia and the Yukon*. Vancouver: Douglas, David and Charles.

Canning, S., and R. Canning. 1999. *Geology of British Columbia: A Journey through Time*. Vancouver: Douglas and McIntyre.

Dalichow, F. 1972. *Agricultural Geography of British Columbia*. Vancouver: Versatile.

Energy, Mines and Resources Canada. N.d. "Earthquakes in Southwest British Columbia." *Geofacts*. Pamphlet. Sidney, BC: Geological Survey of Canada, Pacific Geoscience Centre.

Hare, K., and J. Thomas. 1974. *Climate Canada*. Toronto: John Wiley.

Harris, C. 1997. *The Resettlement of British Columbia: Essays on Colonialism and Geographical Change*. Vancouver: UBC Press.

Jones, R.K., and R. Annas. 1978. "Vegetation." In *The Soil Landscapes of British Columbia*, ed. K.W.G. Valentine, P.N. Sprout, T.E. Baker, and L.M. Lavkulich, 35-45. Victoria: Resource Branch, Ministry of the Environment.

Lavkulich, L.M., and K.W.G. Valentine. 1978. "Soil and Soil Characteristics." In *The Soil Landscapes of British Columbia*, ed. K.W.G. Valentine, P.N. Sprout, T.E. Baker, and L.M. Lavkulich, 49-57. Victoria: Resource Branch, Ministry of the Environment.

Monger, J. 1990. "Continent-Ocean Interactions Built Vancouver's Foundations." *Geos* 4: 7-13.

Rider, J.M. 1978. "Geology, Landforms, and Surficial Materials." In *The Soil Landscapes of British Columbia*, ed. K.W.G. Valentine, P.N. Sprout, T.E. Baker, and L.M. Lavkulich, 11-33. Victoria: Resource Branch, Ministry of the Environment.

Tarbuck, E.J., and F.K. Lutgens. 1992. *The Earth: An Introduction to Physical Geology*, 4th ed. New York: Macmillan.

Welsted, J., J. Everitt, and C. Stadel. 1996. *The Geography of Manitoba: Its Land and Its People*. Winnipeg: University of Manitoba Press.

Wynn, G., and T. Oke, eds. 1992. *Vancouver and Its Region*. Vancouver: UBC Press.

INTERNET

For information about terranes, see www.geop.ubc.ca/Lithoprobe/transect/terrane.html

For climate, see Climate Normals, 1961-1990, www.cmc.ec.gc.ca/climate/normals/E_BC_WMO.HTM

UBC Earth and Ocean Sciences, www.geop.ubc.ca/Lithoprobe/transect/terrane.html

Geophysical Hazards: Living with Risks

3

A geophysical hazard, or natural hazard, can be defined as an assessment of risk from the earth's forces. This implies two components: the jeopardy for people and their property, and the natural forces at work. In hazards research, unless humans or their property are involved in the extreme geophysical event, by definition there is no hazard.

The first component demands an examination of community locations, transportation systems, industrial processes, and human activities generally – in short, the process of decision making in relation to the physical environment. The decision-making process is crucial; conditions and technologies shaping our human landscape change over time, and so does our knowledge about the potential for disaster or catastrophe. Recognizing risks and developing measures to reduce or eliminate them requires decision making at various levels of society, from the individual to government institutions. This province, with its rugged, mountainous landscape, tectonic activity, turbulent rivers, and exposure to many climate regimes, has many geophysical hazards. Unfortunately, lack of knowledge or simply ignoring risks has resulted in a great deal of death and destruction. And as the population grows, the risk of future disasters often increases. Land use planning is an absolute necessity.

The second component involves understanding the physical environment, or forces of the earth, which can be placed into three categories:

- *Tectonic hazard* refers to spreading of the sea floor and the consequent collision, or subduction, of continental and oceanic plates. The risk here is from earthquake activity, volcanic eruption, and tsunamis.
- *Gravitational hazard* refers to the discharge down slope of surface material such as rock, earth, snow, and all manner of debris under the force of gravity. Snow avalanches, rock- and mudslides, debris flows, and debris torrents are all examples of gravitational hazards.
- *Climatic hazard* is evidenced in unusual weather conditions resulting from extreme temperatures, lack of moisture (drought), excess moisture (floods), lightning, hail, and violent winds (hurricanes, typhoons, and tornadoes).

Although categorized separately, hazards can overlap. An earthquake (tectonic hazard), for example, may trigger a landslide (gravitational hazard). A volcanic eruption, such as Mount St Helens in 1980, may cause debris torrents, floods, hurricane-force winds, and may even affect the global climate because of the amount of ash and debris dumped into the atmosphere. The eruption of Mount St Helens was a catastrophic event.

NATURAL HAZARDS MODEL

The following model separates the two components, or systems, in order to examine the dynamics of each in assessing risk and reduction of risk from natural hazards (Figure 3.1). The term "system" implies that complex processes are often involved, and the potential for conflict between the two is reflected by the term "versus" in the model. Each system requires further investigation.

The natural/physical system represents all the physical processes, outlined in the three categories above, that affect the landscape of British Columbia. Chapter 2 examined the factors influencing not only landform processes but also weather and climate. The concept of an "extreme geophysical event" recognizes that the forces of nature do have extremes, referred to as earthquakes, typhoons, floods, and so forth. They may be infrequent, but they are not unnatural. The challenge is to understand when, where, and why these physical processes occur.

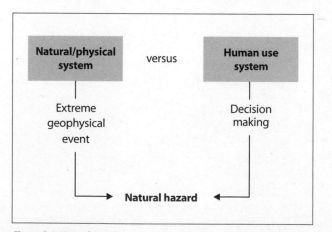

Figure 3.1 Hazards model

Because people or property have to be involved for an event to be categorized as a hazard, the many earthquakes recorded off the west coast of Vancouver Island each year are not a hazard. Similarly, though avalanches occur throughout the mountainous terrain of the province in winter, only a few affect people and consequently fit into the hazard model.

Analyzing the human use system involves analyzing how location decisions are made. Where are homes, commercial structures, institutional buildings, industrial parks, railway lines, highway systems, and recreation facilities located? All have been governed by decision making. Housing subdivisions, for example, are frequently located on flat land as a result of political decisions governing zoning, density, and services available. Economics plays a role in decision making by developers, who in turn involve architects and financial institutions as they decide on the type of buildings and how many will be built in any location.

Decisions about location can be understood from an historical perspective. In British Columbia, river systems served as an effective means of transportation and irrigation, and river flood plains provided fertile soil for agriculture and flat land for housing. Subsequent decisions to build and expand on these initial locations have resulted in whole communities being situated on flood plains. Unfortunately, the decision-making process of the human use system often has little respect for the natural/physical system until the flood occurs.

When the two systems are in conflict during any hazard, it may help to ask which system is to "blame." In other words, when a flood causes loss of life or property damage, is it the fault of the extreme geophysical event or of the human use system in which decisions were made to build in a risky location? The literature on hazards suggests that the natural/physical system is passive (Burton, Kates, and White 1978; Whittow 1980), implying that it is absolutely necessary to gain as much knowledge about our physical environment as possible. In studying river systems, we come to understand the processes for water discharge, how a river creates a flood plain, and where the risk of locating is greatest.

The human use system is active and therefore must take the responsibility for damage and loss of life. Location decisions are usually made from an economic and political perspective, with little consideration of the physical environment. When the flood occurs or an avalanche hits the community, however, there is a tendency to view the extreme geophysical event as the cause of the catastrophe. In fact, these disasters are commonly referred to as "acts of God," and insurance companies normally do not insure against them. In this simplistic way, extreme geophysical events in nature are put into a realm that suggests we know nothing about physical processes and have no responsibility for our location decisions. Governments usually come to the rescue when a catastrophe occurs, compensating individuals for losses. The negative aspect of this process is that compensation is viewed as a right rather than a gift, and responsibility is rarely taken at the individual level. The importance of the hazards model is its ability to help us understand the processes operating in each system, so that decision making will seriously consider the natural/physical system and thus reduce or eliminate risks.

The story of Walhachin, on the South Thompson River between Cache Creek and Kamloops, illustrates the conflict between the two systems. In the early 1900s, American owners of these bench lands, or terraces, sold off ten-acre farms, mainly to British remittance men from England. (The men had commissions in the military and their families had money.) Land in the Thompson River Valley, which was known as Walhachin, was promoted as comparable to that of the Okanagan and of the Wenachee Valley in Washington; by the beginning of the twentieth century both had excellent reputations for fruit farming. Immigration to Walhachin began by 1910, and the immigrant remittance men soon discovered that this was no Okanagan Valley. The area was a desert, and the struggle to obtain water for irrigation was never ending. In those days, it was not possible to pump it from the Thompson River. Over 200 kilometres of flumes were built to carry water from dammed-up streams. Even with these expensive efforts, there was rarely enough water. The soil was alkaline and not conducive to tree fruit agriculture; to make matters worse, the region was subject to frosts severe enough to kill fruit trees. With the advent of the First World War in 1914 nearly all the men from Walhachin signed up with the armed forces. Most never returned, and the community was abandoned by the 1920s (Riis 1973).

The town of Sandon, in the West Kootenay region, portrays a more disastrous story. It was a fairly typical mining town, hastily constructed following the discovery of rich veins of silver in the 1890s. The town was located in the Selkirk Mountains, where peaks can exceed 2,700 metres, six metres of snow in winter is not uncommon, and stream valleys are narrow. When the snow melts in the spring, Carpenter Creek, which winds through the community, becomes a raging torrent. Many of the trees surrounding the community were cut, leaving Sandon vulnerable to avalanches and mudslides.

The hazards were many for Sandon. The town was made of haphazard frame structures built too close together, and when fire broke out, it burned to the ground – the upper town in 1900 and the lower town in 1906. After the fire of 1900, the town was rebuilt, and much of Carpenter Creek was encased with a culvert shaped like an upside-down U, so that a road could be put over the top. Of course, this required constant maintenance so that debris from nearby logging would not build up at the front end of the encasement. Over time, the price of silver declined and so did the mining, and more and more people abandoned the town. Maintenance of the Carpenter Creek encasement was not kept up, and eventually most of the town was destroyed as the creek sought its own course.

Another mining community, Britannia Beach, just north of Vancouver on the way to Squamish, faced its share of natural hazards as well. Unlike Sandon, Britannia Beach was a planned company town. Though the main community was located on the shores of Howe Sound, a number of self-contained mining camps, including Jane Camp, were established in the mountains above the main town. The physical setting had steep, mountainous terrain and heavy rainfall in the winter months, with snow at higher elevations. Geologists had warned of the instability of the slopes above Jane Camp. Unfortunately, no action was taken and a major landslide occurred in 1915, claiming over fifty lives and destroying much of the camp.

The main community of Britannia Beach was unwisely situated on the flood plain of Britannia Creek. The accumulation of debris in the creek as a result of all the logging activity associated with the community and from the mining process caused water to pond behind debris dams. The dams broke with the rains of 1921, and the combined flooding and debris torrent washed away many of the homes located on the flood plain into Howe Sound. With little warning, great destruction occurred and many lives were lost.

These examples are just a few of the many tragic stories that colour the history of British Columbia. Common to each is how the human use system and the natural/physical system combine to create catastrophes. Had people or corporations made decisions based on the physical characteristics of the region, the risks could have been reduced or eliminated. Part of the calamity is that knowledge about the natural/physical system was available, but ignored.

MEASURING EXTREME GEOPHYSICAL EVENTS

There are a number of ways to measure extreme geophysical events, and these become the basis of risk assessment for any location. *Magnitude* measures the intensity of the event and thus describes the potential for damage. A common measure of magnitude is the Richter scale for earthquakes. The waves radiating out from an earthquake are calculated by a logarithmic calibration. A reading of 4.0 or less means some degree of shaking but usually little damage. A reading of 7.0 or higher indicates a catastrophic event. Among the climatic hazards, high winds are measured by various categories of speed. Hurricanes, for example, begin at 121 kilometres per hour and are feared for their destructive power. Floods, another climatic event, are measured by the rising volume of water, usually past a base level. All extreme geophysical events are measured by magnitude in some way.

Frequency refers to how often an event occurs in any location. Lightning-induced forest fires are fairly common in the interior of the province during July and August but much less so in coastal locations, and they are very rare in either location during the winter. Some river systems, such as the Fraser, have a high potential to flood each spring, while others are considerably less likely to do so. As can be observed from these examples, seasons influence the frequency of some extreme geophysical events. This is not the case for tectonic hazards.

Speed of onset describes how rapidly the event occurs and, consequently, how much warning is likely. Earthquakes occur with little or no warning, whereas snowmelt floods may take weeks for the waters to rise and inundate areas.

Duration is the length of time the event lasts. Earthquakes may be over within seconds. When a flood occurs, it may take weeks before water levels recede. Wind storms are often over in hours.

Spatial pattern may be the most important descriptive measure of all because it assesses whether any pattern can be discerned on the landscape where extreme geophysical events occur. In other words, can it be mapped? Spatial patterns remind us that the various measures of assessment should not be isolated. It is of fundamental importance to recognize that most geophysical events have a spatial pattern for which risk can be estimated. Clearly, some locations are at greater risk than others for particular types of natural hazards. Earthquake risk in British Columbia, for example, is much higher in coastal locations than in the interior. If one requires zero risk of earthquake, moving to Saskatoon would accomplish this goal. Of course, this geographical location exposes one to the risk of frigid temperatures in the winter.

Combining all these types of measurement gives some ability to predict extreme geophysical events. We have not refined prediction to the point of being able to state the exact day and time that a particular magnitude of earthquake will occur, although research is being conducted toward that end. The accumulation of magnitude, frequency, and location measures over time gives us statistical data about how often floods, earthquakes, and other events have occurred, and at what intensity, in various locations. Statistical probability then allows us to assess the risk of any location. It is thus essential to the decision-making process when choices of location are made.

RESPONSES TO EXTREME GEOPHYSICAL EVENTS

Individuals are often unaware of the range of hazards and consequently simply react to extreme geophysical events when they occur. Knowing that a hazard can happen is an important beginning to a proper response. Knowing how to respond is the crucial next step. How can we protect our lives and property from natural hazards? Can we prevent disasters from happening?

Sims and Baumann list some of the factors to consider when studying hazards: "The first ... [is] hazard experience; that is, for example, does it make a difference as to what one does or will do about a hurricane if one has been through a hurricane before. A second factor [is] the probability of a hazard's occurrence; that is, does it make a difference in what one will do about hurricanes if the likelihood of their occurring is high. A third factor [is] the extent of economic investment; that is, does how much one has to lose in property, say, crops or buildings, make a difference in how one perceives the threat of a hurricane" (1974, 26).

Sims and Baumann go on to explain that while these three factors appear very reasonable, it assumes that we all act rationally. Unfortunately, this is not the case: "Before the lava cools or the flood waters fully recede, people are back on the volcano's slope and the river's edge, rebuilding" (p. 26). It is very difficult to respond in a rational manner when you are personally involved. Moreover, individuals often have few choices about where they live.

Beyond the individual are the neighbourhood, the community, regional government, provincial government, and federal government. If it is valid to ask how the individual perceives hazards, it is equally valid to ask how these institutions perceive them. Each level must be involved in the decision-making process to reduce and eliminate risks.

Four types of geophysical event are primarily responsible for loss of life and property damage by natural hazard in British Columbia: floods, avalanches, debris flows and torrents, and earthquakes. Each will be discussed in terms of its physical characteristics, the destruction it causes, and the responses by individuals and governments.

Floods
The historical need for and use of water systems in British Columbia has spawned housing, industry, and transportation systems in close proximity. It is therefore unsurprising that floods have caused the greatest amount of property damage in the province (Foster 1987, 48).

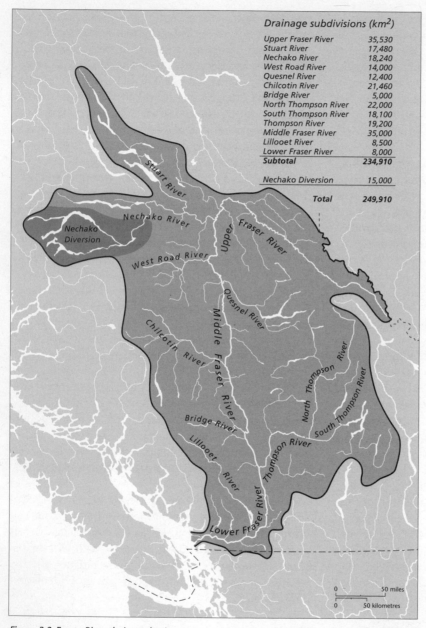

Drainage subdivisions (km²)

Upper Fraser River	35,530
Stuart River	17,480
Nechako River	18,240
West Road River	14,000
Quesnel River	12,400
Chilcotin River	21,460
Bridge River	5,000
North Thompson River	22,000
South Thompson River	18,100
Thompson River	19,200
Middle Fraser River	35,000
Lillooet River	8,500
Lower Fraser River	8,000
Subtotal	**234,910**
Nechako Diversion	15,000
Total	**249,910**

Figure 3.2 Fraser River drainage basin
Source: Modified from Fraser Basin Management Program (1994), 5.

lumbia, Peace, and Skeena, which drain the interior of the province. The volume of water, or discharge, through these systems depends on the size of the drainage basin, the amount of snowpack or accumulation, and spring weather conditions. The Fraser and all its tributaries encompass nearly one-quarter of the land base of British Columbia (Figure 3.2). The system has a large drainage basin and represents a potentially huge volume of water. How much water runs off depends on the amount of snow that has accumulated over the winter and how quickly it melts. The smaller river systems of Vancouver Island, the Queen Charlottes, and the coastal region can produce conditions that result in flash flooding. For these river systems, the rapid rise of water is directly related to intense amounts of rainfall, which can cause the rivers to overflow their banks.

Figure 3.3 uses several graphs to show the range of water discharge for selected rivers in British Columbia. There is a great difference in the volume discharged by the large drainage basin of the Fraser River compared to the small drainage basins of river systems such as the Yakoun River and Cowichan River. Although volume of water is important, however, it is the pattern of discharge that determines flood potential at different times of year. Peak periods of discharge occur in the late spring for both the Fraser and Liard River systems, for exam-

Two types of flooding are associated with the river systems of British Columbia: **snow-melt flooding** and **flash flooding.** Snow-melt flooding, often referred to as spring flooding, occurs on river systems such as the Fraser, Co- ple, producing snow-melt flood conditions in these interior drainage basins. Because large river systems drain so much of the interior of British Columbia, the potential for flooding during spring run-off affects most of

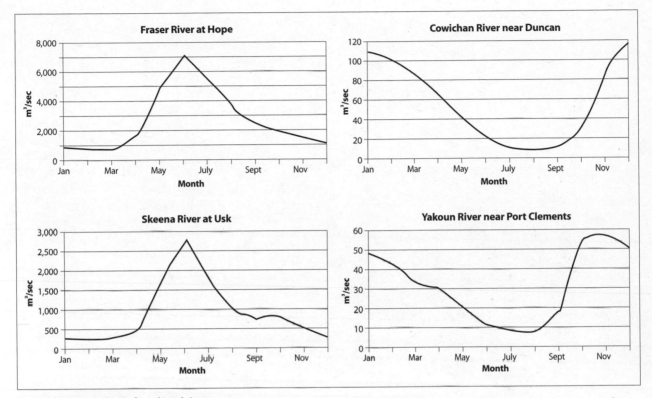

Figure 3.3 River discharge for selected rivers
Source: Data from Environment Canada (1991), 243, 350, 892, 1077.

the province. As the graphs for the Yakoun River in the Queen Charlottes and the Cowichan River on Vancouver Island show, flash flooding usually happens in the winter months, which is when the greatest rates of discharge occur.

Documenting flood events is essential to forming a proper response. The volume of water discharged in any given year is measured in relation to benchmarks for any community. The flood plains affecting most flood-prone communities in British Columbia and the rest of Canada have been mapped. The extent of flooding depends on the level to which the river ultimately rises. The high water "contours" can be mapped to show the areas that will be inundated. Accumulating information about both magnitude and frequency of flooding allows for the production of a map of statistical probability, or what is referred to in Figure 3.4 as a ten-year, fifty-year, or hundred-year flood plain. The fifty-year flood plain

in the example implies the potential of the water rising to that level, or contour, once in fifty years. It is interesting to note that the "Great Flood" in Manitoba in 1997 reached the 500 year flood plain level. In terms of decision making, one can see from Figure 3.4 that the higher in elevation we locate from the river's edge the lower the risk.

The 1948 flood in the Fraser Valley was catastrophic: "This flooding of 22,260 hectares (55,000 acres) left 200 families homeless, caused 10 fatalities and washed out 82 bridges. Together the federal and provincial governments provided approximately 20 million dollars to rehabilitate flood victims and repair and strengthen the dyking system" (Foster 1987, 49). Figure 3.5 gives some idea of the extent of flooding in the Lower Fraser Valley alone. Flooding also occurred throughout the Fraser River drainage system in 1948.

The unusual condition most responsible for the 1948 flood was the prolonged spring season. Snow and cold

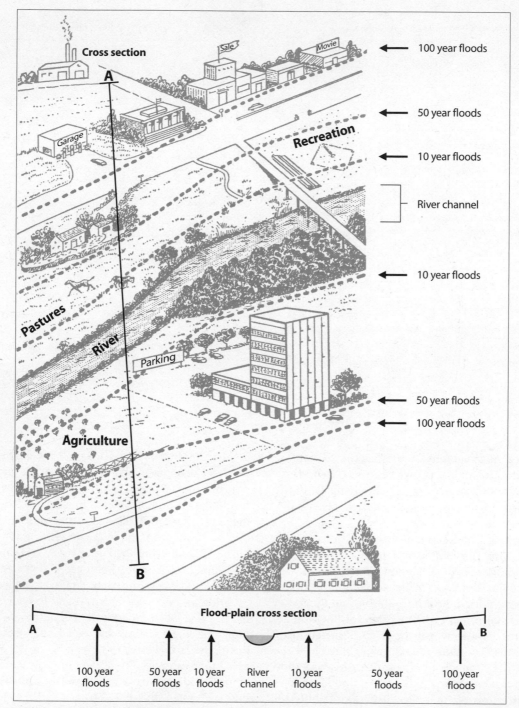

Figure 3.4 Flood plains and land uses
Source: Modified from Tufty (1969), 231, with permission.

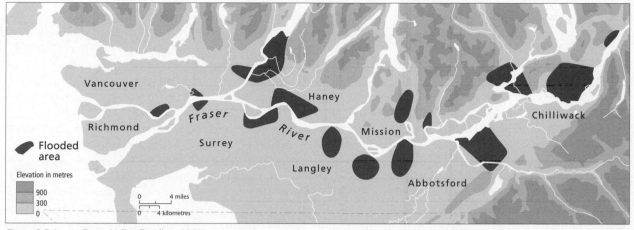

Figure 3.5 Lower Fraser Valley flooding, 1948

weather continued well into April and early May. Then, as many farmers commented, "summer came." The hot weather melted the snow rapidly and the flooding began. Figure 3.6 shows the Fraser River discharge at Hope, measured in cubic metres per second, for 1948 as well as for several other high water years. The 1948 flood caused a great deal of damage and loss of life, but considerably higher flood waters were recorded in 1894. There is little record of the extent of the damage in the earlier flood, however, because few people lived in the Lower Fraser Valley. This information contains two significant facts. First, a flood of the magnitude of 1894 has a statistical probability of recurring at Mission once every 140 years (Fraser Basin Management Program 1994, 27). Second, as the number of people on the flood plain increases, so does the risk. The population of the Lower Fraser Valley has exploded since 1948, and many more homes have been built in the areas most susceptible to flooding. The question becomes: Are we prepared for flood levels that would match or exceed those of 1894? Statistical probability indicates that a flood of that level or higher will recur.

People sometimes get romantic notions about floods, in which rowing to work or school is regarded as an adventure. This myth needs to be dispelled. One of the greatest concerns related to flooding is the spread of diseases. Flood waters carry many pathogens, and each house must be inspected by a health officer after the waters have receded. There is an enormous amount of

damage to home furnishings, and the fine silt that flood water deposits usually renders any motor useless. Moreover, no price tag can be put on pets, photo albums, and many other items lost or destroyed in a flood. In the 1948 flood, the Fraser Valley was cut off from the

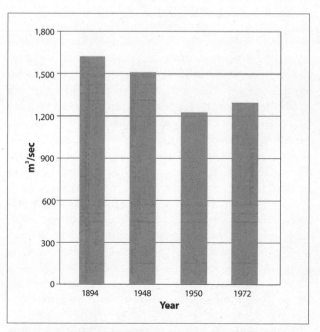

Figure 3.6 Fraser River discharge measured at Hope, 1894-1972
Sources: Data from Fraser Basin Management Program (1994), 27; Environment Canada (1991), 351.

rest of the province for most of June because the waters washed away the railways and roads. The army had to be called in to prevent looting, and the price of staple food products such as bread and milk had to be fixed because of shortages. Floods have a huge cost, and the risk needs to be reduced or avoided.

The most common way to guard against flooding is to build dyking systems. In fact, too much reliance has often been placed on this form of technology. Dykes are subject to erosion, and to remain viable they must be continuously maintained. This is especially true in light of the fact that river drainage systems can change course and, in some cases, dykes are part of the change. As communities in the drainage system grow in population, range land is overgrazed, forests are clear cut, and more pavement is added. The cumulative impact of all these human activities is an increased rate of run-off in the spring, bringing greater pressures on the dyking system.

As more people move into the Lower Fraser Valley and interior communities, dyking systems must be expanded. As more and more kilometres of dykes are built, more and more of the flood waters are contained, resulting in ever greater volumes of water and pressure on downstream dykes. Rivers carry enormous amounts of silt – the Fraser, in particular, has a "muddy" reputation – that are either deposited on the land in a flood or build up the delta at its mouth. When the dyking system is increased, more sediment stays in stream and builds up the bottom of the river floor, thus raising the level of the water and again bringing greater pressure on the downstream dykes. There is no such thing as a permanent dyke. They are subject to erosion and must constantly be reinforced and maintained to reflect the changing dynamics of river systems.

Figure 3.7 shows the variety of potential corrective and preventive measures to reduce or eliminate the risk from flooding. Corrective options are the many short- and long-term responses for people who are affected by flooding. Quite simply, how can people who live in a flood-prone area protect themselves and their property? Immediate responses range from warning systems to sand bagging and evacuation. More long-range approaches involve various means of controlling water, including dams, dykes, or channel improvements such as dredging, and also the recognition that human activi-

ties in the watershed, such as clearcutting, affect water run-off and flooding. Owners of individual homes in the flood plain can invest in flood proofing, thus keeping flood waters from entering basements or ground floors. Urban redevelopment may result in flood proofing of entire high risk sections of the community, or even building on pillars.

Preventive measures keep human activity out of flood plains. Many of these options require various government institutions to establish regulations and barriers. Zoning can be effective in keeping high risk areas as pasture lands, parks, or for uses that exclude housing. Regulation can achieve the same end by not allowing subdivision of flood plains. Building codes and health regulations may also stipulate rules that prevent construction.

There are other options to prevent building in flood plains. A rather subtle form of prevention, for example, is the posting of warning signs. Flood insurance and tax adjustments can be scaled to risk as a preventive measure; the higher the risk, the higher the insurance premium, or the taxes, will be. Development policies may discriminate against construction within a flood plain, or if structures are to be built, a developer must build up the land to an elevation above the 200 year flood level.

It is an interesting exercise to assess the theoretical range of options any community has with respect to reducing and preventing floods and match these options to decisions that have already been made. Public information and education, which is a component of both correction and prevention, is essential and arguably the most important element in all levels of decision making. People at all levels – individual home owners, investors, politicians, and others – must be aware of the options if they are to be implemented.

Figure 3.7 was designed specifically to examine the range of options for flood conditions. As other extreme geophysical events are discussed, however, keep this framework in mind. Reducing or eliminating risk from avalanches, debris flows, and earthquakes has many similar considerations.

Avalanches

Of all the hazards, avalanches have resulted in the greatest loss of life in the province, primarily because railways

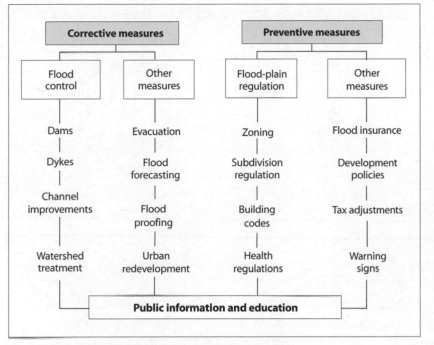

Figure 3.7 Corrective and preventive measures for flood hazards
Sources: Modified from Sewell (1965); Whittow (1980).

have been built through steep mountain passes that have a high incidence of avalanche. While many transportation systems and a few communities are still threatened, the greatest danger today is in the ski industry.

Colin Fraser's book *The Avalanche Enigma* (1966, 79) outlines four interrelated factors that produce avalanches:

1 *Type of snow* varies from hard pellets to the typical "star" shape. Pellets do not bind well and create unstable snow conditions.
2 The *rate of snowfall* is crucial, especially if it is accumulating at 2.5 cm per hour or more. Table 3.1 demonstrates how the accumulation rate of new snow serves as a guide to the stability of conditions.
3 *Terrain,* or slope, is an important factor in avalanche conditions. Slopes greater than 60° are usually too steep for large accumulations of snow but experience frequent, small avalanches instead. Slopes below 30° are usually not steep enough to pose any threat of an avalanche. Snow can build up on slopes between 30°

and 60°, and their terrain is steep enough to be unstable.
4 *Change in temperature* of 6°C per hour or more, either warmer or colder, can create the unstable conditions that result in avalanche.

Avalanche risk should be considered within the context of the slope of the terrain, the rate of snowfall and type of snow that has previously accumulated, and rapid temperature change. Figure 3.8 classifies and illustrates the various forms of avalanche conditions. It should be noted that a dry, airborne-powder avalanche can reach speeds of up to 325 kilometres per hour and destroy practically everything in its path.

Mapping the spatial pattern of our highest risk areas is essential. The road network in British Columbia was mapped for risk after a tragic avalanche destroyed the North Route Café west of Terrace, on 22 January 1974, claiming seven lives. The provincial government then commissioned a task force to assess all highways in British Columbia and recommend measures to reduce the risk from this extreme geophysical event (British Columbia, Ministry of Highways 1974).

Understanding the physical processes involved and accumulating statistics are the first steps toward corrective and preventive measures. Some of the measures used in British Columbia's road and rail systems are: erecting

Table 3.1

New snowfall and avalanche risk

Snowfall (cm)	Risk
15 to 20	Slight avalanche risk
30 to 60	Small avalanche risk, danger for skiers
60 to 90	Moderate avalanche risk, blocked roads and rails
90 to 120	High avalanche risk, structural damage possible
Over 120	Major disaster

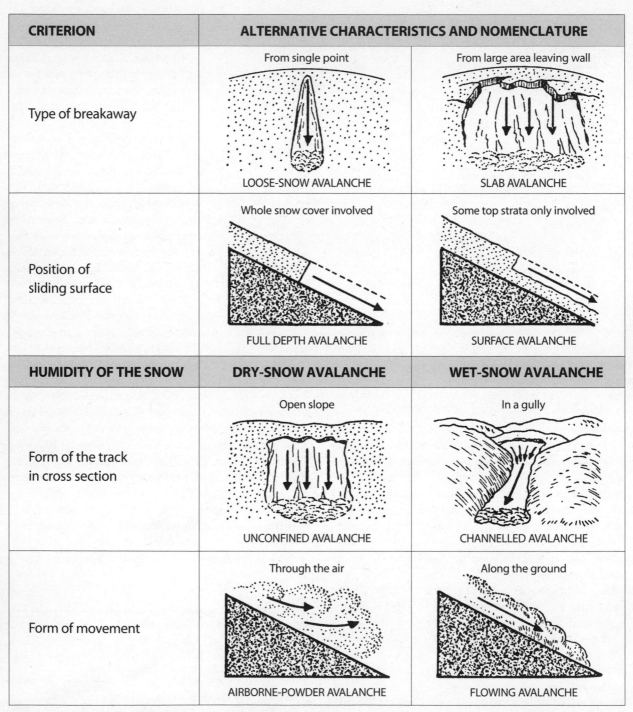

CRITERION	ALTERNATIVE CHARACTERISTICS AND NOMENCLATURE	
Type of breakaway	From single point LOOSE-SNOW AVALANCHE	From large area leaving wall SLAB AVALANCHE
Position of sliding surface	Whole snow cover involved FULL DEPTH AVALANCHE	Some top strata only involved SURFACE AVALANCHE
HUMIDITY OF THE SNOW	**DRY-SNOW AVALANCHE**	**WET-SNOW AVALANCHE**
Form of the track in cross section	Open slope UNCONFINED AVALANCHE	In a gully CHANNELLED AVALANCHE
Form of movement	Through the air AIRBORNE-POWDER AVALANCHE	Along the ground FLOWING AVALANCHE

Figure 3.8 General types of avalanche
Source: Fraser (1966), 52.

roadside warning signs; making fairly extensive use of artillery fire and dynamite to prevent the build-up of large snowpacks; and constructing tunnels, snow fences, snow sheds over roads and rail lines, and berms or catchment basins, which are built in a semi-circle at the bottom of an avalanche path to prevent it from continuing on to a highway. Terrain management is another measure, and involves planting trees to stabilize snow conditions and/or building deflecting barriers. Where buildings are involved, a set of regulatory mechanisms including any or all of the following preventive measures is useful: building codes, restrictive zoning in high risk areas, tax incentives, and prohibitive insurance programs.

As mentioned earlier, the highest avalanche risk today is for skiers and those involved in outdoor snow activities generally. With the growing popularity of backcountry skiing, heli-skiing, and snowmobiling into untracked and often untested snow conditions, the risks are high. Even on managed ski slopes, adventurists who ski out of the prescribed boundaries run a serious risk. The managed, commercial ski slopes employ a host of mitigating measures within their boundaries: warning signs, trained avalanche rescue teams, artillery, closure of high risk runs, and so forth. For the heli-ski industry and other commercial and recreational organizations using the backcountry, the main survival techniques of avalanche preparedness are information, education, and electronic locators to be worn by each participant. Given the significant loss of life during these recreational activities, considerably more attention is required.

Debris Flows and Debris Torrents

Debris flows and torrents are often mistakenly referred to as landslides, mudslides, or lahars (mudslides from volcanic eruption). This is not surprising, as the terms *debris flows* and *debris torrents* are only recent designations for the types of gravitational event in which water, mud, rocks, trees, and all nature of debris flow either slowly or very quickly down a stream bed. This entire mass

can have a very rapid speed of onset, providing little or no warning. The magnitude depends on the slope and topography of the creek bed and the amount of debris carried down the channel (see VanDine and Lister 1983).

Coastal locations are the most susceptible to this hazard because the many streams in these areas usually have their beginning on a steep mountain slope and exit into the ocean (Figure 3.9). These streams have extremely variable discharges that depend mainly on the amount of rainfall, which can be intense and last for weeks. The accumulation of both organic and inorganic matter occurs in mountain stream beds through the normal erosion process and through human activities. The coastal areas have a long history of poor logging practices that have left behind many old roads and bridges collapsed into the waterways, allowing even more debris to build up in stream channels. Small dams can therefore form, with consequent ponding of water. With intense rainfall these dams fail, bringing a huge volume of water mixed with debris down the stream gullies, scouring out the stream bed in the process. The debris is carried downstream by gravity and is deposited on the more gentle slopes – the slopes where people have built their communities and transportation systems.

Nine lives were lost in the Squamish Highway's M Creek disaster in 1981. Here, a debris torrent destroyed the bridge crossing, and unaware travellers plunged into the chasm. Two years later in the same region, a debris torrent at Lions Bay claimed two lives as the debris jumped

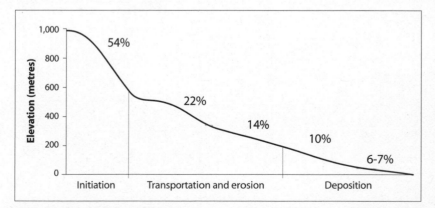

Figure 3.9 Typical debris torrent slope
Source: Modified from Thurber (1983), 8.

Table 3.2

Hazards related to streams on the Squamish Highway

Recorded natural events, 1906-83

Streams	10 September 1906	28 October 1921	16 December 1931	29 December 1933	Fall 1960	Early 1960s	12 October 1962	22 December 1963	18 September 1969	1972	2 November 1972	7 November 1972	15 December 1972	23 May 1973	1976-7	December 1979	28 October 1981	31 October 1981	4 December 1981	6 October 1982	3 December 1982	11 February 1983
Disbrow			U													F						
Unnamed #1																						
Sclufield																						
Montizambert																						
Strid																						
Charles									D		D	D							D			
Turpin																						D
Newman									D										D	F		D
Lone Tree																						
Rundle																						
Harvey									U				F	F					F			
Alberta																					D	D
Magnesia					F		D										D					
M Creek																	D					
Loggers																						
Deeks										F												
Brunswick Pt.																						
Bertram																						
Kallanni							F										F					
Unnamed #7																						
Unnamed #8																						
Furry																F		F				
Unnamed #9																						
Daisy							F															
Thistle																						
Britannia	F	F		F				F														

Note: D = Debris flow; F = Flood; U = Uncertain origin.
Source: Thurber Engineering (1983), 4.

the stream bed and carried away a trailer in which the occupants were sleeping (Needham and Smith 1983). On Vancouver Island, the community of Port Alice has had numerous debris torrents, causing plenty of damage over the years. The risk is high for many communities, and transportation systems and requires both corrective and preventive measures. Table 3.2 reflects the number of natural hazards associated with the streams on the Squamish Highway between 1906 and 1983.

Dyking systems, deflection berms, catchment basins, and stream management have all been deployed as corrective measures to combat this hazard. Along the Squamish Highway, much longer and higher spanning bridges have been installed to allow the debris to flow underneath. Zoning regulations, warning signs, and even insurance programs could act as preventive measures to bring about alternative and safer location decisions.

Earthquakes

The coastal area of British Columbia is an extremely complex and active earthquake zone. From northern California to just north of Vancouver Island, the Juan de Fuca Plate is subducting under the North American Plate. To the north of the Juan de Fuca Plate and in close proximity to the Queen Charlotte Islands, the Pacific Plate slips past the North American Plate, moving in a northerly direction until it reaches Alaska, where it once again subducts under the North American Plate. (See Chapter 2 for an explanation of plate tectonics.) These movements produce earthquakes and in some cases of high magnitude. The whole coast of British Columbia is one of the highest risk earthquake zones in Canada because of these tectonic processes.

The magnitude of earthquakes is usually measured on the Richter scale, on which each number represents a tenfold increase in the intensity of shaking. Table 3.3 matches the Richter scale with potential structural damage and gives some examples. The structural damage suggested here should be accepted with some caution, as the effects of an earthquake depend on many variables, including slope of the land, depth of soil, type of soil, nature of bedrock, building materials, height and design of building, and so on. Brick and stone buildings are less flexible than frame structures, for example, and highrise complexes may not be able to withstand the horizontal motion of an earthquake. Structures located on landfill may be subject to the soil liquefaction and lose their support. This information can be mapped for any community, clearly showing areas of higher and lower risk. (For the Vancouver area see Macdonald and O'Keefe 1996.)

Earthquakes occur frequently off the west coast of Vancouver Island and the Queen Charlottes but fortunately most of these are of a small magnitude and pose little risk. Nevertheless, the west coast has a record of earthquakes of catastrophic magnitude and the potential for a megathrust earthquake of the Juan de Fuca Plate is of great concern (Heaton and Hartzell 1987; Charlwood and Atkinson 1983; Koppel 1989; Mayse 1992). The oceanic plate builds up tremendous pressure to subduct under the continental plate and a megathrust is the rapid release of this pressure, resulting in an earthquake that could exceed 9 on the Richter scale. Evidence through First Nations' oral history, Japanese records of tsunamis, and coastal subsidence records a megathrust earthquake of approximately 9.0 on 26 January 1700 (Internet, Natural Hazards – Earthquakes, 2). The Geological Survey of Canada suggests "that huge subduction earthquakes have struck this coast every 300-800 years" (Internet, Earthquakes in Western Canada, 2).

Table 3.3

Richter scale showing potential structural damage

Potential structural damage	Scale	Examples
Very little if any	1.0	Too numerous to list
	2.0	
	3.0	
	4.0	
	5.0	
Some structures	6.0	
	6.5	Seattle, 1975
Many structures	7.0	Vancouver Island, 1918
	7.3	Powell River/Comox, 1946
Most structures	8.0	Queen Charlotte Islands, 1949
	8.3	San Francisco, 1906
	8.5	Alaska, 1964
All structures	9.0	

Other significant earthquakes include one of magnitude 8.1 that occurred on the Queen Charlotte Islands in 1949 and another calculated at 7.3 that was recorded in the Powell River/Comox area in 1946, causing property damage and one fatality. In June 1997, a 4.3 magnitude earthquake hit the Strait of Georgia region.

Seismic activity also produces related hazards. Landslides, slumping, and destruction of telephone lines, hydro lines, roads, rails, and pipelines may result from earthquake activity and bring their own damage and threat to life. The great earthquake of 1964 in Alaska caused enormous destruction to the port of Valdez and the city of Anchorage, and the ensuing tsunami damaged port communities all the way to California. One of the hardest hit by the tsunami was Port Alberni on Vancouver Island, where millions of dollars in damage occurred.

The frequency, magnitude, and spatial pattern of earthquakes in the coastal region of British Columbia is reason for justifiable concern for the people living there. The 1980s saw considerable effort go into the public information and education side of preparedness, particularly in the public school system, where earthquake drills are common practice. Earthquake preparedness is essential for when the "big one" occurs. The Provincial Emergency Program (PEP 1992) accomplishes this through simulation exercises. Another important measure to reduce the risk is the regulatory mechanism of building codes, which are constantly upgrading building materials and structural design to counteract seismic activity (see CMHC 1989). This often requires retrofitting older buildings such as schools and other institutional, commercial, and residential sites, as well as structures such as dams.

SUMMARY

The study of natural hazards is an interesting and important area of research that combines the skills of both the physical and the human geographer. Far more information is required to understand the physical processes that produce extreme geophysical events and to produce higher degrees of accuracy in predicting where, when, and at what magnitude events will occur. One of the most important elements in risk reduction is public information and education to ensure that informed location decisions are made from the individual to the governmental level. As the population of British Columbia expands so too does the risk. Through hazards research there is considerable hope that many risks can be avoided.

REFERENCES

British Columbia, Ministry of Highways. 1974. *Avalanche Task Force.* Victoria: The Ministry.

Burton, I., R.W. Kates, and G.F. White. 1978. *The Environment as Hazard.* New York: Oxford.

Central Mortgage and Housing Corporation (CMHC). 1989. *Assessment of Earthquake Effects on Residential Buildings and Services in the Greater Vancouver Area.* Ottawa: The Corporation.

Charlwood, R.G., and G.M. Atkinson. 1983. "Earthquake Hazards in British Columbia." *The BC Professional Engineer* (December): 13-16.

Environment Canada. 1991. *Historical Streamflow Summary British Columbia.* Ottawa: Inland Water Directorate, Water Resources Branch, Water Survey of Canada.

Foster, H.D. 1987. "Landforms and Natural Hazards." In *British Columbia: Its Resources and People,* ed. C.N. Forward, 43-63. Western Geographical Series vol. 22. Victoria: University of Victoria.

Fraser, C. 1966. *The Avalanche Enigma.* London: Murray.

Fraser Basin Management Program. 1994. *Review of the Fraser River Flood Control Program.* Vancouver: Fraser Basin Management Board.

Heaton, T.H., and S.H. Hartzell. 1987. "Earthquake Hazards on the Cascadia Subduction Zone." *Science* 236: 162-8.

Koppel, T. 1989. "Earthquake: A Major Quake Is Overdue on the West Coast." *Canadian Geographic* 109 (4): 46-55.

Macdonald, I., and B. O'Keefe. 1996. *Earthquake: Your Chances, Your Options, Your Future.* North Vancouver: Cavendish.

Mayse, S. 1992. *Earthquake: Surviving the Big One.* Edmonton: Lone Pine.

Needham, P., and D. Smith. 1983. "Mudslide Engulfs Trailer: Two Dead at Lions Bay." *Vancouver Sun*, 11 February, A1.

Provincial Emergency Program (PEP). 1992. *British Columbia Earthquake Response Plan*. Victoria: PEP.

Riis, N. 1973. "The Walhachin Myth: A Study of Settlement Abandonment." *BC Studies* no. 17 (Spring): 3-25.

Sewell, W.R.D. 1965. *Water Management and Floods in the Fraser River Basin*. Research Paper no. 100. Chicago: University of Chicago.

Sims, J.H., and D.D. Baumann. 1974. "Human Response to the Hurricane." In *Natural Hazards: Local, National, Global*, ed. G.F. White, 25-30. New York: Oxford.

Thurber Engineering. 1983. *Debris Torrents and Flooding Hazards, Highway 99, Howe Sound*. Vancouver: Thurber Engineering.

Tufty, B., 1969. *1001 Questions Answered about Natural Land Disasters*. New York: Dodd, Mead.

VanDine, D.F., and D.R. Lister. 1983. "Debris Torrents – A New Natural Hazard?" *The BC Professional Engineer* (December): 911

Whittow, J. 1980. *Disasters: The Anatomy of Environmental Hazards*. Markham: Penguin.

INTERNET

Cascadia Megathrust Figures, www.pgc.nrcan.gc.ca/geodyn/megafig.htm#fig2

Climate Normals, 1961-1990, www.cmc.ec.gc.ca/climate/normals/E_BC_WMO.HTM

Earthquake Processes: Cascadia Subduction Zone, www.pgc.nrcan.gc.ca/geodyn/cascadia.htm

Earthquakes in British Columbia – How can geology be used to minimize risk? natural.gov.bc.ca/geosmin/mapinv/surfical/quake/eq6.htm

Earthquakes in Southwest British Columbia, www.seismo.emr.ca/bc.html

Earthquakes in Western Canada, www.pgc.nrcan.gc.ca/seismo/eqinfo/eq-westcan.htm

The Magnitude 8.1 Queen Charlotte Island Earthquake of 1949, www.pgc.nrcan.gc.ca/seismo/hist/1949.htm

Natural Hazards – Earthquakes, nais.ccm.emr.ca/~kramers/hazardnet/b_earthquake/6eqmapw.htm

Natural Hazards – Landslides, nais.ccm.emr.ca/~kramers/hazardnet/i_landslide/lslmapw.htm

Natural Hazards – Tsunami, nais.ccm.emr.ca/~kramers/hazardnet/d_tsunami/tsuintro.htm

Modifying the Landscape: The Arrival of Europeans

4

Aboriginal peoples were the first to discover and settle throughout North America. It is nevertheless useful to understand early European values and settlement patterns prior to discussing First Nations, primarily because the discussion in Chapter 5 focuses on the contrasting value systems of First Nations and European peoples and Native reactions to changes imposed by the British. This chapter therefore focuses on the eighteenth- and nineteenth-century struggles of colonization by the British (first against the Spanish and later against the Americans), and the historical processes that have modified the landscape of British Columbia.

When the British claimed the territory we now know as British Columbia, they were initially motivated by desire to control the valuable fur resources of the northwest coast. The fur resources of the interior of the province, an extension of the fur trade initiated from eastern Canada, also played a role in colonial control. Until the 1850s, Europeans had little impact on the land: fur trade forts were erected, few resources other than furs were exploited, basic transportation systems were developed, and British sovereignty was established.

The discovery of gold by the mid-1800s escalated the settlement process by non-Natives and was responsible for the establishment of the present provincial boundaries. The attitude of those arriving was fundamental to rapid changes to the landscape. The search for gold caused much destruction to streams, salmon habitat, and forests and trespassed on First Nations territory. Permanent and temporary communities were built, transportation systems were developed, and predominantly British institutions evolved. By 1871, the gold rush was over and British Columbia was facing tough economic times. It was then, however, that British Columbia joined Confederation, and with the union came the promise of a national railway.

EARLY EUROPEAN CLAIMS AND INFLUENCE

Europeans "discovered" America, as the textbooks say, and the reference is usually to Columbus landing in the Bahamas and then in Cuba. This was not the first discovery of North America by Europeans, although it may have been the most important in terms of its impact.

Through the technologies of navigation the Europeans discovered new lands, and through superior military firepower they expanded their empires. "Discovery" implied an attitude of ownership that translated into the control and colonization of the new lands, with the expectation of wealth from resources. This quest for wealth expanded mercantilism and eventually became global in scale. Territories were defined and redefined as European nations battled each other in claiming colonies. Hand in hand with this economic, political, and military approach to exploration came other views. Christian missionaries, who were not particularly tolerant of other religions, or of animism, were there to convert the "heathens." European views of settling in a new land included rules regarding civilization, superiority, ownership of land, and use of the land.

The voyages of Columbus, and others who followed, were given credit for the European discovery of North and South America, but their main quest was to find a way to the riches of the Orient. Sailing between Europe and Asia could only be accomplished in one of two directions – either south around Africa and past India, or across the Atlantic and around South America. In either case, the location of what is now British Columbia was very remote and isolated, which partly explains the rather late interest by Europeans.

European interest in the northwest coast of North America arose essentially out of two complementary motivations in the late 1700s. The first, and more complex situation, involved exploiting the resources of colonial territories in exchange for the valued silks, teas, and porcelain of China. The second quest was for the Northwest Passage and the prize money for its discovery.

By the 1700s, the Spanish had claimed most of South America, Central America, and a good portion of present-day southwestern United States, and they had established forts in what is now known as California. The Spanish forts became the bases from which the resources of the colonies, primarily silver, were shipped to China. The Chinese had little use for European goods but were most interested in silver as an exchange item. The continual challenge for Europeans, who were reluctant to part with silver, was to seek other resources that the Chinese would find favourable. The discovery by the late 1700s that sea otter pelts from the Pacific northwest coast were a much desired commodity in China enveloped the northwest coast of North America in a fierce struggle over expansion of colonial territory.

The concept of **territory** combines geographic space with political control, or sovereignty, over that space. Laying claim to foreign territories was accomplished by establishing forts and by documenting, through a ship's log book, the voyages of the early explorers. The Spanish, with their silver trade out of South and Central America, assumed that they had control over the whole west coast of North America and became concerned about the Russians, who were building forts and laying claim to Alaska. The northernmost fort for the Spanish, however, was San Francisco. Spanish vessels had sailed up to the Queen Charlottes and stopped at Nootka Sound on the west side of Vancouver Island, where some initial trading with Aboriginal peoples occurred in 1774. Captain Cook arrived at Nootka Sound in 1778 in his search for the much sought-after Northwest Passage, hoping to gain the £20,000 reward on offer, and in doing so laid claim to this "undiscovered" territory and its valuable fur resources in the name of Britain.

While the Spanish were very secretive about their voyages, Cook's publication of his journals implied that the British had been there first and thus claimed the territory. It was not long before other British vessels came to this region in search of the valuable sea otter pelts, and the conflict over sovereignty ignited.

The Nootka Sound Incident of 1789, in which a British officer was taken at sword point and three British ships were confiscated by the Spanish, brought to a head the rival claims. The skirmish may have been local, but the ramifications were global. The resolution to this territorial conflict required six years of protracted negotiations in Europe. Between 1789 and 1795, Captain Vancouver for the British and Captains Malaspina, Galiano, and Valdés for the Spanish made a number of voyages up and down the coast, surveying the landscape and recording their impressions of the Aboriginal peoples. The many Spanish place names in the area – Texada Island, Galiano Island, and Juan de Fuca Strait are examples from over a hundred names – remain as a legacy of this period (Archer 1981; Gibson 1981).

A map from the journal of Captain Vancouver shows a fairly detailed outline of the northwest coast of North America (Figure 4.1). He was impressed with the rugged topography and highly indented coastline. Interestingly, only the mouth and a portion of the Columbia River, and none of the other large rivers in British Columbia, are shown on the map. British and Spanish place names are evident as well.

As a result of the negotiations Spain ceded the northwest Pacific region to Britain. Transfer to British territorial control, however, did not diminish the intense competition for valuable sea otter pelts. The Russians had the advantage by being first into the region and establishing coastal fur trade forts across Alaska and down the Panhandle. The Americans, or Yankee traders, also aggressively pursued this trade and in doing so threatened British control of the territory. The near extinction of the sea otter by the early 1800s reduced the intensity of territorial conflict.

The overland fur trade, mainly for beaver pelts, began to make its way into British Columbia with Sir Alexander Mackenzie's trek through the Rockies to the coast at Bella Coola in 1793. Like the sea otter trade, the overland fur trade provoked competition, but this time the main rivals were competing commercial empires rather than competing countries. The Hudson's Bay Company and the North West Company struggled for supremacy over the fur-bearing animal trade. The North West Company, headquartered in Montreal, was by far the more aggressive trader, as it established a greater number of forts throughout the Prairies and British Columbia.

As employees of the North West Company, the early explorers of British Columbia – Mackenzie, Simon Fraser, and David Thompson – surveyed the potential for fur trade forts and made contact with indigenous peoples. Company forts were built and the mainland territory of present-day British Columbia was divided into two regions: the Columbia District and New Caledonia (Figure 4.2). The Columbia River could be navigated, whereas the Fraser was far too turbulent, and an overland trail was constructed along its route. By 1821, the bitter rivalry ended with the buyout of the North West Company and merger of the two companies; consequently, the Hudson's Bay Company gained the forts and a monopoly on the fur trade (Harris 1992).

The fur trade across Canada had been established for over 200 years before it reached British Columbia, and with the merger the Hudson's Bay Company was in

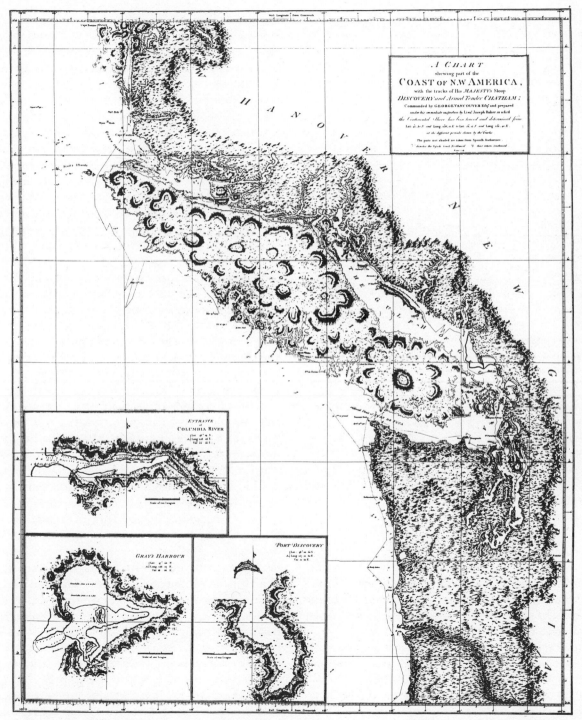

Figure 4.1 Map from the journal of Capt. Vancouver, 1798.

Figure 4.2 Fur trade forts and routes, 1805-46
Source: Modified from Harris (1997), 37.

control of a continent-wide industry, albeit an industry in decline. Furs had been overharvested throughout the country, and fur traders ran into additional problems in British Columbia.

Birch bark canoes and extensive river systems were the main means of transporting furs to the Hudson's Bay forts and to Montreal. Many of the rivers in British Columbia were too turbulent to navigate, however, resulting in many portages. Moreover, for much of the province, there were no birch trees to make canoes. Overland trails and horses became the new, more dependable means of transportation. Some forts, particularly in the southern interior, became horse ranches to answer the need, but this system added extra costs to the transportation of furs.

Fur traders also encountered difficulties with respect to the sheer number of First Nations. Common practice for fur trade companies on the Prairies and elsewhere in Canada was to negotiate with tribal chiefs who, in turn, organized fur trapping. For the most part the territorial boundaries of these Native groups were fairly large, with few chiefs. In British Columbia, this system led to frustration for the trading companies. The geographic boundaries of the region included approximately thirty or forty major ethnic groups, and these could be divided further into hundreds of clans, or houses, each with its own territorial boundaries and hereditary chief (Muckle 1998, 7). Fur brigades made up of thirty or more men, usually French Canadians, replaced Native trappers and were symbolic of this frustration. The brigades would enter an area, do all the trapping, and transport the furs east. The system procured furs, but at a higher cost, and it also trespassed on Aboriginal lands.

As the fur trade struggled to remain economically viable, its other role, of asserting British sovereignty over the land, became important. The head of each fort, the fort factor, was responsible for the British institutions of marriage, law, and order, and was obligated to establish British rule over the land. The question of who had effective control over the land was still problematic, for the territory now referred to as British Columbia "was a bilateral free-trade zone, not a colony" (Harris 1997, 37).

An agreement between Britain and the United States in 1818 made the forty-ninth parallel the border between Canada and the United States, but it included the territory only as far west as the Rocky Mountains. Left still to settle was Oregon Territory – the land from the end of the Alaskan Panhandle (at 54° 40'), east to the divide in the Rockies, and south to the Hudson's Bay Company fur trade forts on the Columbia River (Figure 4.2). American westward and northward expansion, promoted in part by a philosophy of **manifest destiny,** led to a claim for the whole Oregon Territory (Figure 4.3).

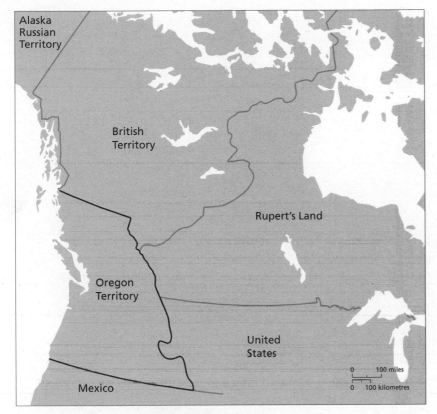

Figure 4.3 American claims to western North America, 1825
Source: Modified from Gentilcore (1993), Plate 21.

but the prospect of finding more gold in the territory was stimulated. A great deal more was found throughout the province, and it was gold, ultimately, that shaped the boundaries of British Columbia and changed the geography of this region permanently.

THE GOLD RUSH: OPENING UP BRITISH COLUMBIA

The Fraser River and Cariboo gold rush, which began in 1858, should be viewed in the context of an earlier gold rush in the San Francisco area in 1849. The San Francisco rush saw over 100,000 miners migrate into the Bay area and resulted in urbanization, roads, and a fair amount of chaos in the competition to find the elusive metal. As gold discoveries waned in northern California, the possibility of becoming rich from new discoveries on the Fraser attracted a great deal of attention. The rush was on as some 25,000 to 30,000 miners pushed into the lower reaches of the Fraser panning for gold. Prior to the gold rush, there may have been a few hundred non-Natives in this region. The arrival of thousands of miners spurred the "opening up" of the region, with significant impact on the landscape.

Neither British Columbia nor the separate colony of Vancouver Island was prepared for this onslaught of miners, many of whom were American. The arrival of so many Americans refuelled concerns about manifest destiny and prospects of border violation. In response, the new colony of British Columbia, which included only the mainland, was created in 1858 (Figure 4.4). Chief Factor James Douglas of the Hudson's Bay Company had been appointed governor of the colony of Vancouver Island in 1851 by the British government. He was concerned for British sovereignty, as well as over the real possibility of war with Aboriginal peoples, whose lands

This conflict was brought to a climax with the US presidential election slogan "54°40' or Fight." The **Oregon Treaty** of 1846 resolved the conflict by extending the forty-ninth parallel from the Rockies to the Pacific. Vancouver Island, which extends below the forty-ninth, remained British. This treaty cut off Hudson's Bay Company access to much of the Columbia River, the key river system used for navigation by the fur trade. The transportation of furs in the remaining British territory would have to rely much more on overland routes.

In 1849, the Hudson's Bay Company was given jurisdiction over the territory and a mandate to establish a colony that included Vancouver Island and the Queen Charlottes. This region became more important with the discovery of gold on the Queen Charlottes in 1850. The gold did not amount to much and attracted few miners,

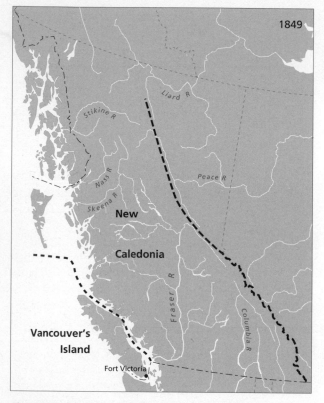

Figure 4.4a Colony of Vancouver Island, created 1849
Source: Modified from Kerr (1975), 32.

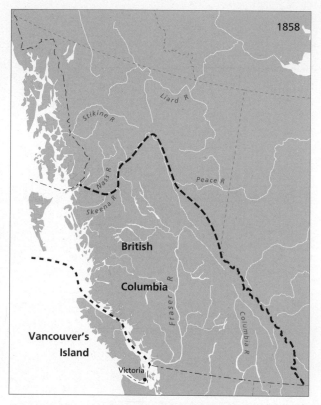

Figure 4.4b Colony of British Columbia, created 1858
Source: Modified from Kerr (1975), 32.

were being trespassed upon and violated. He created a tax on Americans to remind them that they were on foreign land. For the First Nations, he negotiated a number of small treaties on Vancouver Island and established large reserves in the interior. The reserves were not created through treaties, however, and therefore were not accompanied by any formal rights.

As the gold seekers pushed up through the canyons of the Fraser and on into the Cariboo, where even more gold was discovered, new communities such as Barkerville sprang up and a demand arose for transportation routes between Victoria and the interior (Figure 4.5).

Further discoveries of gold, on the Stikine River in 1862, resulted in the creation of the Stickeen Territory and pushed the northern boundaries to the sixty-second parallel. By 1863, the British amalgamated the Stickeen Territory with British Columbia, and with the

discovery of more gold on the Peace River and prospects in the Kootenays pushed the boundaries east to the present-day configuration (Figure 4.6). Nevertheless, establishing boundaries and maintaining them are two separate processes. The United States' purchase of Alaska in 1867 left British Columbia facing American territory on two sides. Also of concern was that many of British Columbia's inhabitants were American. When British Columbia joined Confederation in 1871, the issue of territoriality was finally put to rest: British Columbia was part of Canada.

The gold rush was to have a profound impact on the landscape and on the indigenous peoples of British Columbia. The term **frontier mentality** describes the attitude of people who came to the frontier in search of wealth but had no intention of staying. Actions prompted by this attitude were particularly destructive to the environment. Most miners were single men from other parts of British North America, the United States, Europe, and China, lured by the possibility of striking it rich with gold. **Placer mining,** mining gold in its pure form, was very attractive for these individuals because it required little investment other than a shovel, a gold pan, and hard work. Gold, because of its weight, settles to the bottom of streams or on its banks and sand bars, where it can be recovered quite easily. Over time, the technology of acquiring gold changed, and it was not long before gold pans gave way to small sluices, larger sluices, dredges, and hydraulics, all having a destructive impact on the rivers and streams that were the habitat of salmon, the basic food staple for many First Nations.

Gold mining communities, thrown up in a haphazard fashion with frame structures and at extreme risk to fire, reflected this attitude of impermanence. Similarly, the destruction of the forests for fuel, building supplies, and mining left communities vulnerable to avalanches, mud flows, and flooding. The miners of the Cariboo provided a market for beef, and cattle drives, often originating in the western United States, competed to be the first into communities such as Barkerville in the spring. In the process, the natural grass lands of the south central interior valleys, where the highly nutritious bunch grass was the dominant form of vegetation, were seriously overgrazed. The vegetation that succeeded bunch grass, and is dominant through these dry interior valleys today, is sage brush, which has deep, large roots, and cannot be eaten by cattle.

Not all who were present in British Columbia during this early period had a frontier mentality. Many intended to remain in the region and demonstrated a desire for permanence. Initially, the economic and colonial interests of the Hudson's Bay Company had provided the impetus for the territory to remain British. Individuals such as Governor Douglas, who taxed American miners and established British law and order along with the many other British institutions, represented the attitude of permanence. The Royal Engineers, who provided a military presence, were responsible for planning, surveying, and building structures such as the Cariboo Road, and laying out the town site of New Westminster. The famous "hanging judge," Sir Matthew Baillie Begbie, left the reputation of a lawful British frontier as opposed to

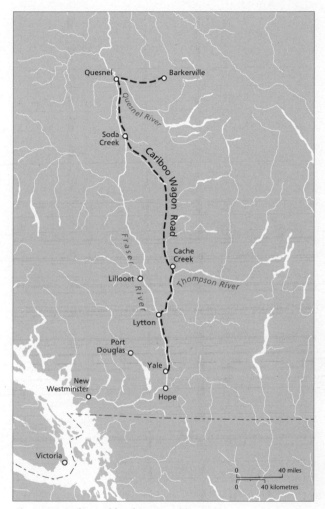

Figure 4.5 Cariboo gold rush communities, 1863

an unlawful American one, although historical geographer Cole Harris (1997) questions the truth of this interpretation.

Before long, families began to migrate to British Columbia intending to call the region home. These families were often British and were imbued with the values of the land from where they came. They began the formidable task of remaking this wild landscape into a more familiar, tamer one.

Differences in attitudes toward, and uses of, the land did not end with the gold rush. Rather, philosophically

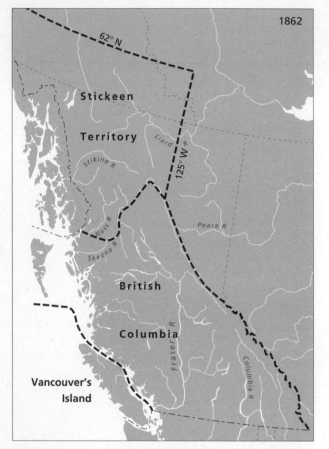
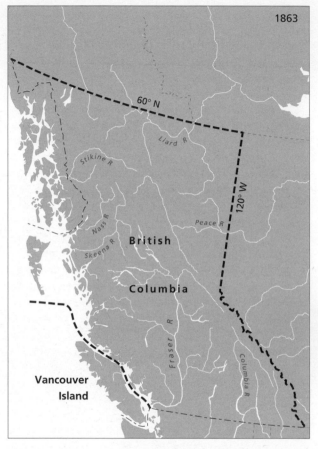

Figure 4.6 Boundary changes to British Columbia, 1862 and 1863
Source: Modified from Nicholson (1954), 32.

opposing views over resource development and management were expressed through political parties and much rivalry.

When British Columbia joined Canada in 1871 the act brought a sense of permanence to its boundaries and an expectation that the wilderness was to be "civilized" with British people and British values. Politically, the move brought two levels of government: federal and provincial. Economically, the province was strengthened as a frontier of expanding industrial capitalism, and the promise of a national railway connected to an international port fuelled land speculation and growth generally. Political scandals prevented the railway from being built in British Columbia until the 1880s, which was

the next "boom" period (Chodos 1973). Provincial dependence on resource and infrastructure development and on an external demand for resources was the beginning of an enduring cycle of boom and bust.

The census of 1881 divided the population of British Columbia by ethnicity and location. Ethnically, the population was classified as "White," "Chinese," "Indian," and "African" (Table 4.1). The population of Africans, at 274, was considered insignificant as most were scattered throughout the mining areas of the Cariboo and Cassiar regions. The white population was concentrated in the southwestern corner of the province, and especially Victoria, although some lived in the mining camps of the Cariboo and Cassiar areas and a few were involved in fishing, forestry, and farming. The Chinese "tended to occupy locations that had been created by whites" (Galois and Harris 1994, 41) and therefore worked the placer mines, in the canneries, in the mills, or on the farms. Those in urban locations tended to be in the service occupations. Aboriginal people were the most numerous and were spatially separated from the other three groups. They experienced the greatest change with the arrival of non-Natives. Previously unknown diseases, a newly acquired problem of alcohol addiction, confinement on reserves, and missionary influences removed them from their traditional lands. By 1881, the once Native-dominated landscape was altered, reshaped, and dominated by a white society.

SUMMARY

The arrival of Europeans to the northwest coast of North America came late in the European phase of colonization. The existing system of global trade, colonization, and limited transportation (by sailing vessel), however, made this geographic region one of the most isolated in the world. Britain was able to secure the coastal territory through negotiations with Spain, while the interior also came under British domination as an extension of the overland fur trade companies operating out of Montreal and later Hudson Bay.

Until the Fraser River and Cariboo gold rushes, European settlement at fur trade posts was mainly temporary, and few resources other than gold were developed. The discovery of gold changed the landscape dramatically. Politically, its boundaries had solidified, but the greatest changes came throughout the southern interior with the building of communities, farming, ranching,

road building, and mining. The frontier mentality led to a great deal of destruction. Conversely, an attitude of permanence recognized the forces of institution building and landscape modification, often in the image of British value systems.

The search for gold had produced a frenzy of activity – permanent settlement, transportation systems, the immigration of many more people, especially Americans and, to an extent, Chinese – that challenged the existing British colonial system governing the territory. By the end of the 1860s, considerably less gold was being discovered, and British Columbia experienced the downward cycle of a bust economy.

Essentially, this historical account ends in 1871, when British Columbia joined Confederation. With this act, British Columbia was no longer a British colony but a separate province with elected members operating within the realm of powers outlined in the British North America Act. Confederation also put to rest the fear of annexation by the United States. Finally, Confederation came with a promise of another economic boom – the building of a national railway connected to an international port. For political reasons, the railway was delayed for ten years.

Table 4.1

1881 census population

Ethnic group	Population
Indians	26,849
Chinese	4,195
Whites	19,069
Africans	274
Total	50,387

Source: Uncorrected nominal returns. Modified from Galois and Harris (1994), 39.

REFERENCES

Archer, C.I. 1981. "The Transient Presence: A Re-appraisal of Spanish Attitudes toward the Northwest Coast in the Eighteenth Century." In *British Columbia: Historical Readings,* ed. W.P. Ward and R.A.J. McDonald, 37-65. Vancouver: Douglas and McIntyre.

Chodos, R. 1973. *The CPR: A Century of Corporate Welfare.* Toronto: James Lorimer.

Galois, R., and C. Harris. 1994. "Recalibrating Society: The Population Geography of British Columbia in 1881." *Canadian Geographer* 38 (1): 37-53.

Gentilcore, R.L., ed. 1993. *Historical Atlas of Canada.* Vol. 2, *The Land Transformed 1800-1891.* Toronto: University of Toronto Press.

Gibson, J.R. 1981. "Bostonians and Muscovites on the Northwest Coast, 1788-1841." In *British Columbia: Historical Readings,* ed. W.P. Ward and R.A.J. McDonald, 66-95. Vancouver: Douglas and McIntyre.

Harris, C. 1992. "The Lower Mainland, 1820-81." In *Vancouver and Its Region,* ed. G. Wynn and T. Oke, 38-68. Vancouver: UBC Press.

–. 1997. *The Resettlement of British Columbia: Essays on Colonialism and Geographical Change.* Vancouver: UBC Press.

Kerr, D.G.G. 1975. *Historical Atlas of Canada,* 3rd ed. Toronto: Thomas Nelson.

Muckle, R.J. 1998. *The First Nations of British Columbia.* Vancouver: UBC Press.

Nicholson, N.L. 1954. *Boundaries of Canada, Its Provinces and Territories.* Ottawa: Queen's Printer, Department of Mines and Technical Surveys.

First Nations and Their Territories:
Reclaiming the Land

5

In British Columbia, "there is general agreement that the term First Nations refers to a group of people who can trace their ancestry to the populations that occupied the land prior to the arrival of Europeans and Americans in the late eighteenth century" (Muckle 1998, 2). The term recognizes that Native peoples have their own languages, customs, and territories and have occupied British Columbia for approximately 10,000 years. Various cultures and their territorial boundaries have changed over those many years, and the initial discussion here examines some of the cultural components of how peoples used the land prior to the arrival of non-Natives.

Europeans made their contact from a number of directions in the 1700s and early 1800s, bringing with them new trade items, technologies, and value systems that included the organization and use of land, a philosophy of mercantilism, and a definition of civilization. The Europeans were few at first but their numbers increased dramatically with the gold rush of 1858. The impact was catastrophic for First Nations. The stories of suppression, assimilation, and loss of land and life were endless and only began to improve slowly after the 1950s. The interrelated roles of the federal government, provincial government, court systems, public opinion, and Aboriginal organizations are examined here in terms of why land is the key issue in Aboriginal affairs and central to the modern treaty system.

Table 5.1

First Nations language divisions

Ethnic division	Language	Dialect
Haida	Haida	Masset
		Skidegate
Tsimshian	Tsimshian	Tsimshian
		Gitxsan
		Nisga'a
Kwakwaka'wakw	Kwakwaka'wakw	Haisla
		Heiltsuk
		Southern Kwakwaka'wakw
Nuu'chah'nulth	Nuu'chah'nulth	Northern Nuu'chah'nulth
		Southern Nuu'chah'nulth
Nuxalk	Nuxalk	
Straits Salish	Comox	
	Pentlactch (extinct)	
	Sechelt	
	Squamish	
	Halq'emeylem	
	Straits Salish	
Interior Salish	Nlaka'pamux	
	Stl'atl'imx	
	Secwepemc	
	Okanagan	Okanagan Lakes (extinct)
Kootenai	Kootenai	
Athapaskan	Tsilhqot'in	
	Dakelh	
	Sekani	
	Tahltan	
	Kaska	
	Dene-thah	
	Dunne-za	
	Tsetsaut (extinct)	
	Nicola (extinct)	
Inland Tlingit	Tlingit	

Sources: Duff (1965), 15; Muckle (1998), 117-22.

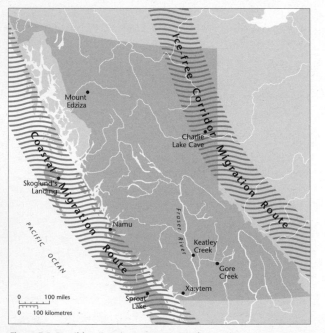

Figure 5.1 Possible migration routes to North America
Source: Muckle (1998), 20.

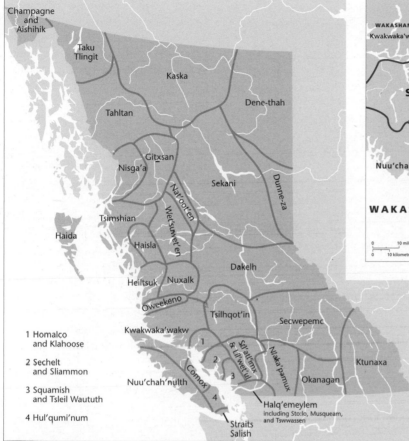

Figure 5.3 First Nations linguistic divisions in southwestern British Columbia
Source: Modified from Harris (1992), 40.

Figure 5.2 Historical territorialization of First Nations in British Columbia
Source: Muckle (1998), 7.

PRE-CONTACT ABORIGINAL SETTLEMENTS

The generally accepted theory of North American settlement is that the first inhabitants came from Siberia via the ice bridge of the Bering Sea some 11,000 to 12,000 years ago, when glaciers from the last ice age were receding. Figure 5.1 indicates two routes into British Columbia and the location of several archaeological sites.

The population and density of Aboriginal peoples varied with geographic location. The highest and most densely populated regions were along the coast and along salmon-bearing streams in the interior. Density was considerably lower in the rest of the interior. Table 5.1 shows the historically different language groups and dialects, and Figure 5.2 matches some of them with their

traditional territory. It should be noted that there is no consensus among either First Nations or anthropologists on the divisions of major ethnic groups or the nations that belong to them. Different writers use different categorizations, territorial boundaries, and spellings. The map showing the linguistic breakdown of Aboriginal peoples in the southwestern corner of British Columbia outlines the complex spatial organization, or territorial boundaries, of one area (Figure 5.3).

The Aboriginal means of production was based on very strong ties to the land and its resources, going beyond the basics of food, clothing, and shelter. Native spiritual beliefs, specifically animism, were intimately tied to the land, as was the social and spatial organization of the various peoples. Territorial boundaries based on language group were further defined in terms of ownership by individual houses or clans. McKee (1996, 4) elaborates on the Native concept of private property: "Private property was vital, and just as sophisticated as that of European nations at the time. Each household possessed

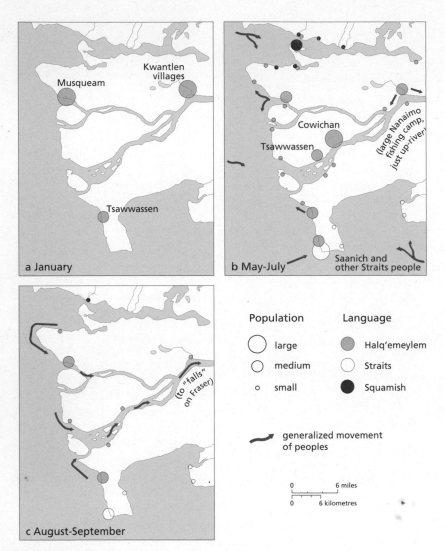

Population
○ large
○ medium
∘ small

Language
⬤ Halq'emeylem
○ Straits
⬤ Squamish

↪ generalized movement of peoples

0   6 miles
0   6 kilometres

Figure 5.4 Semi-nomadic migrations in the Lower Mainland area
Source: Modified from Harris (1997), 39.

in this well-organized migration process.

Today, **Aboriginal rights** are recognized by section 35 of the Canadian Constitution, but it is not clear what those rights include. Ownership or control of territory for each group is recognized as **Aboriginal title.** Aboriginal rights also imply the freedom to pursue a particular mode of production (through use of land, labour, resources, technology, and capital) in combination with forms of political, social, and spiritual organization that ensure survival.

The Fraser River delta sustained many land and water resources – especially salmon – which were available at different times of the year, thus influencing the semi-nomadic movement of various Native groups throughout the year. Figure 5.4 covers the migrations for typical portions of the year, illustrating how many groups travelled beyond their traditional boundaries to procure food. As a consequence, there was considerable overlap of territorial boundaries. This has created some difficulty in the modern treaty process, which begins by attempting to recognize title to specific lands.

Aboriginal societies were based on oral tradition; history, legends, stories, and wisdom were held by the elders and passed on to the next generation. Stories, dances, and songs were important possessions in defining the resources of the land and the territorial limits of the land base. These were communally based societies that functioned through an elaborate division of labour in which each individual had a role in sustaining the group.

First Nations of the coast had complex social, spiritual, and political organizations. An elaborate division

land for village sites and for hunting and food gathering. Specific items and rights were held in common by the members of each household, including canoes, totem poles, ceremonial objects, and rights to hunt and fish in specific waters and harvest particular food species." Native groups were also semi-nomadic, moving from season to season depending on the availability and location of resources. Nevertheless, the same sites were occupied for hundreds and perhaps thousands of years

of labour, sophistication of carvings (on totems, dug-out canoes, and so forth), shamanism, and the potlatch were symbols of well-organized cultural groups within well-defined boundaries. The potlatch was an extremely important political institution practised almost exclusively by coastal First Nations. It was a form of government that organized and legitimized the decisions each nation made: "Potlatch government business is conducted at feasts. The house invites members of other clans to its feasts to witness the decisions taken, and to receive gifts that reinforce co-operation between the various houses. Potlatch feasts are held to commemorate deaths, to pay debts, to resolve disputes, and to confirm positions within the house leadership. They also provide a forum for discussing and resolving community problems. Potlatch feasts are organized according to strict procedures as to invitations, seating, speaking, gift giving, payments, dress and conduct" (Brown 1992, 4).

British Columbia, and particularly coastal British Columbia, had far more linguistic groups and clans, living in much higher densities, than anywhere else in Canada. They developed very sophisticated cultures and elaborate social organizations based on territorial boundaries. Their value systems and their historical use of the land defined a geography that was seriously challenged and changed with the arrival of Europeans.

EUROPEAN CONTACT AND IMPACT

The history of early explorers romanticizes individuals who battled the unknown elements with rather primitive technologies, claiming land for European monarchs. These initial discoveries usually led to further exploration and exploitation of the resources of the land. All development was carried out in the belief that, because they were the first Europeans to a "new" land, they had the absolute right of ownership of the land. This presumption is implicit in the European place names given to the new territory.

This was a **Eurocentric** view, in which exploitation of other people, their land, and their resources was justified even when military force was used. It was also a racist view, in which superiority was based on skin colour. Christian missionary zeal went hand in hand with colonialism as it attempted to convert Native people from their supposedly inferior and heathen beliefs. Eurocentrism encompassed a view of history that defined civilization and proposed a set of civilizing rules. Within British Columbia this meant that one should speak English, adopt Christian values, live within the British common law system, accept the principle of private land ownership, and develop the land for agricultural use. These are largely the values of a capitalist system, and they produced a serious clash with First Nations values. According to Harris (1997) this resulted in a **deterritorialization** of Aboriginal land and way of life and a **reterritorialization** by non-Natives with a capitalist value system. One of the persistent struggles to this day for First Nations is assimilation, or the loss of their cultural values and a whole way of life related to the land. It is often said by Aboriginal peoples that the land is the culture.

Initial European contact was made on the coast via sailors in the late 1700s and was not marked by a great deal of conflict, although conflict did occur. Rather, Native curiosity about the Europeans in ocean-going vessels and interest in trade goods was more common. There evolved a mutual dependence: Europeans wanted Aboriginal peoples to acquire sea otter pelts, which were so valuable in China, and Aboriginal peoples wanted European goods.

The overland fur trade entered British Columbia from eastern Canada in the early 1800s and resulted in a new relationship between Europeans and Aboriginals. It was marked by European frustration over trade agreements and difficulties in transporting furs. Consequently, First Nations were often eclipsed by European fur brigades. The abusive attitudes and actions of Eurocentrism surfaced as fur trading posts were established and non-Natives began to trespass on Native land. (See Chapter 4 for more historical detail.)

Continuous contact between Natives and non-Natives in British Columbia was late in European history, occurring at the end of the eighteenth century and beginning of the nineteenth. And even this contact was minimal until the Fraser River gold rush of 1858. In eastern and central Canada, there was a much longer history of contact and conflict, and a very significant event eventually affected the whole of Canada – the Royal Proclamation of 1763.

Aboriginal Rights, Title, and Treaties

Through the proclamation the British recognized that Aboriginal peoples had occupied the land first and that they had a form of self-government over the territory. Most significantly, the proclamation established a formal process by which Aboriginal peoples received certain rights under British rule. The proclamation stated that First Nations should not be abused, they should have the right to continue governing themselves, and they should be compensated for the use of their land (Tennant 1990; Frideres 1988). It is this last point particularly that defined a new geographic relationship.

Aboriginal title refers to the right a tribe or band has to a given block of land it used and defined as its territory historically. Because these territories were not static,

their boundaries overlapped. The Royal Proclamation recognized Aboriginal title and the treaty process, implemented throughout much of Canada, extinguished, or ceded, Aboriginal title in exchange for compensation. Compensation was usually made in terms of small cash settlements, material goods, and reserve lands. It also recognized that Aboriginal peoples could continue to hunt, fish, and trap as they always had.

Figure 5.5 shows the various historical treaties throughout Canada. Only two applied to British Columbia. First, Europeans acquired land on southern Vancouver Island through the Douglas Treaties. Second, Treaty Number 8, signed in 1899, was precipitated by the Yukon gold rush and included the Peace River region of the province. Here, reserves were created on the basis of "one

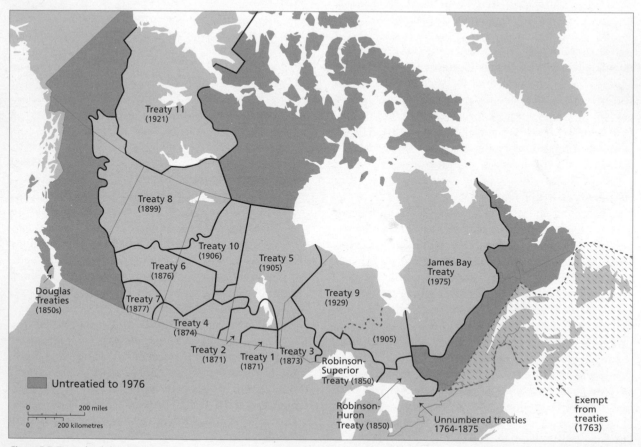

Figure 5.5 Treaty land divisions in Canada

square mile (1.61 square kilometres) for each family of five. Initial cash payments were given to the chiefs and councillors of the various tribes, and provision was made for education, farm stock, implements, and ammunition" (McKee 1996, 22). It should be noted that few Aboriginal peoples could read, write, or speak English at this time, and most believed these arrangements to be peace treaties, not treaties relinquishing their right to land. Throughout the rest of British Columbia, the land was simply taken and occupied by non-Natives; no treaties were signed and no compensation was given.

James Douglas, the chief factor for the Hudson's Bay Company, became governor of the colony of Vancouver Island in 1851. He was concerned about the potential for conflict between the few Europeans who had settled in the region and the Native population, which greatly outnumbered them. By 1854, he had arranged fourteen small "purchases" of land from the local Aboriginal peoples on southern Vancouver Island, which became known as the Douglas Treaties. The gold rush of 1858 saw some 25,000 to 30,000 non-Natives enter British territory, mainly through Fort Victoria, and then push into the lower reaches of the Fraser River searching for gold. Douglas was then appointed governor of the mainland colony of British Columbia, as well as Vancouver Island, to deal with this potentially explosive situation. Historical documents show that through his term as governor, which lasted to 1864, he upheld the spirit of the Royal Proclamation and had every intention of accomplishing more treaties, but Britain was unwilling to supply funding (Cumming and Mickenberg, 1981). Douglas also created reserves without treaties and extended rights of settlement to Aboriginal peoples similar to those extended to Europeans.

From the beginning of non-Native land settlement in British Columbia, the British had both experience with Native peoples (especially in the South Pacific and eastern Canada) and a clear view of how to deal with them. Their intention was to give them a western education, convert them to Christianity, and integrate them into the economy. It was a policy of assimilation, and it denied historical rights to the land. Douglas retired in 1864 and a new regime, with radically different views on First Nations, came into power. Joseph Trutch, the new Chief Commissioner of Lands and Works, remarked that "'the Indian was an obstruction to settlement and progress.' Indeed, in a report on Indian claims to certain lands which had been laid out as reserves for them on Douglas' instructions, Trutch commented: 'The Indians have really no right to the lands they claim, nor are they of any actual value or utility to them, and I cannot see why they should either retain these lands to the prejudice of the general interests of the Colony, or to be allowed to make a market of them either to the Government or to individuals'" (Cumming and Mickenberg 1981, 193).

With Trutch came an attitude that reversed the initial direction of Douglas in upholding the requirements of the Royal Proclamation. Aboriginal title was denied, and there was no intention of compensation. This condescending view and policy toward First Nations in British Columbia occurred prior to the province entering Confederation in 1871 and would persist in subsequent provincial governments for the next 120 years.

Negotiations over Confederation had much more to do with providing a railway than with understanding and addressing First Nations issues in British Columbia, though the federal government became responsible for Aboriginal affairs, taking on what is known as a **fiduciary trust.** This duty meant that the federal government had the legal obligation to look after the best interests of First Nations, which has been interpreted to include establishing reserves. This responsibility spoke more of European interests than of Aboriginal interests. The federal government, confronted with demands for more land by First Nations, often acquiesced to existing provincial policies that instead denied them any rights to the land. Only after a near war with Aboriginal peoples in the late 1870s was the provincial government prepared, reluctantly, to increase the number of reserves. The size of reserves then became a crucial issue in negotiations.

Wherever the federal government was involved in treaties across Canada, reserve size was frequently based on 160 acres per family. In some cases, as in Treaty 8, one square mile (640 acres) was allocated to a family of five. The quarter section (160 acres) was consistent with the Homestead Act, however, which allowed European immigrants to claim a holding of this size. Creating a reserve under this formula simply involved calculating the number of families and multiplying by 160 acres. In

British Columbia, because the provincial government was opposed to increasing reserve land and because there were numerous bands to be considered, the federal government recommended eighty acres per family. The provincial government proposed ten acres per family. The two governments then agreed to ten to twenty acres per family, which is why so many "postage stamp" sized reserves exist in the province today.

Assimilation and Decimation

The impact on First Nations went far beyond the racism and discrimination described above. The term "four plagues" describes some of the key factors in deterritorialization – a new geography in which Aboriginal peoples were no longer able to use the land as they had historically. The four plagues were diseases, alcohol, the reserve system, and the role of missionaries.

Diseases took a huge toll on First Nations. Smallpox epidemics claimed the greatest number of lives, although measles, influenza, tuberculosis, and venereal diseases accounted for many deaths. Alcoholism, although a disease, is noted separately because it was deliberately offered to First Nations as a commodity of trade. The consequence to societies that had no experience of this drug was debilitating and responsible for further deaths.

The reserve system included only some of the historical village sites and a fraction of the traditional territory, or Aboriginal title. The system of deliberately confining reserves to small, fixed plots of land seriously restricted the semi-nomadic hunting and gathering activities of Aboriginals, which were the foundation of their economy. For societies in which land was integral to the way of life, reserves imposed a major alteration and deterritorialization.

Missionaries encouraged the reserve system because it rendered Aboriginal societies less mobile and thus easier to reach. In partnership with government, missionaries also became responsible for education. From today's perspective, the residential school is seen as one of the most devastating institutions imposed on First Nations. Children were separated from their parents and forced to attend schools where their Native languages were forbidden and the educational values were questionable. Recently reported stories of sexual abuse within residential schools are horrific. This social upheaval led to the breakdown of the family unit in Native society.

One simple way of defining a culture is through its language. The language represents a medium of expression for that culture, or the values that the culture holds. Envision a circle as representing all the words in the English language and a separate circle representing all the words in another language (Figure 5.6). Even though the two circles may be the same size, they do not overlap each other perfectly when one tries to compare the meanings of the words and phrases of each language. Anyone who speaks two or more languages knows this well, since language represents ideas and values that may not exist in the other language. To take a language away, which is what the residential school system did, is a form of cultural genocide. Unfortunately, a number of First Nations languages are now extinct and others are in jeopardy.

All of these changes were devastating to First Nations. Table 5.2 illustrates just how much of an impact disease and other causes of death had on the various populations. Cole Harris (1997, 26) documents how deadly diseases such as smallpox had already diffused overland from the Gulf of Mexico to the coast of British Columbia prior to European arrival on the west coast

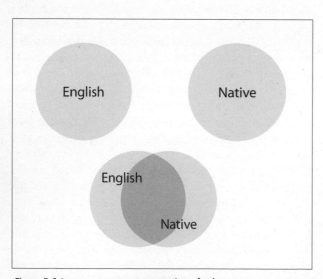

Figure 5.6 Language as a representation of culture

Table 5.2

Aboriginal populations for selected years

Ethnic division	1835	1885	Low	(year)	1963
Haida	6,000	800	588	(1915)	1,224
Tsimshian	8,500	4,500	3,550	(1895)	6,475
Kwakwaka'wakw	10,700	3,000	1,854	(1929)	4,304
Nuu'chah'nulth	7,500	3,500	1,605	(1939)	2,899
Nuxalk	2,000	450	249	(1929)	536
Straits Salish	12,000	5,525	4,120	(1915)	8,495
Interior Salish	13,500	5,800	5,348	(1890)	9,512
Kootenai	1,000	625	381	(1939)	443
Athapaskan and Inland Tlingit	8,800	3,750	3,716	(1895)	6,912
Provincial total	70,000	28,000	22,605	(1929)	40,800

Source: Duff (1965), 32.

by sailing vessels in the late 1700s. It should be recognized that the 1835 population figures in Table 5.2 are an estimate, and further, that diseases had already eliminated some of the population. Harris suggests that perhaps 90 to 95 percent of the overall Aboriginal population perished (p. 30).

The young and the old were the most vulnerable to disease. Because Native societies were oral, the death of elders meant that their wisdom was disappearing with them. It meant that the elaborate division of labour necessary for communal societies was shattered. With so many dying, being pushed onto reserves, having their children taken away to residential schools, and being forced to exist under the rules of assimilation, Native societies were in chaos.

Many policies attempted to assimilate First Nations besides the missionary school system. The Indian Act of 1876 became the main instrument of assimilation and alienation from the land. It imposed a system of bands with an elected chief and council, thus abolishing the hereditary system. With the appointment of a non-Native Indian agent to relay the needs of each band to Ottawa, the new system was also paternalistic. When potlatches were outlawed in 1884, First Nations lost their chief political institution.

Enfranchisement was another divisive tactic, and essentially it offered the choice of becoming non-Indian. Aboriginal peoples were often persuaded to leave the reserves in exchange for certain rights. They could sign a document that gave them the right to vote, to enter liquor stores, to receive a small sum of money, and, in the early days, to receive a parcel of land. This process created the designations of non-status and status Indian, with profound implications for future generations. A non-status Indian had no rights to the reserve and its governance. If a status male were to marry a non-status female their children would be able to retain their status, but until 1985 the children of a status female married to a non-status male were not entitled to status.

Many other restrictions occurred: dances and ceremonies were prohibited, a status card had to be carried at all times off the reserve, and in 1927 non-Native lawyers were not allowed to represent First Nations bands on any land claim issues (Tennant 1990). In many ways British Columbia, along with the federal government, had instituted an apartheid system.

The McKenna-McBride Commission (1912-16) reveals the frustration for the First Nations in dealing with both levels of government in British Columbia over reserve lands. John McKenna represented the federal government and Richard McBride, the premier of the province, represented BC interests. The intention of the commission was to sit down at the table and finalize the size of reserve lands once and for all. Its official mandate included both the expansion and the reduction of reserves, provided that the bands consented. This consent clause was ignored, and "laws were later passed stating that Indian consent was not really necessary after all" (Gunn 1976, 5). The following documents the extent of the cut-off lands: "Cut-offs were recommended by the McKenna-McBride Commission on 35 reserves amounting to a total of 36,000 acres. All but one of these reserves are located outside the urbanized Vancouver area. The one reserve within the Vancouver area is the Capilano lands of the Squamish Indians. A 130 acre piece of land was removed from the reserve on the north shore of Burrard

Inlet near the mouth of the Capilano River (some reports claim 132 acres)" (Gunn 1976, 9).

Eighty thousand acres were added to other reserves, but the lands were of marginal value. The Vancouver area example above is the present-day location of the West Vancouver Park Royal Shopping Centre, where the land is valuable indeed. The McKenna-McBride Commission was only one process that resulted in lands being cut off from already small reserves. Reserve lands were lost to any number of transportation developments: highways, railways, hydroelectric transmission lines, and so forth. Lands adjacent to or within municipal boundaries were other targets for annexation. All these losses from existing reserves are referred to as cut-off lands in British Columbia.

The long litany of discrimination, assimilation, and basic injustice has not happened without the resistance of Aboriginal peoples. There was a near war in 1876 when the BC government took the position that it would not set more land aside for reserves, and more reserves were created. The story of the Nisga'a, who escorted non-Native surveyors off their land by gunpoint in 1887, illustrates not only resistance but also a clear view of Aboriginal title. Paul Tennant (1990) documents Native organizations that came together after the Indian Act (1876) to fight common injustices. One of the strengths of those organizations, with a leadership brought up under the missionary school system, was a common language – English.

DEVELOPMENTS AFTER THE SECOND WORLD WAR: NEW RIGHTS AND NEW THREATS

After the Second World War, new attitudes toward discrimination against all visible minorities resulted in changes to old policies. Status Indians were allowed to vote by 1949, and by 1951 the bans on the potlatch, dances, and ceremonies were lifted. Residential schools were phased out in the 1960s. By 1988, bands were able to levy taxes on the leased portions of their lands and, by 1991, British Columbia finally elected a provincial government that recognized Aboriginal title.

Bill C-31 was passed in 1985 in an attempt to correct the problems that status and non-status designations caused. This Bill now legislates four categories of First Nations:

1 Status with band membership – Indians who have the right to both registered status and band membership;
2 Status only – Indians who have the right to be registered without the automatic right to band membership;
3 Non-status band members – Indians eligible to be registered under a band list in accordance with the Band Citizenship Code but who do not have the right to registered status;
4 Non-status Indians – Indians who are still not entitled to be registered (Joseph 1991, 69).

Some bands benefited by increased membership while others saw the Act as another top-down set of rules imposed by Ottawa. A more serious concern is expressed by Susan Joseph (1991, p. 69): "The first generation cut-off clause dictates that only the first generation descendants of an individual are entitled to be registered. Second and succeeding generations will never be allowed status, nor will they be allowed to pass a right to status on to their children."

The transition to fewer discriminating policies and racist attitudes was not smooth. The 1960s, in particular, brought other, more serious threats to First Nations. This was the decade of megaprojects, during which the provincial government encouraged large, and often foreign, capital investments in resource development. The expansion of the forest and mining industries, combined with major infrastructure developments of hydroelectric dams, railways, and roads, put enormous pressures on the land. Many First Nations bands felt threatened by the assault on traditional lands that had never been surrendered. Court action and in some cases road blocks and threats of violence resulted.

With the rapid economic growth of the 1950s and 1960s, a major struggle for First Nations was to get federal and provincial governments to recognize Aboriginal title and Aboriginal rights. Both levels of government were reluctant because of the economic implications. Consequently Aboriginal peoples turned to the court system and public opinion to persuade government to change its mind. The various actors involved – courts, governments, and public – are shown in Figure 5.7. The lines between public opinion and the two levels of

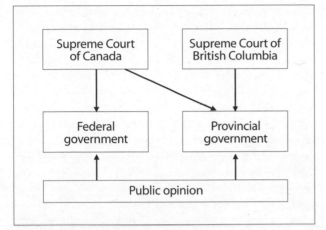

Figure 5.7 Interrelationship of decision makers

government represent the connection between the public and the federal and provincial governments it elects, indicating the importance of public opinion in government decisions. Individuals or groups have the right to challenge existing laws and conditions through the courts, and the judgment of the court becomes law until an appeal or new challenge occurs. The diagram recognizes the power of the Supreme Court of Canada to interpret laws in ways that affect both the provincial and federal governments. The 1969 White Paper forced First Nations people to become more familiar with these interconnections.

In the 1960s, the federal government under Prime Minister Pierre Trudeau was struggling to come to terms with the Indian Act and growing resistance by Aboriginal peoples. The resulting 1969 federal White Paper proposed radical changes: the Indian Act would be abolished, reserve lands would be divided, and special status would end. In other words, there would be no recognition of Aboriginal title, no compensation for past injustices or expropriation of land, and no special status to recognize the inherent right of Aboriginal self-government. In short, the White Paper intended to turn all First Nations into "ordinary Canadians" – the ultimate in assimilation. Harold Cardinal, as president of the Indian Association of Alberta, commented: "For the Indian to survive, says the government in effect, he must become a good little brown white man. The Americans to the south of us used to have a saying: 'the only good

Indian is a dead Indian'. The MacDonald-Chrétien doctrine would amend this but slightly to, 'The only good Indian is a non-Indian'" (1969, 1).

The reaction by Aboriginal peoples across Canada was to form their largest national organization ever to reject the proposed policy. The federal government did withdraw the White Paper, but change in its policy was achieved through a decision by the Supreme Court of Canada. The *Calder* case in 1973 is famous because the Nisga'a Tribal Council took the federal government to court, asserting that their Aboriginal title had never been surrendered, or ceded. The Supreme Court ruled "that aboriginal title is rooted in the long-time occupation, possession and use of traditional territories. As such, title existed at the time of original contact with Europeans, regardless of whether or not Europeans recognized it" (British Columbia, Ministry of Aboriginal Affairs 1994). The court split evenly between six judges over the question of whether Aboriginal title still existed, and a seventh judge dismissed the case on a technicality. This became a landmark decision because its recognition of Aboriginal title produced an abrupt change in federal policy. The federal government began a process of negotiating **comprehensive claims**.

Comprehensive claims were, and are, conducted on unsurrendered lands within Canada. These are lands not previously subject to treaty. Comprehensive claims give rise to modern treaties in which negotiations can include land, resource management rights, cash compensation, education opportunities, economic development, and other rights and compensation, but initially not self-government. It took constitutional debates, military stand-offs, and much confrontation before self-government was included as a possibility in 1995. Comprehensive claims are distinct from **specific claims**, which are for individual injustices done to specific bands. Cut-off lands, for example, are matters for specific claims.

With the about-face of the federal government, the flood gates opened for treaty negotiations. Nevertheless, claims did not proceed rapidly. By 1997, all except one were settled in the Northwest Territories and Yukon, where provincial governments were not involved. In British Columbia, the provincial government stood fast until the 1990s on its position that Aboriginal title did not exist, making negotiations extremely frustrating.

In this province, the conflict between resource exploitation on traditional territories continued, resulting in road blocks and court cases. The Meares Island decision in 1985 by the British Columbia Court of Appeal upheld an injunction to prohibit MacMillan Bloedel from logging the area until a land claim by the Nuu'chah'nulth people was settled. Logging interests were blocked again at the Stein Valley near Boston Bar, and the logging of south Moresby Island in the Queen Charlottes was so contentious that most of the disputed land became a federal park. Industrial capitalism at the frontier was being blocked, and corporate profits, along with resource revenues to the provincial government, were being curtailed. Resource allocations for the future were uncertain.

The *Sparrow* decision (1990) put into question historical Aboriginal entitlement to fish versus commercial entitlement to salmon through the regulatory system of licences and run openings. Essentially, the Supreme Court of Canada overturned an earlier lower court decision to convict an elder of the Musqueam Nation of illegal fishing, ruling that the Constitution Act provided "a strong measure of protection" for Aboriginal rights to fisheries. The decision stated that Aboriginal people "must be given priority to fish for food over other user groups." The federal government, which manages the fishing industry, interpreted the court decision to mean that a Native-only commercial fishery had priority over the non-Native commercial fishery. In an industry already suffering competing interests and declining stocks, the United and Allied Fishermen's Union launched an appeal. The 1996 appeal decision reiterated that Aboriginal people had an unextinguished right to fish for food and that this right must be broadly interpreted.

Many of the court cases of the 1980s and early 1990s appeared to be upholding the terms, conditions, and concept of Aboriginal title, but the *Delgamuukw* case set aside this pattern. Thirty-five Gitxsan and thirteen Wet'suwet'en hereditary chiefs went to the Supreme Court of British Columbia in 1984 to claim compensation for the loss of land and resources on their traditional lands. They also claimed the inherent right of Aboriginal self-government over these lands. Chief Justice Allan McEachern did not accept traditional oral descriptions as evidence of historical land use and consequently ruled in 1991 that Aboriginal title had been extinguished at the time of Confederation. The news release by the Gitxsan and Wet'suwet'en in reaction to the decision shows their frustration: "This judge is attempting to push back justice for native people in Canada at least 20 years. This has been done by either ignoring or rejecting legal gains made in the last few years by native people all over the country. This judge goes as far as saying on page 300 of the document that native people, especially the Gitksan and Wet'suwet'en, are economically marginalized because they live on reserves and that the only hope of change can come if they leave and assimilate into the mainstream. This is the 1969 White Paper, a policy long rejected by successive federal regimes" (1991).

Needless to say they appealed this decision, and in 1993 the BC Court of Appeal reversed much of the former judgment and recognized Aboriginal title, though it also ruled that those rights were non-exclusive. In December 1997, the Supreme Court of Canada ordered a new trial and ruled on several aspects of the *Delgamuukw* case, making it absolutely clear that Aboriginal title means ownership of the land. There was no recommendation on compensation though. For the Gitxsan and Wet'suwet'en, hope for justice lies in the courts still. As in the *Calder* case, the *Delgamuukw* decision could have national implications.

In the 1980s, the court cases, blockades, and public opinion were turning against the provincial government's entrenched position. The public was becoming aware of and concerned about the environmental damage brought on by resource exploitation and was beginning to make the link between treatment of the land and treatment of the people who were attempting to stop the exploitation. The public was demanding fair treatment of First Nations so that these issues would not continue for the next generations.

Premier Bill Vander Zalm created a Ministry of Native Affairs in 1990 in response to the pressure. Its mandate was to investigate Aboriginal title, although ironically it was made clear that the provincial government did not recognize Aboriginal title. Interestingly though, it made the statement that if Aboriginal title did exist, then the federal government would be 100 percent responsible for compensation. This ministry, and its mandate, did little to satisfy British Columbians.

Finally, in 1991, a change in government led to recognition of Aboriginal title. The British Columbia Claims Task Force was appointed and its recommendation to enter into a treaty process was accepted (British Columbia, Ministry of Aboriginal Affairs 1991). Bill 22 established the BC Treaty Commission in 1993. Modern treaties had finally come to British Columbia. It is worth noting that had the New Democratic Party not won the election in 1996, the treaty process would have been dissolved by all other political parties as stated in their party platforms.

MODERN TREATIES AND THE FUTURE

The BC treaty negotiations, or **modern treaty** system, is designed to overcome the lack of treaties and compensation for all bands and tribal councils in British Columbia. This system established a compensation formula between the federal and provincial governments in which the federal government is mainly responsible for the cash compensation portion of any settlement and the province is responsible for land and resource allocations (McKee 1996; *BC Studies* 1998-9). The negotiations themselves have a six-stage process:

Stage 1: Statement of intent by the First Nation to negotiate a treaty
Stage 2: Treaty negotiations preparation
Stage 3: Negotiation of a framework agreement
Stage 4: Negotiation of an agreement-in-principle
Stage 5: Negotiations to finalize a treaty
Stage 6: Treaty implementation

Throughout the province, various bands and tribal councils are at different stages of the process, and approximately one-third of the bands by 1999 have not engaged in the process at all. An update about the bands and tribal councils that have signed on and which stage each group has reached can be found on the British Columbia Treaty Commission website (under Status of Negotiations). The rights of third parties are a major concern from a non-Native perspective. Third parties include all the users of land within any claim area: resource companies, farmers, other private land holders, local governments, and organizations that have a vested interest. Simply put, third parties are concerned that their

interests will be jeopardized by a treaty. The mandate of the treaty process is to protect the right of continued use by third parties; bands and tribal councils receive cash and access to resources in compensation.

A report by consulting company Grant Thorton makes the following statement: "Completing treaties to settle Aboriginal land claims will bring a net financial benefit of between $3.8 and $4.7 billion to British Columbia over the next 40 years" (Minister of Aboriginal Affairs, 24 March 1999). The importance of settling the land claims issue and bringing finality to land use and ownership is obviously in the provincial government's best interest, though the benefits will take many years to materialize.

Although it follows a similar process, the 1996 Nisga'a agreement-in-principle is separate from the BC Treaty Commission guidelines and rules because it has been in progress with the federal government since 1976. It is an interesting document, and while it claims not to set a precedent, the various issues it addresses could establish the basis of future treaty negotiations for many bands and tribal groups throughout British Columbia. The provincial government gave its approval to this agreement-in-principle and accepted it as a treaty in 1999. The treaty now awaits ratification by the federal government.

Figure 5.8 shows the traditional lands, or Aboriginal title, of the Nisga'a, based on the Nass River system. The agreement-in-principle allocates 8 percent of the traditional area as a land base for the Nisga'a peoples, and it includes both surface and subsurface rights. This land base is centred on the existing communities. The proposed Nisga'a land will no longer be reserve land, the Indian Act will no longer apply, and the tax exemption status on reserve lands will disappear. There will be a cash compensation for the other lands of $190 million and a further $11.5 million for the Nisga'a to become involved in the commercial salmon fishery. The agreement-in-principle also includes "an annual treaty-entitlement of salmon, which will, on average, comprise approximately 18 percent of the Canadian Nass River total allowable catch. In addition, the Nisga'a will receive an allocation of sockeye and pink salmon for commercial purposes under a harvesting agreement outside the treaty" (*Nisga'a Treaty Negotiations* 1996, 3). The document gives the Nisga'a responsibility within their territory for wildlife

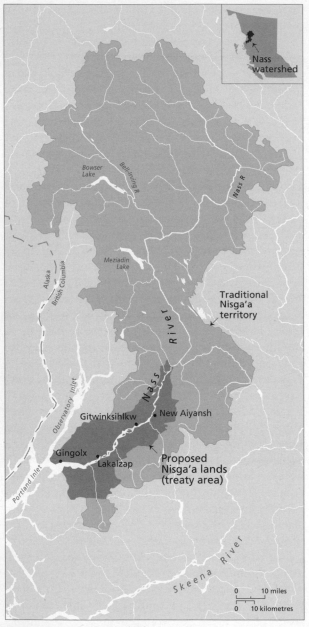

Figure 5.8 Nisga'a Aboriginal title
Source: Modified from *Nisga'a Treaty Negotiations: Agreement-in-Principle* (1996), appendix G.

management as well as for establishing environmental standards. There is also provision for self-government and the establishment of justiciability, which means the right to establish policing and court systems.

Ratification of the Nisga'a treaty will put an end to the long story of injustices and illegal use of Nisga'a lands. It should also end the road blockades and confrontations over land use, and the Nisga'a will be much closer to equality with the rest of Canadian society.

Not all Aboriginal groups have signed on to the treaty process, and incidents such as a blockade at Gustafsen Lake in 1995 that resulted in an armed confrontation for much of that summer reflect separate and hostile views. These confrontations stem from an overwhelming frustration at the government's inability to deal with issues facing Native peoples, such as Aboriginal title, inadequate housing, lack of employment opportunities, the inordinately high suicide rate among Native youth, and continued resource exploitation where treaty negotiations have not been finalized (Willems-Braun 1997). Millions of dollars can be raised to subsidize resource industries operating on contentious, untreatied lands or to mobilize an army to put down a Gustafsen Lake uprising, but dollars needed to improve a band's conditions receives little attention. The delay in gaining each of the steps in the treaty process is being viewed by groups such as the Sechelts as not negotiating in good faith, and consequently they threatened litigation in 1998. This strategy may have some merit, as in April 1999 the Sechelt Nation became the first band to reach Stage 6 of the modern treaties.

From the perspective of First Nations involved in negotiations, the treaty process is long overdue in righting the wrongs of the past. Use of the land and its resources, management rights, cash compensation, and self-government are all ways to establish a solid foundation for preserving cultural values and an economic base for the future. Examples of First Nations strategies for economic development exist across Canada (Anderson and Bone, 1995). The specific location of the Aboriginal group, its access to resources, and how its development plans fit into the global economic arena must all be considered. Successful treaties, self-government, and First Nations economic ventures will be an advantage to all in British Columbia.

SUMMARY

The story of First Nations is one of semi-nomadic peoples who used the land and its resources for long periods before Europeans arrived in British Columbia. This traditional geography was drastically interrupted with the arrival of Europeans, and a new geography of predominantly British and capitalist values was imposed on the land. The actions of the newcomers, both intentional and unintentional, had a catastrophic impact on First Nations. Their numbers were reduced to perhaps 5 percent of former days, and their way of life was altered forever with the creation of reserves, residential schools, and a never ending set of policies governing who they were and how they lived.

Despite these injustices, Aboriginal peoples did not give up the struggle for their culture and their land. It was not until the early 1990s that the process of land claims was seriously considered in British Columbia, and modern treaties are close to being finalized for a number of bands. Although it will not cure all ills, land claim settlement with accompanying self-government will establish individual dignity and an economic base for the future.

REFERENCES

Anderson, R.B., and R.M. Bone. 1995. "First Nations Economic Development: A Contingency Perspective." *Canadian Geographer* 39 (2): 120-30.

BC Studies. 1998-9. Special issue, *The Nisga'a Treaty*. No. 120.

British Columbia, Ministry of Aboriginal Affairs. 1991. *The Report of the British Columbia Claims Task Force*. 28 June, Victoria.

–. 1994. "Landmark Court Cases." Victoria: Communications Branch.

Brown, D. 1992. "Aboriginal Rights." Educational report for the Carrier-Sekani Tribal Council, Prince George.

Cardinal, H. 1969. *The Unjust Society*. Edmonton: Hurtig.

Cumming, P., and N. Mickenberg. 1981. "Native Rights in Canada: British Columbia." In *British Columbia: Historical Readings*, ed. W.P. Ward and R.A.J. McDonald, 184-211. Vancouver: Douglas and McIntyre.

Duff, W. 1965. *The Indian History of British Columbia*. Victoria: Royal British Columbia Museum.

Frideres, J.S. 1988. *Native Peoples in Canada: Contemporary Conflicts*, 3rd ed. Scarborough, ON: Prentice-Hall.

Gunn, A. 1976. "The Lost Lands." *The Province* (Vancouver), 20 January, 5.

Harris, C. 1992. "The Lower Mainland, 1820-81." In *Vancouver and Its Region*, ed. G. Wynn and T. Oke, 38-68. Vancouver: UBC Press.

–. 1997. *The Resettlement of British Columbia: Essays on Colonialism and Geographical Change*. Vancouver: UBC Press.

Joseph, S. 1991. "Assimilation Tools: Then and Now." *BC Studies* no. 89 (Spring): 64-81.

McKee, C. 1996. *Treaty Talks in British Columbia: Negotiating a Mutually Beneficial Future*. Vancouver: UBC Press.

Muckle, R.J. 1998. *The First Nations of British Columbia*. Vancouver: UBC Press.

Nisga'a Treaty Negotiations: Agreement-in-Principle. 1996. Issued jointly by the Government of Canada, the Province of British Columbia, and the Nisga'a Tribal Council. 16 February, Victoria.

Tennant, P. 1990. *Aboriginal Peoples and Politics: The Indian Land Question in British Columbia, 1849-1989*. Vancouver: UBC Press.

Willems-Braun, B. 1997. "Colonial Vestiges: Representing Forest Landscapes on Canada's West Coast." In *Troubles in the Rainforest: British Columbia's Forest Economy in Transition*, ed. T.J. Barnes and R. Hayter, 99-127. Canadian Western Geographical Series vol. 33. Victoria: Western Geographical Press.

INTERNET

Aboriginal Rights Coalition of British Columbia, vvv.com/~arcbc/

British Columbia, Ministry of Aboriginal Affairs, www.aaf.gov.bc.ca/aaf

British Columbia Treaty Commission, "Status of Negotiations," www.bctreaty.net/files/status.html

6 The Geography of Racism: The Spatial Diffusion of Asians

The Chinese were the first Asians to arrive in British Columbia and may even have arrived before Europeans. There has been some speculation that Buddhist monks reached the shores of western North America in approximately AD 458 and later in AD 594 (Caley 1983; Lai 1987). Certainly the Chinese were involved in the initial building of the British Nootka Sound Fort in 1788. It was the gold rush of 1858, however, that resulted in significant immigration of the Chinese to British Columbia. It was also the beginning of attitudes and policies of racism and discrimination that increased in intolerance to the point of hostility and were applied to all Asians in the province.

Spatial diffusion is the movement over space and through time of people, goods, and ideas. The forces or influences that assist in this movement are the *carriers,* and the forces that oppose or block the diffusion process are referred to as *barriers.* The spatial diffusion of people, the focus of this chapter, is known as relocation diffusion. "Push" and "pull" factors assist in assessing the movement of people. Push factors include the various forces – political, economic, religious, and so forth – that sway, or even force, people from their present location. Pull factors represent the many influences that attract people to a specific location. Through this concept we can assess the relocation of Chinese, Japanese, and East Indians to British Columbia at a number of geographic scales. Both push and pull factors were involved in the move from their home countries, and once they arrived and "settled," other factors pushed and pulled these people throughout the province and Canada. Each group had its own unique push and pull factors, although racism was a common force. The following periods are used:

- Chinese immigration and diffusion from 1858 to 1907
- Japanese immigration and diffusion to 1907
- East Indian immigration and diffusion to 1907
- Chinese immigration and diffusion from 1908 to 1947
- Japanese immigration and diffusion from 1908 to 1941
- East Indian immigration and diffusion from 1908 to 1947

- Japanese relocation diffusion from 1942 to 1949
- Asian immigration and diffusion to the 1990s

CHINESE IMMIGRATION AND DIFFUSION FROM 1858 TO 1907

The Fraser River gold rush was the second major gold rush in North America, the first being in the Bay area of California in 1849. As Chapter 4 described, many of the early miners came to British Columbia from the San Francisco area, and some of those were Chinese. Other Chinese came directly from China. Statistics were not particularly accurate for those early days, but David Lai suggests that there were about 1,000 Chinese here by 1862, and 1,759 by 1868 (Lai 1987, 337 and 339). There may have been more, considering that the north half of Barkerville was known as "Chinatown" and Barkerville was a community of around 4,000 at its peak in 1865. Indeed, there may have been as many as 6,000 or 7,000 Chinese in the province in the early 1860s (Con et al. 1982, 14).

The lure of gold was clearly a "pull" factor, and the possibility of becoming rich and returning to China provided an obvious incentive for relocation. The actual decision to leave family and friends to go to a very remote, foreign, and not entirely friendly land does deserve some analysis and consideration of the conditions in China at this time.

China generally, and southern China specifically, can be described as in a state of chaos throughout the second half of the nineteenth century. The fixed amount of agricultural land simply could not support the number of people. In many cases the subdivision of family lands to sons was no longer possible as the plots were too small to support even one family. As a result, second, third, and following sons became landless and left to gain income by selling their labour. At the overall political level, the xenophobic Ch'ing dynasty was being seriously challenged by Europeans, who used their military superiority to introduce opium to southern China and to demand trade. To make matters considerably worse, the Taiping Rebellion in southern China from 1850 to 1864 claimed an estimated 30 million lives (Hucker 1975). Needless to say, all this provided some incentive to leave the country. The decision to go was facilitated by a family, or lineage, system that often pooled the meagre family resources to send landless sons to the new country with the expectation that they would remit any income. The "push" factors were powerful indeed.

The Chinese who came to British Columbia in those early days were mainly Cantonese-speaking single men from Guangdong province in southeastern China. They intended to earn whatever they could in this strange and foreign land and send surplus income to their families in China, always with the intention of returning themselves with sufficient money to afford land and support a family.

Within British Columbia another spatial diffusion process was to occur, and this one would be influenced initially by economic opportunity. Many Chinese became prospectors and followed the trails up the Fraser and into the Cariboo. Others became involved in the service industries in the communities of Victoria, Nanaimo, New Westminster, Lillooet, Quesnel, and Barkerville. As the Cariboo gold finds waned, new gold interests opened up on the Stikine, Peace, and Columbia Rivers. The Chinese located in these regions also.

European attitudes of superiority were inherent in colonization, treating all other visible ethnic groups as inferior. British Columbia was no exception. Added to this base of racism and discrimination was the economic nature of the new British colony. It depended on external demand for its natural resources, a condition that led to "boom" and "bust" times. During the "bust" periods, racism and discrimination tended to increase because of intense competition for scarce employment. After the "boom" period of the gold rush, the Chinese were desperate for employment and often without the option of returning to China. They were willing to work for very low pay. White labour versus Chinese labour became a significant part of the growing bitter racial conflict by the late 1860s.

Institutional racism was, in many ways, the most serious form of racism because it resulted in legislation that discriminated on the basis of race. It expressed the attitudes of the dominant white population through their elected politicians. In British Columbia, this sentiment came early. In 1872, the new provincial government proposed charging a **head tax** of fifty dollars on all incoming Chinese to limit immigration. Immigration came

Table 6.1

Chinese occupations in British Columbia, 1884

Occupation	Number of persons	Population percentage
Labourers		
Railroad	2,900	27.6
Store employees	302	2.9
Ditch diggers	156	1.2
Others	296	2.8
Farming, logging, and mining		
Miners (mostly gold)	1,709	16.3
Coal miners	727	6.9
Farm labourers	686	6.5
Wood cutters	230	2.2
Fuel cutters	147	1.4
Vegetable gardeners	114	1.1
Production processes		
Fishers	700	6.7
Sawmill workers	267	2.5
Boat builders	130	1.2
Brick workers	85	0.8
Service and recreation		
Cooks and servants	279	2.7
Launderers	156	1.5
Barbers	71	0.7
Prostitutes	70	0.7
Managerial		
Merchants	120	1.1
Restaurant keepers	11	0.1
Sales		
Peddlers	67	0.6
Occupations not stated		
Married women	55	0.5
Girls	33	0.3
Boys under 17	529	5.0
New arrivals	602	5.7
Professional and technical		
Doctors	42	0.4
School teachers	8	0.1
Total	10,492	100.0

Note: Total does not equal 100 percent, due to rounding.
Source: Lai (1975), 16.

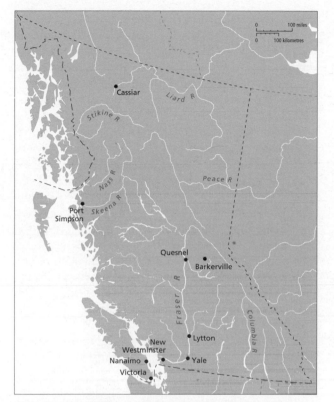

Figure 6.1 Main locations of Chinese in British Columbia, 1881
Source: Modified from Galois and Harris (1994), 42.

ment opportunities (mining and service industries) and new employment opportunities (railway construction).

The main construction years in British Columbia were between 1881 and 1885. The railway line was divided into many individual contracts, and each contract meant an increase in demand for labour in a province that had a limited population of white workers. This provided the CPR with the rationale for importing Chinese labour. Andrew Onderdonk, the contractor for the CPR, brought in 15,701 Chinese from the United States and China over these years, though many either went into other professions or left the province after their short contracts were up. They were paid half the wages of white labourers, and many were promised return passage to China (Chodos 1973; Gunn 1975; Li 1988; Lai 1987). This promise was never fulfilled.

As the contracts ended, a serious condition of surplus labour and unemployment occurred. The Americans anticipated the crisis and compounded the problem by trapping the Chinese in British Columbia. The US Exclusion Act, passed in 1882, would not allow the Chinese in British Columbia access to the United States. These events resulted in further diffusion of the Chinese

under federal jurisdiction, however, and the federal government rejected this proposal. In 1878, the provincial government excluded the Chinese from any provincial works, and thus began the institutional barriers to employment. Anderson (1989) documents well the historical roots of anti-Asiatic sentiments and the many forms of institutional racism practised in Canada, in British Columbia, and in Vancouver.

The building of the Canadian Pacific Railway (CPR) was another "boom" time and promoted a major influx of Chinese immigration – and a racial backlash. The census of 1881 probably underestimates the number of Chinese in British Columbia. Table 6.1 shows the range of employment, with CPR construction workers heading the list. Figure 6.1 shows the spatial distribution of the Chinese throughout the province. The main influence on distribution was a combination of old employ-

Table 6.2

Chinese immigration and emigration, 1886-1900

	Entries	Exits	Net immigration	Estimated population
1886	212	829	-617	11,400
1887	124	734	-610	10,800
1888	290	868	-578	10,100
1889	892	1,322	-430	9,600
1890	1,166	1,671	-505	9,100
1891	2,125	1,617	508	9,129
1892	3,282	2,168	1,114	9,110
1893	2,258	1,277	891	9,800
1894	2,109	666	1,443	10,400
1895	1,462	473	989	11,000
1896	1,786	696	1,089	11,500
1897	2,471	768	1,703	12,200
1898	2,192	802	1,390	12,600
1899	4,402	859	3,543	13,500
1900	4,257	1,102	3,155	15,000

Source: Con et al. (1982), 296.

throughout the province and, in some cases, back to China (Table 6.2).

With the significant increase in Chinese immigrants, many of whom were unemployed and viewed as a threat, more racial legislation was enacted. A provincial law passed in 1885 denied Chinese the vote, and the provincial government was able to convince the federal government to impose a fifty-dollar head tax on all Chinese immigrating to British Columbia. An examination of Table 6.2 indicates that the combination of high unemployment and a head tax had the desired effect of limiting immigration, but only in the very short run.

By the 1890s, the number of Chinese paying the head tax had increased considerably. Consequently, it was raised to $100 in 1901 and $500 in 1904, a great amount of money in those days. Again the tax reduced immigration – only 8 arrived in 1905, and 22 in 1906. By 1908, the number was up to 1,482, and another 7,078 came in 1913 (Lai 1987, 345). The increases reflected the continuing conditions of chaos and lack of economic opportunity in China. British Columbia, for all its racism and discrimination, provided the Chinese with a substantially better economic environment than did China.

Racism often led to Chinese people being placed in hazardous work situations as well as being subject to physical violence and property damage. The many deaths from unsafe working conditions in the building of the CPR is well documented (Burton 1970). Riots occurred in a land-clearing contract for the Brighouse Estate in Richmond in 1887, and strikes took place at the coal mines of Nanaimo over the hiring of Chinese. By far the most publicized violence was the Labour Day riot of 1907. Here anti-Asiatic feeling was whipped up by fire-and-brimstone speakers calling for a white Canada. Most of the establishments of Vancouver's Pender Street Chinatown were damaged as a result, as well as many of the buildings in "Little Tokyo," the Japanese district on Powell Street (Adachi 1976).

While many British Columbians were unaffected by the presence of Asians in the province, newspapers, politicians, and special interest groups were quick to stereotype and blame Asians for many economic problems. Nevertheless, they were appreciated by those who hired them and profited by their hard work.

JAPANESE IMMIGRATION AND DIFFUSION TO 1907

To know the story of the Japanese in British Columbia requires some understanding of Japanese history and international relations. Japanese arrived in the province much later than the Chinese, had different "push" and "pull" factors, and had different forms of institutional racism applied to them.

In the Tokugawa Period (1600-1867), Japan was closed to western trade and influence. The shogun, or leader of Japan, not only rejected outside influences but also prohibited Japanese people from leaving the country. Indeed, the size of fishing boats was limited by law specifically to prevent people from building ocean-going vessels. First the United States and then Europe used military superiority to impose unequal trade treaties on Japan. Japan responded with a revolution that overthrew the shogun system and installed an emperor, beginning the period referred to as the Meiji Restoration (1868-1912).

Unlike China, Japan adopted western technology, science, a constitution, and industrial development almost overnight, which resulted in major restructuring of the society and economy. Agricultural workers were displaced, the population grew rapidly, the rigid class structure diminished, and taxes were very high, setting the stage for revolts and emigration. By 1885, emigration was encouraged.

Industrialization required raw materials such as iron ore, coal, and other mineral resources, of which Japan had very little. In a quest for these materials, Japan embarked on its own imperialism by exerting control over Taiwan and Korea and having unconcealed designs on mineral-rich Manchuria by the 1890s. Russia also had designs on Manchuria and sent troops in to secure the rail line that ran from Russian territory through Manchuria to Port Arthur. Russian and Japanese troops soon clashed, creating tension between the two countries.

Russia was attempting to become a world military force in the 1890s and early 1900s by building up a modern naval fleet and army on the European side of the continent. This posed a threat to British naval interests. Japan's movement toward Manchuria was viewed by Russia as a threat to its port of Vladivostok. In an attempt to control Russia, Britain and Japan signed an alliance in

1902. Ultimately war broke out between Russia and Japan (1904-5). Japan's decisive victory on both land and sea elevated its global status to a modern military state, ended the unequal trade treaties imposed earlier, and cemented relations between Japan and Britain.

The Anglo-Japanese Alliance meant that the British colony of Canada could not impose restrictive head taxes on the Japanese as they had on the Chinese. Nevertheless, the Japanese were denied the vote in 1895, and disenfranchisement meant restrictions on employment opportunities for both immigrant Japanese and future generations born in Canada.

The 1901 census records show only 4,738 Japanese in British Columbia. Most were poor, single men lured by the opportunities of fishing or employment in the coal mines and lumber mills. Few at this time could afford land for farming. Besides, most intended to work for a few years only and then return to Japan (Adachi 1976). Few Japanese emigrated to British Columbia, or anywhere, between 1901 and 1904 because of the impending war with Russia. Once the war was over, however, immigration to British Columbia was renewed at approximately 2,000 per year. Events in 1907 increased Japanese immigration to British Columbia, and racism and discrimination increased as a consequence. White British Columbians demanded that immigration be halted.

Hawaii was the most popular destination for emigrating Japanese. Between 1885 and 1907, approximately 180,000 Japanese made their way to Hawaii. Many returned to Japan after accumulating some income, but many others went from Hawaii to the United States and some to British Columbia. In 1907, the United States passed an act that prohibited the Japanese from leaving Hawaii to go the United States. As a result, British Columbia received a large increase in Japanese immigrants – over 7,000 that year (Adachi 1976). This coincided with a substantial increase in Chinese and East Indian immigration at the same time. Rumours of the new Grand Trunk Pacific Railway demanding even more Asian labour, along with a plan by the CPR to bring in a thousand Japanese for an irrigation project in southwestern Alberta, added fuel to anti-Asiatic feelings and provoked the Labour Day race riot of 1907.

The Chinese boarded up their establishments on Pender Street in response to the mob of rioters. The Japanese organized themselves into a resistance force, and a physical confrontation occurred that ended the riot (Adachi 1976, 74). Following the riot, the Chinese went on strike, and eventually the City of Vancouver paid $26,990 in compensation for the damages (Li 1988, 32). The event made headlines around the world and became an embarrassment to Canada because of the Anglo-Japanese Alliance.

EAST INDIAN IMMIGRATION AND DIFFUSION TO 1907
The emigration of people from India to British Columbia was almost exclusively Sikh prior to 1960. The Sikhs had their own unique struggles within India that produced the "push" factors for emigration. Particularly distinctive here was that the Sikhs, unlike the Chinese or Japanese, came from another colony of Britain. They, like British Columbians, were British subjects. This held little merit with the majority of British Columbians, however, who did not want to see any Asians in Canada.

Sikhism evolved out of Hinduism, but not without the confrontation and violence of religious and political separation. India's colonial period was marked by resistance to British imperialism on the part of the two main religious groups, Hindus and Muslims. The Sikhs sided with the British and as a consequence were rewarded with land in the Punjab region of northern India for their "homeland." By the 1900s, the chaos had reached extreme levels; the Punjab region was being heavily taxed, religious and political conflicts with India continued, famines struck, and so did major outbreaks of diseases (Sandhu 1972).

Sikh troops visiting the province in 1897 brought back the word that British Columbia was opening up and that there was a demand for labour in the lumber industry, a field that Sikh workers were familiar with in the Punjab. There were few Sikhs in British Columbia by 1904 – perhaps a 100 or less – but 387 came in 1905, 2,124 in 1906, and 2,623 in 1907 (Minhas 1994). The attraction of employment in the lumber industry of British Columbia had been discovered.

Repeating the pattern of the Chinese and Japanese immigration, most of the Sikhs were poor, single men and most became employed in the sawmills of the Lower Mainland. Spatial diffusion occurred throughout British Columbia as employment opportunities on Vancouver Island and in the interior became available for land clearing, road building, and other labour-intensive activities.

By 1907, the number of Sikhs in British Columbia had gained the attention of, and alarmed, the dominant anti-Asiatic white society. Yet because Sikhs were British subjects, the range of racially discriminating laws was limited. They were denied the vote in 1907, and disenfranchisement put a ban on working in the professions. The riot of 1907 did not directly involve Sikhs, but it spelled out very clearly the sentiment of exclusion that dictated that Asian immigration should be curbed. The only question was how to curb the immigration of British subjects.

CHINESE IMMIGRATION AND DIFFUSION FROM 1908 TO 1947

The "solution" to halting Chinese immigration, the $500 head tax, was not working. In fact, it was becoming a windfall in revenues for the Canadian government (Table 6.3). Chinese immigration lessened during the First World War because considerably fewer ships were travelling between China and British Columbia, but the numbers increased significantly when shipping resumed as the war ended (Figure 6.2). In 1919 alone, 4,066 people arrived. As Chinese immigration to British Columbia climbed, so too did the demand for Chinese exclusion.

The Chinese were employed in the service industries and resided in the Chinatowns of Vancouver, Victoria, New Westminster, Cumberland, Nanaimo, and many of the old mining and railway communities of the interior. Some were still employed in mining, some in the lumber mills, a great many in canneries, and agriculture became an important way of life for others. The Chinese were beginning to see British Columbia as home, and this was

Table 6.3

Revenue from head taxes on Chinese entering British Columbia

	Number paying head tax	Number exempt	Revenue ($)
1886-94	12,197	264	624,679
1895-1904	32,457	430	2,374,400
1905-14	27,578	4,458	13,845,977
1915-24	10,147	2,807	5,678,865
Total	82,379	7,959	22,523,921

Source: Li (1988), 38.

of grave concern to the anti-Asiatic sentiments of many. The year 1921 saw a federal election, and all the candidates from British Columbia, regardless of political party, ran on platforms to end Chinese immigration (Lai 1987, 347). There were approximately 24,000 Chinese in British Columbia by this time, forming a significant minority group. The Chinese Exclusion Act, passed in 1923, had a profound impact on immigration and on Chinese people already in the province. Approximately twelve Chinese immigrated between 1924 and 1946.

The act prohibited Chinese from re-entering Canada if they left for a period of more than two years, and it denied entry to Canada to wives and children left in

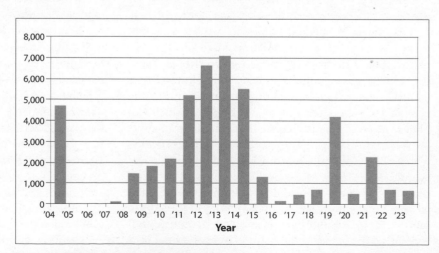

Figure 6.2 Chinese immigration, 1904-23
Sources: Modified from Lai (1987); Urquhart (1965).

China by men who had already emigrated (Lai 1987, 348). The new rules created serious hardship. A number of Chinese returned to China permanently; others began to spread across Canada. Many settled in Toronto, but the largest concentration still formed Vancouver's Chinatown.

JAPANESE IMMIGRATION AND DIFFUSION FROM 1908 TO 1941

The embarrassment of the 1907 race riot resulted in Canada's labour minister being sent to Japan to make amends and to work out immigration limits. The Gentleman's Agreement of 1908 created four categories of Japanese immigrants: A, prior Canadian residents; B, domestic servants for Japanese residents; C, contract labourers; and D, agricultural labourers. A limit of 400 per year was set for classes B and D but no restrictions on classes A and C.

A new "class" of Japanese immigrant was not restricted by this agreement – wives. Marriages were arranged by parents in Japan. The single men in British Columbia recognized that they were here to stay and requested their parents in Japan to find suitable wives for them. This became known as the "picture bride" era. It became very common for relatives in Japan to send a picture of a prospective bride to British Columbia, and if she met with approval, to send the woman over on the next vessel. Wedding ceremonies, often in Christian churches, would then be conducted.

This process caused considerable racist sentiment in British Columbia on two counts. First, the concept of an arranged marriage appeared immoral to British, Christian morals. Second, the natural consequence of many of these marriages was children, increasing an Asian population that was perceived as unassimilable and undesirable.

Powell Street became the commercial centre for the Japanese, and the enclave of Steveston in south Richmond preserved the language, religion, and customs. The main employment in Steveston was fishing, which the Japanese began to dominate from a fairly early time. By 1901, for example, the Japanese held 1,958 fishing licences out of a total of 4,722 issued. With two people to a boat, or licence, over 4,000 Japanese were engaged in the fishing industry (Adachi 1976, 47). Fishing interests also led to the diffusion of the Japanese to the central and north coast fishing grounds of Rivers Inlet and up to the Skeena River region. Forestry, mining, and labouring jobs attracted others throughout British Columbia. The majority of the Japanese, however, remained in the Lower Mainland region.

The demand for exclusion of the Japanese continued even though a number of Japanese living in this country fought for Canada in the First World War, and some died. Those who survived were promised the vote, but opposition was so strong that they did not receive it until 1931 (Adachi 1976, 155). Japanese domination of the fishing industry, a source of much concern, became the target of the Duff Commission in 1922: Japanese fishing licences were reduced by 40 percent. With over 1,200 fishers driven from employment the Japanese focused on farming. The often marginal lands of the Fraser Valley became most popular; others went to farm the Comox Valley, the Okanagan, and Cariboo.

By 1922, the alliance forged between Britain and Japan had expired and so, apparently, had the need for treating the Japanese in British Columbia with diplomacy. In 1923, class B and D immigrants were restricted to 150 per year, and by 1928 all classes, including "picture brides," were restricted to that number. As with the Chinese, the door to Canada for the Japanese was nearly closed.

The impact of not being able to vote became more and more restrictive, particularly for the *nisei*, or second generation Japanese, who were born here. Educated in British Columbia's schools and, in some cases, graduating from the University of British Columbia, the professions were out of reach for them because they were not on the voters' list. By 1936, the Japanese in British Columbia organized and lobbied Ottawa for the federal vote. Unfortunately, most of the 1930s was influenced by the Depression. As well, on the international scene Japan was providing a military challenge to China and, indirectly, to Britain and the western world. The backlash affected all Japanese, even if they had been born in Canada. No franchise was granted. In fact, as the Second World War began in 1939, Japanese Canadians who volunteered for military duty were rejected and all Japanese over the age of fifteen were forced to be fingerprinted and registered.

The bombing of Pearl Harbor on 7 December 1941 by the Japanese and the capture of Hong Kong on Christmas Day, resulting in Canadian casualties and hundreds of prisoners of war, solidified the Pacific west coast's worst fears – imminent invasion by Japan. All Japanese in western North America become suspect and classified as enemy aliens. This fear had a devastating impact on the Japanese in British Columbia, who were forcibly moved from the coast and lost goods and property.

EAST INDIAN IMMIGRATION AND DIFFUSION FROM 1908 TO 1947

The Canadian authorities were well aware that Sikhs were British subjects and that they belonged to a minority group seeking independence in India, a conflict they did not want to spill over into Canada. They were also amply aware of the anti-Asiatic feelings of those British Columbians who wanted all Asians banned. The new rules established for Sikh immigration, in 1908, were as effective as the Chinese Exclusion Act. Two key rules were put in place. First, immigrants from India had to have $200 in their possession when entering Canada, a great sum of money for the mainly poor peasants immigrating to find employment. Second, immigrants had to arrive by continuous passage, but with shipping technology at this time no vessels travelled directly between India and British Columbia. Because of these restrictions, Sikhs were no longer able to enter Canada legally. Many left the province, some going to the United States and others returning to India.

These discriminatory rules were not without challenges, the most publicized being the *Komagata Maru* incident in 1914. This Japanese ship carrying approximately 400 Sikhs arrived in Vancouver on 23 May but did not comply with the continuous passage rule or, in most cases, the $200 rule. Refused docking privileges, food, and water, the *Komagata Maru* remained in Vancouver's harbour for eight weeks while protracted legal negotiations resulted in riots and bitterness. The federal government finally provided food and water, but few of the passengers were allowed to stay in British Columbia and the *Komagata Maru* was escorted out of Canadian waters by a navy gun boat. By the end of the First World War, only a thousand or so Sikhs remained in Canada. The door was effectively closed on this Asian group.

JAPANESE RELOCATION AND DIFFUSION FROM 1942 TO 1949

Most Japanese in British Columbia lost everything. They were suspected of spying and consorting with the enemy; their fishing boats were confiscated along with automobiles and radios. All were auctioned off at fire sale prices. Japanese-language newspapers and schools were outlawed, and many employers laid off their Japanese employees. The racist politicians of British Columbia used the threat of a Japanese invasion as a club to incarcerate all Japanese. Mass evacuation from the coast of British Columbia was seen as the "solution" to the Japanese problem.

A hundred-mile demarcation line from the coast of British Columbia was established and all Japanese were to be relocated east of it (Figure 6.3). Initially, the most suspect were rounded up and sent to a prisoner-of-war camp in northern Ontario. Over 2,000 men were sent to road construction camps in the interior, while families were rounded up in the cattle exhibition area of the Pacific National Exhibition (PNE) grounds in Vancouver. Most relocation was to nearly bankrupt ghost towns in the Kootenays – Sandon, Greenwood, New Denver, Slocan Valley, and Kaslo – as well as to Tashme, a name derived from the first two letters of Taylor, Shirras, and Meade, members of a security commission created under the War Measures Act. Tashme was only twenty-three

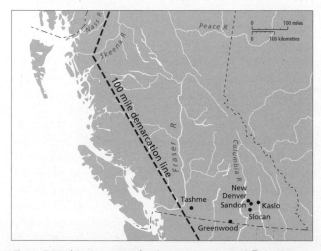

Figure 6.3 Relocation camps for the Japanese in British Columbia

Table 6.4

Japanese evacuees, 31 October 1942

	Number of evacuees
Detention camps	
Greenwood	1,177
Slocan Valley	4,814
Sandon	933
Kaslo	964
Tashme	2,636
New Denver	1,505
Subtotal	12,029
Road construction camps	
Blue River/Yellowhead	258
Revelstoke/Sicamous	346
Hope/Princeton	296
Schreiber	32
Black Spur	13
Subtotal	945[a]
Sugar beet projects	
Alberta	2,588
Manitoba	1,053
Ontario (males only)	350
Subtotal	3,991
Other permits and situations	2,520
Total	19,485

a   Between March and June 1942, 2,161 Japanese were placed in road construction
    camps. Many were allowed to have their families join them by October 1942.
Source: Adachi (1976), 415.

kilometres from Hope and became the site of the Hope-Princeton road construction project, as 2,300 people spent three years building it in supervised confinement (Evenden and Anderson 1973, 42). The first winter was exceptionally cold and uncomfortable, with two families per uninsulated cabin. Insulation and weather proofing were added for the next winters.

The war effort produced a shortage of labour everywhere, and politicians saw the Japanese as pools of labour not only for road building but also for the harvest of sugar beets on the Prairies and in Ontario. Several thousand Japanese relocated to these destinations, although Alberta expressed concern over inheriting British Columbia's so-called Asian problems (Table 6.4; Adachi 1976, 415).

As the war progressed, the Japanese settled into the detention camps in a more permanent way. The Canadian government recognized that the Japanese still had equity in the buildings and property from which they had been evicted on the coast. These possessions became the next round for the auctioneers, and in the process the Japanese lost most of what they had acquired up until 1941. Homes, farms, and commercial enterprises were all sold off.

As the war drew to an end, provincial politicians began to pressure the federal government to prevent the camps from becoming permanent locations. By the end of 1944, the decision was made to close the detention camps, and the Japanese were given a choice. They could relocate east of the Rockies, since the coast was still out of bounds, or be repatriated back to Japan. Rejected by Canada, many Japanese initially opted to be deported to Japan; when it came time to leave, however, only about 4,000 actually went back to war-torn Japan. Those who remained underwent another relocation process, with the greatest number taking up residence in Ontario, mostly in Toronto.

Finally, in 1949, four years after the end of the war, the Japanese were allowed to return to the coast of British Columbia to live, and they were given the vote. One of the tragedies of this period, as can be seen in Table 6.5, is that the majority of Japanese incarcerated were born in Canada.

ASIAN IMMIGRATION AND DIFFUSION TO THE 1990s
The end of the Second World War ushered in a period of affluence and a change of thinking about minority groups in Canada. A new acceptance led to renewed

Table 6.5

National status of Japanese Canadians in 1941

	Number	Percentage
Japanese nationals	5,564	25.2
Naturalized Canadians	3,223	14.6
Born Canadians	13,309	60.2
Total	22,096	100.0

Source: Adachi (1976), 414.

immigration and the franchise. The Chinese Exclusion Act was repealed in 1947 and limited immigration again resumed. For East Indians, the restrictive rules on continuous passage ended in 1947, at which time they also received the vote. The Japanese were allowed back to the coast and received the vote in 1949. These were important changes in breaking down institutional racism, but a quota system for immigrants by race was in place, making the immigration process restrictive and racist still.

International pressures removed racial discrimination as a policy of immigration by 1962, but the non-racist point system that currently operates was not put in place until 1967. The mix of immigrants changed considerably in both numbers and background with the modifications to the immigration rules. Many more Chinese and East Indians, especially, began to immigrate to Canada. The class of new immigrants also changed substantially. These newcomers were no longer dominated by poor farmers, fishers, and labourers as in the old days, nor did they have to confine themselves to the Lower Mainland of British Columbia.

In terms of spatial diffusion throughout Canada, it is interesting to see that, while each of these groups initially established themselves in British Columbia, they spread out across Canada, particularly after the Second World War, and relocated mainly to Ontario. The Chinese, Japanese, and Sikhs have very different histories in British Columbia but share a common trait of suffering from racism, distrust, and discrimination. Ontario stands out as being far more receptive to racial differences, and consequently, more Chinese and Japanese live in Ontario than in British Columbia.

For the Chinese, the residential segregation that created the many Chinatowns throughout British Columbia has largely disappeared from the landscape. The exceptions are Vancouver and Victoria, where Chinatowns exist mainly as tourist centres and as the location of some Chinese institutions, goods, and services but provide little in the way of housing.

The East Indian population is still largely Sikh and is now greater than 300,000 in Canada. They too have diffused across the country, with about 100,000 each in the Toronto area and the Lower Mainland of British Columbia.

For the Japanese in Canada, 1989 was an important year as Prime Minister Brian Mulroney issued an official apology for their treatment during and after the Second World War. Moreover, a compensation package of approximately $21,000 was offered to every Japanese Canadian born or residing in Canada prior to 1948. A reflective message, however, is offered by Sunahara (1981, 169): "In a country that prides itself on its democratic tradition, it is sobering to note that everything done to the Japanese Canadians was, and still is, legal under Canadian law."

REFERENCES

Adachi, K. 1976. *The Enemy That Never Was.* Toronto: McClelland and Stewart.

Anderson, K.J. 1989. "Cultural Hegemony and the Race-Definition Process in Chinatown, Vancouver: 1880-1980" *Environment* 6 (2): 127-49.

Berton, P. 1970. *The National Dream: The Great Railway 1871-1881.* Toronto: McClelland and Stewart.

Caley, P. 1983. "Canada's Chinese Columbus." *The Beaver* (Spring): 4-11.

Chodos, R. 1973. *The CPR: A Century of Corporate Welfare.* Toronto: James Lorimer.

Con, H., R. Con, G. Johnson, E. Wickbers, and W.E. Willmot. 1982. *From China to Canada.* Toronto: McClelland and Stewart.

Evenden, L.J., and I.D. Anderson. 1973. "The Presence of a Past Community: Tashme, British Columbia." In *Peoples of the Living Land: Geography of Cultural Diversity in British Columbia,* ed. J.V. Minghi, 41-66. BC Geographical Series no. 15. Vancouver: Tantalus.

Galois, R., and C. Harris. 1994. "Recalibrating Society: The Population Geography of British Columbia in 1881." *Canadian Geographer* 38 (1): 37-53.

Gunn, A. 1975. "Our History of Ethnic Prejudice." *The Province* (Vancouver), 18 November, 5.

Hucker, C.O. 1975. *China to 1850: A Short History.* Stanford: Stanford University Press.

Lai, D.C. 1987. "Chinese Communities." In *British Columbia: Its Resources and People,* ed. C.N. Forward, 335-57. Western Geographical Series vol. 22. Victoria: University of Victoria.

–. 1975. "Home Country and Clan Origins of Overseas Chinese in Canada in the Early 1880s." *BC Studies* no. 27 (Spring): 3-29.

Li, P.S. 1988. *The Chinese in Canada.* Toronto: Oxford.

Minhas, M.S. 1994. *The Sikh Canadians.* Edmonton: Reidmore Books.

Sandhu, K.S. 1972. "Indian Immigration and Racial Prejudice in British Columbia: Some Preliminary Observations." In *Peoples of the Living Land: Geography of Cultural Diversity in British Columbia,* ed. J.V. Minghi, 29-40. BC Geographical Series no. 15. Vancouver: Tantalus.

Sunahara, A.G. 1981. *The Politics of Racism.* Toronto: James Lorimer.

Urquhart, M.C., ed. 1965. *Historical Statistics of Canada.* Toronto: Macmillan.

FILMS

Lerman, Jeanette, director. 1975. *Enemy Alien!* 25 mins. Wolf Koenis, producer. National Film Board.

Resource Management
in a Changing Global Economy

7

Resources have been, and continue to be, the strength of British Columbia. The value of these resources has fluctuated over time and new resources have been discovered, bringing changes to regional economies and sometimes conflict. This chapter begins by defining resources and describing several categories of use, or models, for examining them. Resources are managed at the three levels of government: federal, provincial, and municipal. The responsibilities of the three may overlap and cause conflict.

Not only is there a political dimension to resource management but there is a host of economic considerations, in terms of the way in which resources are used, or valued. These have changed over time and are influenced by the interrelated and shifting roles of labour, technology, capital, land, and other factors affecting the production of goods and services. Here again, all levels of government play a significant role. British Columbia's reliance on resources and on the changing factors of production and markets has left its mark of boom and bust economies. The global economy of today produces spatial differences in British Columbia in terms of what resources are used, how resources are used, and how people of British Columbia make a living.

DEFINING RESOURCES AND CATEGORIZING THEIR USES

A simple definition of a **resource** is any naturally occurring substance that is of value to society. The key to this definition is not in naming an endless number of substances but in understanding their function in society. Implied is that resources are culturally defined. Herring roe, for example, is a significant resource in this province, bringing in several millions of dollars to BC fishers even though very little roe is consumed in British Columbia. It is highly valued in Japan, however, where it is considered a delicacy. Different cultural groups value materials, or resources, in very different ways.

Also implied in the definition is that the value of a resource can change. One only needs to consider the value of coal from the beginning of the twentieth century to now. It used to be the most important energy source for cooking and heating homes, running locomotives, and firing steam-driven engines in industry. By the 1950s and 1960s, petroleum, in the form of diesel fuel and gasoline, was replacing the transportation uses

of coal, while electricity and natural gas were replacing its residential and industrial uses. These were not good times for coal miners in the Fernie area of southeastern British Columbia. Then, in the 1970s, new technologies and demands by new markets increased the value of coal. Japan became the main purchaser of BC coal, using coking coal in the iron and steel industry and thermal coal in creating electricity. The new communities of Elkford and Sparwood in the Fernie area and Tumbler Ridge in the Peace River area symbolize new interests and employment related to the metamorphosis of this resource.

The same resources may have more than one value. Fresh water, for example, has numerous values. Anglers, swimmers, boaters, environmentalists, and others value fresh water in its natural setting from very different perspectives, although all these views may be complementary. The use of fresh water for hydroelectric energy production or as a system to discharge effluent from industry or domestic sewage is more contentious. Competing values are the essence of the need for resource management.

Resources can be categorized in a number of ways that help in understanding the dynamics of their use and management. One way is to separate renewable from non-renewable resources. Renewable resources are living, or biotic, resources, which can be further divided into plants and animals. The various species of trees, agricultural crops, and all other plant life, along with domestic and wild species of animals, and humans are all classified as renewable resources. A cautionary note is required with this classification: these resources are renewable *only if they are properly managed*. The extinction of plants and animals is of concern throughout the world. Noted Canadian geneticist David Suzuki has enlightened many Canadians about the concerns of the threat to biodiversity due to decimation of species and reduction in resource options.

Non-renewable resources are non-living, or abiotic, such as minerals and fossil fuels. Non-renewable implies that there is a fixed or finite amount. In practical terms, it is not a simple matter to determine the fixed amount of a non-renewable resource such as oil or natural gas, much less gold or copper. More frequently we view these resources within a more specific, regional context and make projections about how long a particular mine

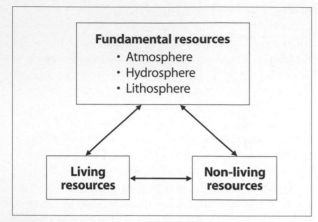

Figure 7.1 Ecosystem model of resource use

or oil well will produce until it is exhausted. Even then, these estimates are often only good for a particular time and a particular technology. The concept of a finite amount of any resource is important to the management process.

Figure 7.1 is an ecosystem model of resource use, illustrating that our use of resources affects the environment and the use of other resources. The term *fundamental resources* refers to air (atmosphere), water (hydrosphere), and land (lithosphere). The mix of these three essential spheres, along with energy from the sun, produces photosynthesis. They are thus fundamental to life and are therefore known as life-giving resources.

The dynamics of the ecosystem model are revealed by looking at the example of how people rely on and use more and more fossil fuels for energy and materials. Using these non-living resources results in ever greater amounts of waste (pollution) such as sulphur dioxide, which is given off to the atmosphere, one of the fundamental resources. In the atmosphere, sulphur dioxide (SO_2) combines with water vapour (H_2O) and falls as acid precipitation (H_2SO_4), which affects the land (lithosphere) and the resources of the land, including living resources. Increased acidity from acid rain can have a serious impact on drinking water, forests, agriculture, and fish living in the streams and lakes (hydrosphere). The ecosystem model demonstrates the cyclical nature of resource use, thereby recognizing environmental impact as a "cost" in the production of any given commod-

ity. This assessment incorporates the real cost of commodities and raises the concept of sustainability of resources.

In the field of resource management it is essential to view the whole process of how resources are developed into products. A struggle in the past that continues to this day is recognizing and assessing the impact and costs associated with resource development: the packaging that increases the costs for solid wastes, the effluents from industries discharged into water, and the potential environmental effects, including human health, of using products such as herbicides and pesticides. What level of effluent or discharge is "safe," and who pays the costs of clean up? Considering that many of the consequences to human health and the environment take a great deal of time to assess, it is not surprising that these negative factors have been labelled as externalities to the production process. Corporations have therefore not had to include environmental repercussions as a cost of production.

Sustainability of resources raises the fundamental question of the rate of resource use. Will resources be maintained for the future? Managing to prevent overuse and depletion of either renewable or non-renewable resources is a difficult task, particularly when increasing populations create a corresponding increase in demand for products. Public awareness and participation in programs such as the "three Rs" – reduce, reuse, and recycle – forge a link between consumerism, waste, and sustainability.

Subsequent chapters examine individual resources that have been and continue to be important to British Columbia. A common theme is to see how resource use has developed and changed. As conditions of society change, so do the need for and use of resources. Technologies have changed, ways of making a living have changed, understanding of and attitudes toward the environment have changed, and British Columbia has become much more urban. Change provokes new conflicts over resource use. Traditional forestry practices, for example, run headlong into other valued resources of the forest such as old-growth preservation, wildlife, water quality, and tourism. More and more, all resources require management, and the various levels of government are responsible.

MANAGING OUR RESOURCES:
THE ROLE OF GOVERNMENTS

The federal, provincial, and municipal levels of government have the responsibility for managing resources. The distribution of powers between the federal and provincial governments was initially laid out in the British North America (BNA) Act in 1867 (renamed the Constitution Act of 1867) and now the Constitution Act (1982). The conditions and responsibilities of municipalities and regional districts in British Columbia are established by the Municipal Act under the provincial government. It should be kept in mind that the responsibilities of each entity are constantly being challenged and changed.

The federal government is responsible for relationships with other countries and for things external to Canada. This includes all foreign trade. Controversial issues such as the sale of CANDU reactors to Korea or fresh water to the United States are in the hands of Ottawa managers. The federal government also has jurisdiction over oceans and the resources of the oceans. For British Columbia this includes not only any potential offshore oil and gas reserves but also, and most important, the fishing industry. Chapter 9 describes how the management of fish stocks, salmon especially, has been fraught with difficulty. Aboriginal affairs are another federal government responsibility, and Chapter 5 describes the modern treaty process, which in many ways is about the management of resources. Most of the main infrastructure in British Columbia – ports, railways, the Trans-Canada Highway, pipelines, and airports – has been established by the federal government. As well as being crucial to the movement of resources, this infrastructure represents a substantial use of resources in its construction and operation. Many other areas of federal responsibility influence resource use, ranging from research and development funding to health, education, and social welfare.

The provincial government's responsibilities and involvement in managing resources are very broad. The following is a partial listing of the relevant ministries:

Ministry of Aboriginal Affairs
Ministry of Advanced Education, Training, and Technology

Ministry of Agriculture and Food
Ministry of Education
Ministry of Employment and Investment
Ministry of Energy and Mines
Ministry of Environment, Lands, and Parks
Ministry of Fisheries
Ministry of Forests
Ministry of Health
Ministry of Labour
Ministry of Small Business, Tourism, and Culture
Ministry of Transportation and Highways

Clearly there is some overlap between provincial and federal government jurisdiction. One of the reasons salmon stocks are so difficult to manage, for example, is that salmon spend critical periods in freshwater streams before migrating to the ocean. Although the freshwater streams are regulated under the federal Fisheries Act, the provincial government has allowed other resource developments that damage the quality of salmon-bearing streams, such as dams, forest activity, mining, urban development, and the discharge of effluent from any number of sources. Similarly, the modern treaty negotiations require agreement between provincial and federal levels of government as well as the First Nations.

Local government in British Columbia is made up of incorporated municipalities and, since 1965, regional districts. Their jurisdiction over land and other resources is subordinate to provincial authorization. Local government plays the most significant role in resource management through the process of land use zoning. Zoning influences where people work, live, and engage in recreation. Communities also build infrastructure that is essential to tourism, such as parks and recreation facilities. Again, overlap between local and senior levels of government can lead to conflicts.

In the mid- to late 1980s, fish farming took on the appearance of a "gold rush," as many companies wanted to raise salmon in netpens. Most of this activity occurred on the Sunshine Coast, mainly in the Sechelt Inlet. A classic multijurisdictional struggle to manage this new resource developed. Because the netpens were located in salt water, the federal Department of Fisheries and Oceans was involved, but as it was necessary to anchor the netpens to the foreshore, the provincial Ministry of

Lands became involved with foreshore leases. At the local government level, the Sunshine Coast Regional District expressed concerns over many resource use conflicts, from pollution to the privatization of the foreshore. These issues were in their backyard, but they were without the jurisdiction or political power to resolve the conflicts. Eventually all three levels of government were able to co-operate to make decisions on the resource uses in Sechelt Inlet.

Political managers are responsible for setting the "rules" for resource development. Governments establish permits for resource development, including rates of harvest and allowable levels of discharge of wastes into the environment. The various ministries listed above do much more than set the rules; they promote and encourage resource development through infrastructure projects, laws, and, in some cases, direct subsidization. The role of government can be powerful and is usually justified in terms of creating employment and gaining taxation revenues. Another way of viewing resources and resource management is from the perspective of economic viability. How resources have been used to develop the economy of British Columbia is best seen from an historical perspective that situates the province with reference to the rest of Canada and of the world.

RESOURCES, DEVELOPMENT, AND THE BC ECONOMY

Staples Theory

The Canadian economic historian Harold Innis used **staples theory** to describe the development of Canada (Watkins 1963; Barnes and Hayter 1997). His theory is based on the exploitation of resources, or staples, in Canada. Its main assumption is the existence of an external demand for these resources. Innis identified five resources in the development of Canada: fish, furs, timber, wheat, and minerals. These five represent both an historical and an east-to-west development. Minerals, the last resource on the list, are the exception to the directional trend. The first resource was cod, taken off the coast of Newfoundland beginning in the late 1400s. Next, the continental resources were developed from east to west with the fur trade, followed by timber and agriculture. Minerals tended to be discovered sporadically from region to region. Through these resources and the

economic activities tied to them, the Canadian economy developed.

The terms *backward, forward,* and *final demand linkages* are used in staples theory to describe the types of economic activity, including implications for employment, associated with each staple. Backward linkage refers to all the conditions necessary to export a resource. The most important backward linkage for any resource is the collection of transportation systems because it can influence so many other economic activities. When the country was first developed, this meant building and running port facilities with warehouses, boat repairs, and all the employment related to loading resources onto ships for export. Over time ports were connected to canals, railways, and road systems. Backward linkages include the employment created from building these facilities as well as the construction and manufacturing of rails, boats, trains, or any of the inputs to export a resource.

Forward linkage is the process of adding value to any resource through further processing or manufacturing prior to export. In British Columbia, there has always been concern about the export of raw logs versus the forward linkage of milling the logs into dimension lumber. Obviously, if the sawmilling activity occurs here, then more jobs are created in British Columbia. Much higher value-added to wood can be gained by manufacturing furniture, doors, or even musical instruments, with greater benefits to employment and to government revenues. Once a resource is tagged for export, there is usually a community involved, whether at the port or at the location of resource extraction. Over time, and with the accumulation of backward and forward linkages, some of these communities have become major centres.

Final demand linkages are defined as the demand for production of goods and services for the local or domestic market. In smaller communities, consumer goods may have to be imported. Nevertheless, as the population increases, the ability to reach thresholds for local production also increases.

The accumulation of all these linkages is referred to as the multiplier effect. A pulpmill, for example, locates in a community, adding 500 workers (forward linkage). These workers in turn need housing, food, and many other necessities and luxuries, and so the local economy grows to fulfill this increased demand (final demand linkages). One more pulpmill in the province may be all it takes to stimulate the manufacture of pulp machine components (backward linkages). All this economic activity can mean considerably more than 500 people working. To be kept in mind as well though is the main assumption of this theory, that there is an external demand for the resource. Periods of recession and depression are reminders that external demand has not always been sustained. When the mill or mine shuts down, the multiplier effect works in reverse and many more, apart from those in the mill or mine, face unemployment. When it is the only industry, closure of a mill or mine may put an entire community in jeopardy.

British Columbia's history of dependence on external demand for its resources continues today. Table 7.1 shows that the bulk of BC exports are resource-based products. Staples theory is therefore a useful framework to assess the impact that various forward, backward, and final demand linkages have in terms of employment and social benefits. This is not necessarily easy to determine,

however, because the technologies of resource processing change along with accessibility, competition, markets, and so forth.

Fordism and Post-Fordism
One of the most fundamental changes to resource processing occurred early in the twentieth century when Henry Ford originated the assembly line process. Since then the term **Fordism** has been used to define mass production of standardized goods, usually at a centralized assembly plant. Assembly line technology transforms resources into consumer goods. In British Columbia, Fordist methods were employed by the resource industries in the canning of salmon, the concentration of minerals, the manufacture of two-by-fours and other dimensional lumber, and the production of pulp and later paper and plywood.

After the Second World War, a period of prosperity created the greatest ever demand for consumer goods. The principles of Fordism were refined, and an increased demand for the forest and mineral resources of British

Table 7.1

Merchandise trade of British Columbia, 1994

	Exports			Imports		
	$ billions (1)	% of total BC (2)	% of total Canadian (3)	$ billions (4)	% of total BC (5)	% of total Canadian (6)
Resource-based products						
Agriculture and fishing	1.53	6.7	8.7	2.16	11.9	17.2
Energy	2.2	9.6	10.1	.22	1.2	3.1
Forestry	13.99	61.3	44.9	.43	2.4	23.8
Industrial materials	2.74	12.0	7.0	2.77	15.2	7.2
Subtotal	20.46	89.7	18.6	5.58	30.8	9.3
Other products						
Machinery and equipment	1.49	6.5	3.5	5.72	31.4	8.7
Automotive production	0.37	1.6	0.6	3.78	20.8	7.9
Consumer goods	0.3	1.3	5.1	2.66	14.6	11.3
Special transactions (trade)	0.18	0.8	1.9	.47	2.6	9.5
Subtotal	2.34	10.3	2.0	12.63	69.4	8.9
Total	22.81	100.0	10.1	18.19	100.0	9.0

Note: Columns do not always add to total, due to rounding.
Source: Wilkinson (1997), 138.

Columbia resulted in a major expansion of these industries in particular. Investments in manufacturing plants, however, tended to be made in remote locations. Many single-resource communities were created and became dependent on the particular resource.

The 1960s and 1970s saw the rise of oligopoly conditions, in which large corporations with little competition began to control the resource sectors of British Columbia and elsewhere. They employed new scientific management principles that included capital-intensive technologies to produce ever greater amounts of goods from resources, but these new machines and techniques began to replace labour. The old, labour-intensive ways of processing resources, from sawmilling to mineral extraction, were fast disappearing.

The struggle for provincial governments and to a lesser degree for the federal government was to encourage – usually by providing the infrastructure – corporate investment in the resource sector to produce more employment. As long as new plants were being opened, the potential multiplier effect of new employment resulted, and greater revenues from resource taxation accrued to government. Once the plant and equipment were in place, however, new capital-intensive technologies eventually reduced the number of workers required. More and more it became the government's role to become involved in establishing a social safety net, along with educational retraining for displaced workers. It should be remembered, though, that up until the early 1970s almost the whole country experienced good economic times with plenty of job opportunities, and governments had money to spend.

Unions were caught in an even greater dilemma. The new technologies often necessitated increased productivity to remain competitive, making it difficult for unions to oppose the equipment that reduced their rank-and-file members. The trade-off for unions was to ensure better working conditions and wages – for those workers who remained.

Figure 7.2 illustrates the interlocking relationship of various groups within British Columbia; change in one sphere will influence change in the others. For example, the introduction of an automated greenchain (an assembly line of freshly cut dimension lumber) at a sawmill immediately increases the productivity of the mill but

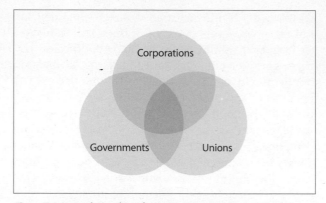

Figure 7.2 Interrelationship of corporations, governments, and unions

leaves many workers seeking (un)employment insurance or retraining and fewer rank-and-file members in the International Wood Workers Union.

British Columbia and the Global Economy

Figure 7.3 illustrates the condition of British Columbia's resource-based and resource-dependent economy, in which our resources achieve only low value-added processing (e.g., pulp, lumber, mineral concentrates). The corporations that own the resource industries in British Columbia reside in heartland areas beyond its borders such as southern Ontario, California, Britain, or Japan. A corporation decides where to locate its head office and research facilities and where to develop new products and implement new technologies. The related high-paying and often professional jobs are not located in the hinterland. Similarly, the production of high value-added products – mainly consumer goods produced from semi-processed resources – and the suppliers of transportation, goods, and services are also located outside British Columbia. Moreover, because resource industries within the province are extraprovincially controlled, the profits generated are exported, and the province depends on imported goods, services, and high value-added consumer goods.

Thomas Gunton uses this model to express concern over encouraging large foreign corporations to invest in the resource sector of the province: "Overall ... a staple region such as B.C. which is dominated by externally controlled firms interested in obtaining a secure supply of resources for externally located operations will not develop strong regional linkages. Instead the economy will develop only lower order processing and service activities tied to the regional market. It will lose out on much of the 'footloose' employment such as head office management, research and development and higher order processing which will be located outside the region simply because the firm is externally controlled" (1982, 10). Gunton suggests that British Columbia should develop far more linkages with its resource base.

Barnes et al. (1992, 178-80) confirm these trends and suggest that British Columbia has developed in four unique ways in comparison to the rest of Canada. First, industries employing Fordism were decentralized into the resource frontier. Second, these Fordist industries gained the fewest forward and backward linkages for British Columbia of any province in Canada. Third, the provincial government played the most facilitative role of any government in Canada in providing infrastructure and attracting corporations. Fourth, British Columbia has experienced the greatest boom and bust cycles of any region in Canada.

In good economic times – in British Columbia from the end of the Second World War to the early 1970s –

the changes and impact on workers were handled by relatively wealthy federal and provincial government programs. Establishing community colleges with training programs and a major expansion of the social safety net were accomplishments during this period.

Global conditions changed, and these changes had serious implications for economies that relied on resource development. By the 1970s, the world was moving into **post-Fordism.** Early in the decade, currencies that had been pegged to fixed gold rates became destabilized and the energy crisis created global unease. Increased pressure for global trade, along with major economic recessions from 1981 to 1986, in the early 1990s, and again in the late 1990s, indicated that the traditional economy was not working.

Two trends transformed the production process and caused a major restructuring of resource-based economies such as in British Columbia. **Time-space convergence,** or time-space collapse, refers to the ability of transportation technologies to shrink the world in terms of patterns of movement. The advent of supertankers, jet air-cargo carriers, coaxial cables, satellites, telephones, fax machines, Internet, e-mail, and other technologies has resulted in an incredible conquering of distance, permitting a global scale of organization in the movement of goods and information.

The second trend is the fragmentation of the production process, whereby the phases of a given process are divided and assigned to various locations in the most cost-efficient manner. Two questions apply here: What are the components of any good or service and where are the cheapest places in the world to produce these components?

Time-space convergence and process fragmentation have resulted in the advent of the multinational or transnational corporation: "IBM, for example, one of the truly giant multinationals, actually has no single plant outside the United States in which a complete product is manufactured. Each phase of the

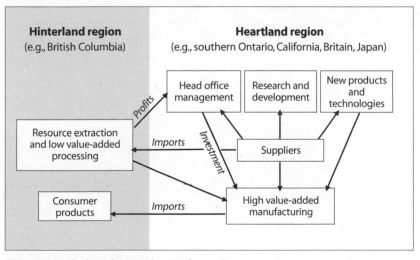

Figure 7.3 Staple dependency and external control
Source: Gunton (1982), 14. Reproduced with permission.

production process – ranging from management control and raw material production, through simple and more complex component manufacture, to research, design, and final assembly – is usually located in a different place, with the market perhaps in still another country" (Galois and Mabin 1987, 10). Another result is the international division of labour. Labour is the least mobile global factor, and "cheap" labour regions such as Mexico and China have emerged. Globalization of production has shaken the world order of nations in terms of the production of goods and services.

The global marketplace is in a state of uncertainty, particularly with respect to traditional production processes that produce standardized, one-size-fits-all materials. Post-Fordism brings with it the concept of flexible specialization, by which goods can be produced to fit individual consumer demands. This is often accomplished by contracting out to small firms throughout the world. The term *restructuring* is often heard these days and refers to the ways in which resources and technologies are combined to produce goods and services in the global marketplace. Restructuring also has a somewhat negative connotation: when corporations, or whole economies, restructure it implies major changes, including employment loss and uncertainty for many.

A serious recession from 1981 to 1986 caused many traditional Fordist operations to restructure in British Columbia. MacMillan Bloedel's sawmill in Chemainus on Vancouver Island is a typical story of the impact and conditions of restructuring. This was a fairly large mill, producing dimension lumber mainly for the American housing market. That market was in a seriously depressed state in the early 1980s. The mill closed in 1982, laying off 642 workers, a major blow to the town's economy. A

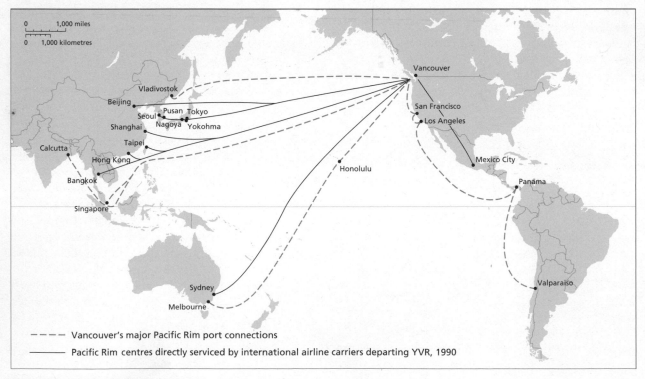

Figure 7.4 Vancouver and the Pacific Rim
Source: Modified from Barnes et al. (1992), 191.

new mill opened in 1985, organized on the principles of flexible specialization. It employed specialized equipment to cut lumber to fit the market demands of Japan and Europe as well as the United States, but only 145 employees were required. Meanwhile, the company contracted out a number of the functions that it had previously run under the old Fordist system, such as a dry kiln operation (Barnes et al. 1992; Rees and Hayter 1996). For the community of Chemainus, these were tough times marked by high unemployment and outmigration in search of jobs elsewhere. As a response, Chemainus turned to the tourism industry, where it was promoted as the "Little Town That Did." The history of Chemainus was depicted in large murals on the walls of the downtown stores, successfully attracting tourists. New, service-oriented jobs, often at minimum wage, along with contract employment in fields paying much less than before were part of the restructuring.

Restructuring of the BC economy was not restricted to the primary resource industries. The service sector, whose products are often knowledge and business transactions, has been subject to both new technologies and flexible specialization. One only needs to consider banking and the automated banking machine (ABM), or the number of firms using contracted message services. Both large and small firms have experienced layoffs, or what the media have commonly referred to as "downsizing" and "re-engineering," as a result of restructuring. Consequently, all communities, whether small resource-based ones or large metropolitan centres such as Vancouver, are affected by these global economic systems (Norcliffe 1994).

Not all these changes are negative. As we enter the twenty-first century, referred to as the "Pacific century," Asian countries are becoming the new leaders in industrial growth, development, and investment. The Pacific Rim nations include over 50 percent of the world's population, and British Columbia is well positioned to engage in these markets (Figure 7.4).

BC experience in resource production and in infrastructure and urban development have become potential knowledge-based exports. British Columbia is politically stable and comes complete with modern banking and financial institutions, excellent educational facilities, and a highly educated population: attributes that attract immigration and investment. As well, this province still has many resource-based commodities required throughout the world.

The economic sector of British Columbia is experiencing a spatial differentiation, or two geographies. The new knowledge- and information-based employment and economy mentioned above is located mainly in a rather small, triangular geographic region encompassing the area from Victoria to Nanaimo and, most important, the Greater Vancouver area. This region contains over 60 percent of the provincial population and is rapidly growing. Over 2 million people lived in the Lower Mainland alone in 1999, and the figure is expected to surpass 3 million by 2021 (GVRD 1996, 44). Ley, Hiebert, and Pratt (1992), Hutton (1997), Binkley (1997), and others suggest that the growth is linked to Vancouver's

Table 7.2

Manufacturing establishments in metropolitan Vancouver, by Standard Industrial Classification (SIC), 1994

Manufacturing group	Listings	Percentage
Food and kindred products	399	5.9
Tobacco	1	0.0
Textile mill products	76	1.1
Apparel and other textile products	336	4.9
Lumber and wood products	700	10.3
Furniture and fixtures	245	3.6
Paper and allied products	116	1.7
Printing and publishing	1,093	16.1
Chemical and allied products	257	3.8
Petroleum and coal products	21	0.3
Rubber and plastic products	184	2.7
Leather and leather products	33	0.5
Sand, clay and glass products	248	3.7
Primary metal industries	119	1.8
Fabricated metal products	565	8.3
Machinery, except electrical	523	7.7
Electrical and electronic equipment	288	4.2
Transportation equipment	849	12.5
Instruments and related products	145	2.1
Miscellaneous manufacturing	593	8.7
Total	6,791	100.0

Source: Hutton (1997), 242.

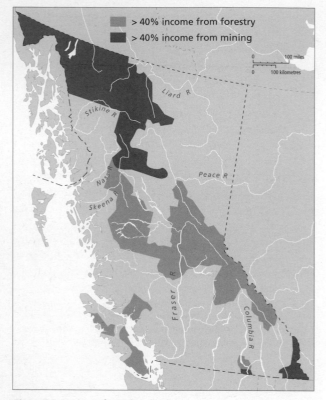

Figure 7.5 Regions of British Columbia dependent on forestry and mining
Source: Modified from British Columbia Round Table on the Environment and Economy (1993).

transformation into a cosmopolitan city that is becoming unhinged, or less dependent on, staples from its provincial hinterland. The growth is also symbolic of increasing economic opportunity and development of manufacturing industries. Table 7.2 shows the number of companies in Vancouver that are listed under the Standard Industrial Classification of Census Canada. Much of the rest of the province forms the second geography, being heavily dependent on resource extraction and processing. Figure 7.5 shows regions that have a great reliance on mining or forestry only.

The BC economy relies on a combination of the older Fordist production systems and the new post-Fordist ways of doing business. The economic structure is thus much more complex than it used to be, and there are many more variables responsible for economic growth.

The development of human resources and knowledge, British Columbia's favourable location for Asia-Pacific trade, a greater involvement in the tourism industry, and the encouragement of new investment locally and from abroad are among the key components. Globalization has also changed the role of government as manager of resources: "Governments, facing pressures to privatize and deregulate and liberalize, increasingly find themselves with diminished control over the processes and strategies of production, except insofar as they can represent the vested interests of their corporations in the development of international rules of corporate behavior and/or privilege" (Wilkinson 1997, 131). Unfortunately, economic and population growth is not evenly distributed throughout British Columbia.

SUMMARY

The early settlement and development of British Columbia was very much tied to resources such as gold and other minerals, forests, fish, energy, and some agriculture; all produced commodities mainly for an external market. The values and uses of these resources have changed over time and so have production processes, corporations, governments, and markets. Categorizing resources as renewable or non-renewable, or using an ecosystem model of resource interaction, recognizes potential limitations and conflicts and the need for resource management. The three levels of government, often with overlapping responsibilities, do not have an easy task as they weigh the many political, economic, and environmental issues associated with resource exploitation.

Staples theory helps to unravel the many economic links between resources and related development. The multiplier effect of increased employment opportunities results from gaining more and more linkages. This occurs provided that the main assumption, the existence of an external demand, holds true. If this fails, the multiplier effect may work in reverse.

The global economic system is not static, nor are the technologies in transforming resources. Fordism was adopted by industrial firms in British Columbia, where employment and development expanded well into the 1970s. Changing economic conditions from the 1970s on saw the creation of multinational corporations that were able to organize the production of goods and services on a global scale, with a global division of labour. Post-Fordism is marked by uncertainty in resource production, restructuring for many firms, and also new opportunities, particularly with respect to the economies of the Pacific Rim.

British Columbia still has resources, and they are especially important to the employment and economy of the interior and the north end of Vancouver Island. The core area of the province, Victoria/Nanaimo/Vancouver, is capturing new employment opportunities based on knowledge and information. British Columbia has essentially two economies, producing even greater challenges for the managers of our resources.

REFERENCES

Barnes, T.J., and R. Hayter, eds. 1997. *Troubles in the Rainforest: British Columbia's Forest Economy in Transition*. Canadian Western Geographical Series vol. 33. Victoria: Western Geographical Press.

Barnes, T.J., D.W. Edgington, K.G. Denike, and T.G. McGee. 1992. "Vancouver, the Province, and the Pacific Rim." In *Vancouver and Its Region*, ed. G. Wynn and T. Oke, 171-99. Vancouver: UBC Press.

Binkley, C.S. 1997. "A Crossroad in the Forest: The Path to a Sustainable Forest Sector in British Columbia." In *Troubles in the Rainforest: British Columbia's Forest Economy in Transition*, ed. T.J. Barnes and R. Hayter, 15-35. Canadian Western Geographical Series vol. 33. Victoria: Western Geographical Press.

British Columbia Round Table on the Environment and Economy. 1993. *Strategic Directions for Community Sustainability*. Victoria: The Round Table.

Galois, R.M., and A. Mabin. 1987. "Canada, the United States, and the World-System: The Metropolis-Hinterland Paradox." In *Heartland and Hinterland: A Regional Geography of Canada*, 2nd ed., ed. L.D. McCann, 39-67. Scarborough, ON: Prentice-Hall.

Greater Vancouver Regional District (GVRD). 1996. *Creating Our Future ... Steps to a More Livable Region*. Vancouver: Strategic Planning Department.

Gunton, T. 1982. *Resources, Regional Development and Provincial Policy: A Case Study of British Columbia*. No. 7. Ottawa: Canadian Centre for Policy Alternatives.

Hutton, T.A. 1997. "Vancouver as a Control Centre for British Columbia's Resource Hinterland: Aspects of Linkage and Divergence in a Provincial Staple Economy." In *Troubles in the Rainforest: British Columbia's Forest Economy in Transition*, ed. T.J. Barnes and R. Hayter, 233-61. Canadian Western Geographical Series vol. 33. Victoria: Western Geographical Press.

Ley, D., D. Hiebert, and G. Pratt. 1992. "Time to Grow Up? From Urban Village to World City, 1966-91." In *Vancouver and Its Region*, ed. G. Wynn and T. Oke, 234-66. Vancouver: UBC Press.

Norcliffe, G. 1994. "Regional Labour Market Adjustments in a Period of Structural Transformation: An Assessment of the Canadian Case." *Canadian Geographer* 38 (1): 2-17.

Rees, K., and R. Hayter. 1996. "Enterprise Strategies in Wood Manufacturing, Vancouver." *Canadian Geographer* 40 (3): 203-19.

Watkins, M.H. 1963. "A Staple Theory of Economic Growth." *Canadian Journal of Economics and Political Science* 29 (2): 141-58.

Wilkinson, B.W. 1997. "Globalization of Canada's Resource Sector: An Innisian Perspective." In *Troubles in the Rainforest: British Columbia's Forest Economy in Transition*, ed. T.J. Barnes and R. Hayter, 131-47. Canadian Western Geographical Series vol. 33. Victoria: Western Geographical Press.

8 Forestry: The Dominant Export Industry

Commercial forestry has a long history in British Columbia, although there are major differences between the coastal and interior forest industries. Dominant tree species and their physical characteristics, accessibility to the resource and to the market all differ. The coastal forest industry predominated, in terms of the volume of wood cut for export, until changes in corporate organization and technology in the 1960s that allowed the interior forests to fulfill growing demand. Technological changes have had a major impact on all aspects of forestry throughout its history, influencing how the resource is harvested, at what rate, which forests are cut, where mills are located, and how wood is processed and transported, as well as employment, taxation, and the economic well-being of the various regions and the province as a whole.

The provincial government is the manager of this resource and as such has implemented a number of forms of **tenure,** or legal contract, to allow private corporations to harvest forest lands. Types of tenure, tree species, location, transportation, and end use all become part of the complex equation that determines revenues and costs of timber harvesting. With the provincial government in control of 96 percent of the forest land in the province, these revenues are essential to the tax base of British Columbia.

Harvesting and managing the forests has also brought a series of concerns and issues. Vicky Husband of the Sierra Legal Defence Fund notes that "it took 85 years to cut the first half of what we have cut in British Columbia, and only 15 years for the last half" (Husband 1995). The rate of harvesting raises the question whether there will be forests for the future, while harvesting techniques raise issues about the environment and other values of our forests. Simply put, should trees be viewed as fibre only? The various types of tenure provoke concerns about the role and domination of large corporations, about how to attain higher value-added for timber resources, and about privatization. Meanwhile, new treaty negotiations may result in new forms of forest tenure and management for some regions of British Columbia.

COAST VERSUS INTERIOR: A BRIEF HISTORY

The distinction between the coast and the interior is largely based on physiographic and climatic differences,

which in turn result in different tree species domination. The coastal region encompasses the whole coastal strip from the Coast Mountains west to include Vancouver Island and the Queen Charlottes (Figure 8.1). As Chapter 2 discussed, this region is characterized by wet, mild winters and more moderate summer weather conditions than in the interior. The dominant tree species is western hemlock, although large stands of western red cedar and Douglas fir (below 51° latitude) are also characteristic. The coastal trees grow to an enormous size, yielding a huge volume of wood per hectare.

In the interior, the climate is more extreme in both temperature and precipitation. The dominant tree species are lodgepole pine and spruce. These trees are not nearly as large as those on the coast and they grow farther apart, so the volume of wood per hectare is considerably less. That this region is seven times larger than the coastal forest region is a compensating factor.

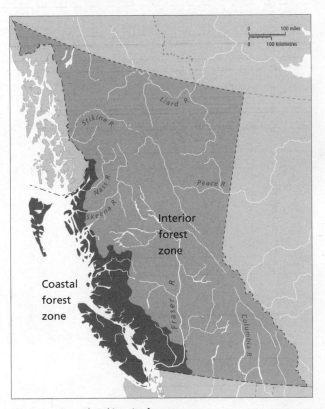

Figure 8.1 Coastal and interior forest zones

Logging for commercial purposes dates back to the 1820s and '30s, mainly for spars for sailing ships. With the California gold rush of 1849, the industry expanded into lumber production (Hayter and Galois 1991). Major lumber production for export accelerated in the 1860s as a number of sawmills were established on southern Vancouver Island and around Burrard Inlet on the mainland. The great advantage of the coastal region, besides very large trees, was its proximity to the ocean for transportation. Logs were dragged to the tideline and floated to sawmills, and the lumber was loaded onto ships and exported to foreign markets. With this advantage, and the growing demand for forest products, the coastal forest industry became dominant.

The early commercial forest industry in the interior served a largely local demand. As the gold rush led prospectors into the Cariboo region in the 1860s, for example, there was a demand for wood for heating and cooking fuel as well as for housing and commercial construction, including pit props, sluices, and flumes. Once the gold claims were exhausted, the community was often abandoned and the demand for forest products ceased. Over time, with the discovery of silver, coal, and other minerals, new mines sprang up in various locations throughout the interior of British Columbia, sparking renewed commercial forest activity.

The completion of the Canadian Pacific Railway (CPR) by 1886 provided a more stable, expanded market in the Prairies and central Canada for the interior mills located on the railway line in places such as Lytton, Kamloops, and Revelstoke. Later, railway developments saw the Kettle Valley Line, Grand Trunk Pacific (now CNR), and Pacific Great Eastern (now BCR) make inroads into different regions of British Columbia and slowly open up forestry activities. Figure 8.2 shows the difference between the coast and interior volumes of wood cut. Transportation developments were important to the interior forest industry, but it took a host of other changes before the interior began to outproduce the coast.

CHANGING TECHNOLOGIES IN A CHANGING INDUSTRY

On the coast, transporting such huge logs to the sawmills was the major problem initially. The first solution was to have teams of oxen drag the logs over skid roads to the tideline. There were sixteen to twenty oxen per team, and the skid roads were made of small logs, often greased with fish oil to reduce friction, and placed on a cleared stretch of ground leading to the ocean. The whole logging and transporting process was labour intensive and slow moving. This produced a linear pattern of logging because it was more efficient to move farther along the coastline than to move inland. Mills such as Sue Moody's, located on the shores of present-day North Vancouver, utilized the logs of Howe Sound and the Sunshine Coast by the 1890s. It was much easier to use the ocean to tow logs to sawmills fifteen to twenty kilometres away than to drag them over skid roads for a kilometre or more.

Different species of trees have different end uses and markets, although there is certainly some overlap. Western red cedar has a very straight grain; it is light and water resistant, making it ideal for

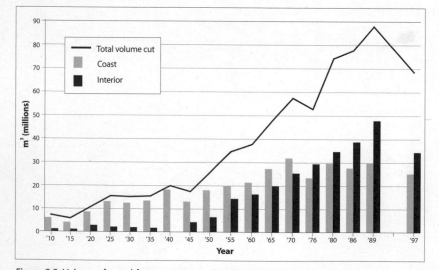

Figure 8.2 Volume of wood from coast versus interior, 1910-97
Sources: Data from Barker (1977); Council of Forest Industries (1994); British Columbia, Ministry of Finance and Corporate Relations (1998)

roofing and siding. The demand for cedar shakes and shingles increased in the 1890s and led to the introduction of two other means of transportation: horses and flumes. The enormous cedar trees could be cut up and split into four-foot (1.2 metre) shingle and shake bolts in the bush. These much smaller sizes could then be stacked on sleds and, because cedar is relatively light, horses could do the job. Horses are considerably faster than oxen, and more manageable, which reduced the time of a round trip and pushed back the linear pattern.

Flumes produced a new pattern of logging. These wooden troughs supported on trestles sloped gradually from the mountain side to the ocean, giving access to cedar stands farther inland. Water from creeks was diverted down the wooden flumes, and in this way the shake bolts were floated down to the ocean, where they would be gathered for the shake-and-shingle mills.

By the early twentieth century, mechanical power in the form of the steam engine began to replace animals and flumes, gaining new efficiencies in harvesting and producing another linear pattern of logging. Logging railways were constructed throughout Vancouver Island and the southern portion of the mainland, often using river and stream valleys. Railways were significant in penetrating considerable distances inland from the coast.

Another important innovation was the steam donkey, or donkey engine. This was a powerful, steam-driven winch that used cables attached to pulleys at the top of a spar tree. Spar trees were created by cutting off all the branches and the top of a tree located adjacent to the rail line. Pulleys were then attached to this tall pole to facilitate cables running from the donkey engine to the fallen timbers. This system would then haul logs from a wide perimeter to a central location, where it could then lift them on to railcars.

With all these changes in transportation technology, the common element was to get the logs to water. On the coast, this usually meant bundling them into log booms (sometimes called rafts) in the ocean to float the logs to the mill. Also, particularly in the interior, rivers and lakes were used to transport logs and became an important factor in the location of mills.

The linear pattern was largely broken with the introduction of the combustion engine. Bulldozers could use switch-back roads to gain access to many areas that the railways could not, and few forests were so isolated that they could not be reached by road. Trucking became the main means of transporting the forest resource by the 1940s, and truck logging is still with us today.

The combustion engine was used in other ways to gain efficiencies in logging. The double-bladed axe and Swede saw gave way to the chainsaw. The donkey engine was replaced by the mechanical spar pole, which could be driven to the logging site. Hydraulics were used to hoist the telescoping steel pole into place, and a combustion engine was used to winch the logs. In the interior, where logs are smaller in diameter, the feller-buncher became the tool of choice by the 1980s. This machine, with metal tracks, cuts the tree, limbs it, and stacks it onto a logging truck – all with one operator.

As the most accessible forests of the coast were depleted, it became necessary to go farther and farther afield and higher up the mountain sides to supply the bulk of mills located on the southern coast of British Columbia. New methods of booming logs were required. The simple rafts of the old days were not sufficient to withstand the rough weather and ocean conditions that were common in towing logs from regions such as the Queen Charlottes. The Davis raft used a fairly successful bundling technique to prevent raft break-up, but self-loading and self-dumping barges, which were faster and posed

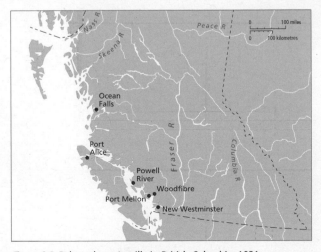

Figure 8.3 Pulp-and-paper mills in British Columbia, 1931
Source: Modified from Farley (1979), 61.

even less possibility of log loss, replaced it (Hardwick 1960, 4).

In areas too remote for road-based logging or where it is desirable to avoid building roads, helicopter logging and balloon logging are able to get the forest resource to a central location. To employ helicopters in such work processes is expensive, so the value of species being harvested has to be high.

Technology has changed throughout the entire industry. The mills themselves have been transformed from labour-intensive to capital-intensive production plants, employing fewer and fewer workers while increasing production. Initially, lumber was cut by hand, or if a stream were nearby a water wheel would be used. The steam engine transformed many of these processes, and by the beginning of the First World War, electric motors ran the sawmill blades. All aspects of milling – the booming grounds, debarking logs, sawmilling, grading, and the greenchain – underwent automation. Hydraulic barkers, new bandsaws with laser directional beams, and the automated greenchain (the assembly line of freshly cut lumber) had restructured sawmilling by the 1980s.

Lumber has been the main end use product for the forest industry of British Columbia in the past and today, but shakes and shingles are important products as well. By the early 1900s, the technology to turn wood fibre into pulp and paper economically resulted in the location of pulpmills in coastal British Columbia to fulfill a global demand. These early locations expanded

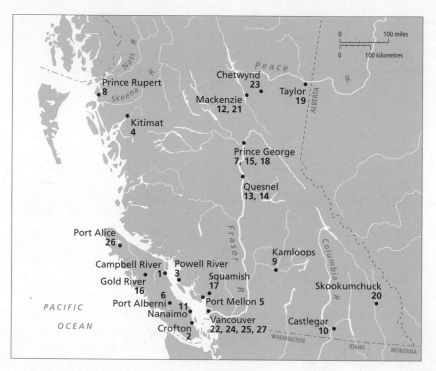

Company	Location	Pulp	Paper
		(1,000 tonnes)	
1 Fletcher Challenge Canada	Duncan Bay	814	576
2 Fletcher Challenge Canada	Crofton	740	411
3 MacMillan Bloedel	Powell River	633	449
4 Eurocan Pulp and Paper	Kitimat	455	455
5 Howe Sound Pulp and Paper	Port Mellon	500	193
6 MacMillan Bloedel	Port Alberni	288	381
7 Northwood Pulp	Prince George	518	–
8 Skeena Cellulose	Prince Rupert	449	–
9 Weyerhaeuser	Kamloops	447	–
10 Celgar Pulp	Castlegar	414	–
11 Harmac Pacific	Cedar	386	–
12 Finlay Forest Industries	Mackenzie	190	190
13 Cariboo Pulp and Paper	Quesnel	323	–
14 Quesnel River Pulp	Quesnel	314	–
15 Canfor Corporation	Prince George	273	–
16 Avenor	Gold River	259	–
17 Western Pulp	Woodfibre	257	–
18 Canfor Corporation	Prince George	255	95
19 Fibreco Pulp	Taylor	231	–
20 Crestbrook Forest Industries	Skookumchuck	224	–
21 Fletcher Challenge Canada	Mackenzie	207	–
22 Island Paper Mills	New Westminster	–	173
23 Louisiana Pacific	Chetwynd	173	–
24 Newstech Recycling	Coquitlam	166	–
25 Crown Packaging	Burnaby	–	164
26 Western Pulp	Port Alice	162	–
27 Scott Paper	New Westminster	31	78
	Total	8,708	3,164

Figure 8.4 Pulp-and-paper mills in British Columbia ranked by size, 1997
Source: Modified from British Columbia, Ministry of Finance and Corporate Relations (1998), 90.

production, and by 1931 Powell River and Ocean Falls were the major producers of pulp (Figure 8.3). The existing mills continued to expand and by 1951 new coastal pulpmills began to appear in locations such as Prince Rupert, Duncan Bay, Port Alberni, Harmac, and Victoria. The locations were often chosen for their proximity to sawmills with their useful wood by-products, which could be used in the production of pulp. As global demand escalated for virtually all wood products by the 1960s, new and more efficient technologies of pulp-and-paper production turned to the largely untouched forests of the interior.

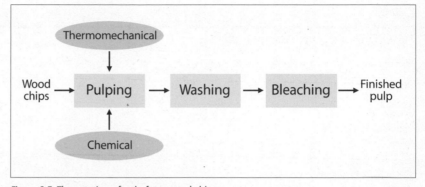

Figure 8.5 The creation of pulp from wood chips
Source: Modified from British Columbia, Ministry of Forests (1994), 3.

Multinational corporations, often foreign, were encouraged to invest in the forest industry – and they did. These corporations believed in consolidation and integration of all components of forestry. Weyerhaeuser, for example, an American company, established a new, state-of-the-art pulpmill in Kamloops in the 1960s, but its operations were much broader than the production of pulp. The company purchased a number of sawmills, thus consolidating many forestry functions under one corporate umbrella. Integration allows a corporation, which owns the rights to the raw material, much more control over its end uses. Owning sawmills in conjunction with pulpmills makes a great deal of economic sense, as the "wastes" from sawmilling can be chipped up for pulp or used as hog fuel for much needed electrical and steam production at the pulpmill. Figure 8.4 shows the number of pulp-and-paper mills in British Columbia in 1996.

During the late 1960s and 1970s, new investments focused mainly on the interior, where some new communities such as Mackenzie were created. It was not long before the volume of wood cut surpassed that of the coast (Figure 8.2). The old mills on the coast were now under serious pressure to compete, and new investments were required to keep them competitive. In the case of Ocean Falls, the old mill was uncompetitive and was initially "rescued" by the provincial government in the early 1970s. A change in government by the mid-1970s resulted in closure of the pulpmill, and the community became a ghost town. New investments also resulted in the consolidation and integration of the forest industry on the coast; old mills were upgraded, a new mill was built in Kitimat, and another on the west coast of Vancouver Island, around which the community of Gold River was created.

The technologies of making pulp have changed considerably from the exclusively chemical processes common before the Second World War to the thermomechanical processes that began in the 1970s. Figure 8.5 is a simplified flow diagram of the conversion of wood to pulp. The thermomechanical means of creating pulp has the advantage of being able to use a wider variety of wood types, as well as sawdust from sawmills, but it requires a great deal of steam and heating, which add energy costs. The chemical processes break the wood fibres down to produce an even stronger pulp. In either method, the last stage – bleaching – is often the most contentious because it requires chlorine. Chlorine releases dioxins and furans in the waste water, thus producing toxic levels of pollutants.

The types of pulp have also expanded to meet the global demand for a growing range of paper products. The Port Mellon Mill on Howe Sound, for example, which is partly owned by Oji of Japan, produces a higher quality of newsprint than is required for the North American market. Pulp products for the European market must be chlorine free. The MacMillan Bloedel mill at Port Alberni has installed a Nexgen plant, using pulp from hardwoods such as cottonwood plus clay to produce glossy paper.

Table 8.1

Value of forest product exports, 1995-7

Commodity	1995	%	1996	%	1997	%
Lumber (softwood)	7,314	43.4	7,736	51.3	7,812	52.8
Pulp	5,540	32.9	3,407	22.6	3,099	21.0
Newsprint	1,553	9.2	1,284	8.5	836	5.7
Paper and paperboard	898	5.3	851	5.6	970	6.6
Plywood (softwood)	262	1.6	585	3.9	831	5.6
Cedar shakes and shingles	209	1.2	222	1.5	241	1.6
Selected value-added	186	1.1	230	1.5	254	1.7
Other	886	5.3	758	5.0	744	5.0
Total exports	16,848	100.0	15,073	100.0	14,787	100.0

Note: Columns do not always total 100 percent, due to rounding.
Source: BC Stats: Quick Facts, 1995, 1996, and 1997.

Considerably more pulp than paper is produced in British Columbia, and the paper is mainly newsprint.

Table 8.1 compares exports by dollar values, showing that lumber is first, followed by pulp and newsprint. A comparison of 1995, 1996, and 1997 shows some of the vulnerability of markets; pulp and newsprint production has seen a considerable reduction. A number of other wood products, however, are important to our export market: paper and paperboard, shakes and shingles, plywood, and value-added wood products such as doors and window frames among them.

With multinational corporations bringing in huge investments, the consolidation and integration of the forest industry brought with it a number of concerns. The new capital-intensive equipment resulted in new efficiencies to harvest trees and produce forest products, but once the major expansion of new mills was over,

Table 8.2

Employment per 1,000 cubic metres of wood harvested

	Jobs/1,000 m³	Timber cut (millions m³)
1950	2.3	19.8
1960	1.9	34.0
1970	1.4	54.7
1980	1.3	76.8
1986	0.8	78.9

Source: Hammond (1991), 78.

the number of workers required decreased (Table 8.2). Clearcutting forests, while cost effective, systematically eliminates old growth and produces many environmental conflicts. In some regions, it also challenges Aboriginal title and recent treaty negotiations. Tied to these concerns is the whole question of sustainability of the forests for the future.

PROVINCIAL MANAGEMENT OF THE FOREST RESOURCE

Historically, the forests were regarded as an inexhaustible resource. Even though it was decided in 1865 to retain crown ownership of forest land, some of the best coastal forests were bartered away for tax revenues and railway development (Barker 1977, 88). Unfortunately, little was done in terms of reforestation or addressing the issues of sustainability until relatively recently.

Tenure is a legal right, granted to private companies, to harvest provincial forest lands, and it produces revenues for the province from the forests. From 1870 to 1905, forest companies paid lease fees and were granted blocks of forest land in perpetuity. A timber licensing system also operated from 1888 to 1909, which not only alienated crown forest lands but also allowed the untaxed transfer of licences. Only once the timber was harvested was a licence holder subject to tax. The folly of this system was that it produced a great deal of speculation and consumption of forest land (Haig-Brown 1961). Leasing and licensing operated side by side during this period.

Crown grants were another means whereby provincial forests came under the control of private interests. Many of these grants took the form of cutting rights given to private railway companies as encouragement to construct new lines. One of the most famous, or infamous, railway grants was the Esquimalt and Nanaimo Railway (E&N) charter. It placed one-quarter of Vancouver Island's forests in private hands. This was some of the best forest land in British Columbia because of its sixty-year growing cycle.

Whether land leases, timber licences, or crown grants, these forms of tenure privatized forest land in British Columbia. And while the actual amount of land in these tenures amounted to only about 4 percent of the land base of the province, it accounted for most of the wood harvested until the 1940s and still accounts for a significant amount today (Figure 8.6). These private forest lands are some of the best and most productive in the province.

Contemporary forms of tenure come from the recommendations of the Sloan Commission of 1945. The commission resulted in amendments to the Forest Act in 1947 and two forms of tenure were developed: tree farm licences (TFLs) and public sustained yield units (PSYUs). The TFL is an area-based tenure that allows a private forest company to obtain a relatively long-term renewable licence (initially twenty-one years and now twenty-five) to a block of forest land. The company is responsible for road building and reforestation within the licensed area. TFLs of the 1950s and '60s were mainly in the coastal forest region, where the large companies dominated, and with this form of tenure they retained a great deal of control over the land. As the industry expanded to the interior, so too did the TFL form of tenure. By 1997, there were thirty-one TFLs in the province.

PSYUs are a volume-based tenure in which the Ministry of Forests plays the lead role. The ministry establishes the volume of wood to be harvested by private compa-

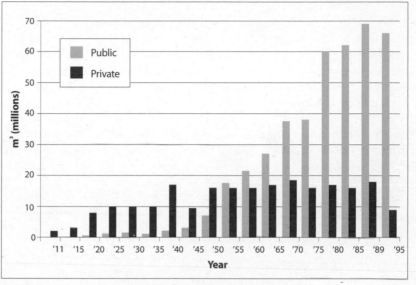

Figure 8.6 Public versus private volume of wood harvested, 1911-95
Sources: Data from Hammond (1991), 76; Canadian Council of Forest Ministers (1997), 8.

nies under a series of licences and harvesting contracts, and is responsible for building the roads and for reforestation. In the 1950s and '60s, this form of tenure applied mainly to the interior, where there were few large forestry companies. Edgell (1987) explains the rationale for these two new forms of tenure under the 1947 changes to the Forest Act. Their purpose was to:

1 ensure industrial access to guaranteed long-term timber supplies
2 stimulate capital investment in processing plants and therefore assure economic stability and development and
3 bring forests under "sustained yield management" (p. 113).

Certainly the concept of sustained yield management was long overdue, and the commission was the turning point. It was finally realized that the forests would not last forever if they were not replanted and better managed than they had been in the past. The two forms of tenure eventually raised the question of who the better manager was, the company with the TFL or the ministry with the PSYU.

Further amendments to the Forest Act occurred in 1978. All lands held under the old timber licences (a term that by now included all historical tenures except crown grants), once harvested, would revert to the crown, and in this way the historically alienated forest land would be clawed back and in the control of the province. Under this amendment, the thirty-one TFLs were extended to twenty-five-year terms. The biggest change was to reorganize provincial forests into thirty-six management regions, referred to as timber supply areas (TSAs) (Figure 8.7). Licences and harvesting contracts under the PSYU tenure were replaced by other licences. The forest licence (FL) was the most common tenure and gave the holder the right to harvest a stated volume of timber each year (British Columbia, Ministry of Forests 1988, 3). The FL retained the volume-based conditions of the PSYU but with a fifteen-year renewable term. Other forms of tenure that replaced the PSYU included the timber sale licence, the small business forest enterprise program, the woodlot licence, and several pulpwood agreements (in the interior only). TFLs and FLs represent 84 percent of the annual allowable cut (AAC) in the province and are thus the dominant forms of tenure (British Columbia, Ministry of Forests 1988, 4). The **annual allowable cut** is the rate of harvest, or volume of wood allowed to be harvested, for each tenure in each timber supply area.

As production increased from the 1960s on, particularly in the interior of the province, new questions about tenure surfaced as the provincial government tried to reduce its bureaucracy and permit privatization. The financial blow to the industry and to the economy of British Columbia brought on by the recession of 1981 to 1986 triggered a move to reduce bureaucracy. This was achieved by significant reductions in ministry staffing, which then raised questions about adequate assessment and monitoring of the forest resource. A 1982 provincial Ministry of Forests policy proposed permitting forest licences and timber sale licences, both of which are for relatively short duration and under considerable government control, to be converted to tree farm licences, which, in the public's view, were much closer to privatization because

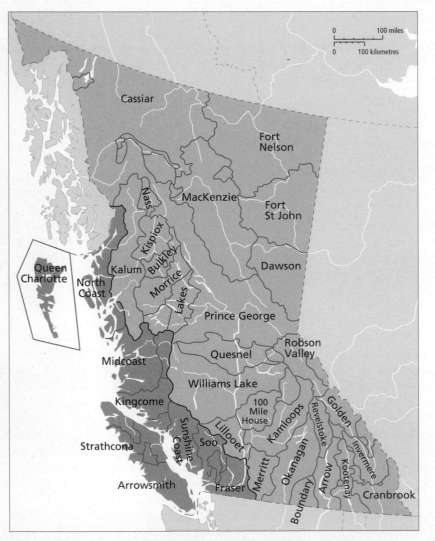

Figure 8.7 Timber supply areas in coastal and interior zones

of the twenty-five-year renewable conditions of this tenure. By 1988, the necessary legislation was in place to allow this conversion.

The reaction by the public, which by this time was both aware of and concerned about logging practices and the apparent lack of management, forced an end to this initiative. With tenure off the agenda temporarily and a new provincial government in 1991, the public expected that the government would come to terms with many of the outstanding issues of this industry.

Forestry produces provincial revenues primarily through taxes, some direct and others indirect. Stumpage is the most important source of direct revenues and the most complex. It is a direct tax placed on logs, but it varies depending on the species, size and quality, and end use. Wood for pulp, for example, can come from a number of species of trees, and since quality is not an issue because the wood is for chipping, the stumpage rate is low. Large Douglas firs, used for lumber, bring in considerably more stumpage.

Besides the market value of the wood, stumpage takes into consideration physical conditions such as difficulty in gaining access to timber and volume of wood available per hectare. Old growth stands of trees on the coast have a much higher volume of wood per hectare than those of the interior; consequently, more roads and more effort are required in interior forests for the same volume harvested on the coast. Stumpage rates are therefore lower to reflect the increased costs and extra labour required.

Finally, the stumpage rate varies with the type of tenure. The tree farm licence, by which a company is responsible for inventory and basically all aspects of managing an area, carries a lower stumpage rate than do volume-based tenures. The volume-based licence holders (for example, those with forest licences) have fewer responsibilities and costs.

Calculating stumpage involves a great many variables and, not surprisingly, several controversial issues. An ongoing battle over stumpage rates and types of tenure has been waged with the United States. A good deal of BC lumber, over 50 percent each year, is sold for housing and other wood products in the United States and is therefore in direct competition with Washington and Oregon lumber producers. American lumber companies complain that BC stumpage rates, inherent in long-term tenures, lead to unfair competition. A number of court cases have resulted, and in 1986 the United States imposed a 15 percent countervail tax for all BC lumber going across the border. By 1990, British Columbia officially increased its stumpage rate by 15 percent and the countervail tax was dropped. Stumpage rates have been increased again since then, causing concern to BC companies but satisfying American producers. The 1996 Canada-United States Softwood Lumber Agreement establishes a quota of softwood lumber from each lumber-producing province in Canada.

Stumpage accounts for about 90 percent of the direct forestry revenues to the province, clearly the most important revenue source from this industry. The remainder is made up of royalties, scaling fees, direct grants by the federal government to the provincial Ministry of Forests mainly for silviculture, and several other charges.

Various indirect revenues go to all levels of government, mainly in the form of taxes on the industry and its employees. Because they are gaining an income, employees pay income and property taxes and many other taxes for consumer goods. The industry uses a great deal of fuel and energy, and taxes are paid on these resources. Companies are also charged land taxes and business licences. The large consumption of water, by pulp-and-paper mills especially, results in a substantial water tax revenue to the province. All these indirect taxes are important sources of government revenue.

The province also incurs costs with respect to the industry. The Ministry of Forests bureaucracy is responsible for administrating the industry. Engineering and building forestry roads, fighting forest fires, and battling pests and diseases that destroy forests all take away from revenues. Reforestation, the goal of silviculture, is another expense but is essential to providing future forests.

It is necessary to distinguish, as the Forest Act does, between basic silviculture and intensive silviculture. Basic silviculture includes the research and development required to ensure the regeneration of healthy forests. Collecting seeds, growing seedlings in similar climate and soils to their eventual location, conducting controlled burns, or scarification, of clearcut areas, and then planting the seedlings are all part of basic silviculture. Table 8.3 shows the rather dramatic increase in planting

Table 8.3

Reforestation of seedlings, 1960-93

	Number of seedlings (millions)
1960	6
1975	65
1989	210
1990	250
1993	215

Source: Council of Forest Industries (1994), 45.

of seedlings by the 1980s and '90s. This increase was necessary because of the backlog of clearcut areas not previously restocked. Another factor in basic silviculture has been the survival rate of seedlings. In 1982, only 54 percent survived; by 1992, 87 percent were surviving (Council of Forest Industries 1994, 45).

In the mid-1980s, the mayors of Vancouver Island used a "carrot patch" analogy to try to persuade the provincial and federal governments to spend money on intensive silviculture. They said that basic silviculture is analogous to preparing the garden and planting rows of carrots in the spring. If you do not tend the garden, however, you cannot expect a bountiful crop in the fall. Probably you will have a bunch of small, overcrowded carrots, all competing for limited nutrients. Replanting seedlings in the forest will result in the same disaster if they are not tended.

Table 8.4 illustrates very clearly the impact of residual spacing to weed out competition (the equivalent of thinning the small carrots), adding fertilizer, and then conducting commercial thinning (harvesting some of the trees earlier in the cycle, the equivalent of picking the carrot crop in stages). Limbing the lower branches is another intensive silviculture practice. As the table shows, it is feasible to double or triple the volume of wood per hectare. A subsidiary benefit is that the larger sized trees can be used for higher value-added wood products, bringing in even more direct revenues to the government. Unfortunately, most attention and money spent on reforestation has focused on basic silviculture. The time required for intensive silviculture, unlike for carrots, is over sixty years for the best forest lands and over eighty years for average forest lands. Governments, from whom this commitment is required, are elected for only four or five years.

Trevor Jones (1983) suggests that the costs of and revenues from logging should be assessed for each region. Are the direct revenues from logging a region such as the Stein Valley, for example, higher than the cost of building roads and bridges, administration, reforestation, and so one? Jones uses the benefit-cost approach to warn the provincial government that "if the Forest Service's costs exceed its direct revenues for a logging operation on Crown land, then the operation is subsidized. This is not necessarily bad, because the forest company may make a profit, employment is created, and the Crown will obtain indirect revenues such as personal and corporate income taxes ... However, the subsidy has to be paid for through the indirect revenues, which are normally used to pay for society's general needs – highways, health facilities, education and other provincial services; external affairs, armed forces, coast guard and other federal services" (1983, 6).

The film *The Fall-down Effect* is worth viewing. Economist Mike Halleran gives a good visual description of the history of logging, technological change, and the conflicts that are now shrinking the forest base and making less wood available for the future. Figure 8.8 summarizes the fall-down effect. The dotted line beginning in the 1980s and leading into the future represents the increase in volume of wood per hectare through basic

Table 8.4

Potential yield through intensive silviculture practices

Forest application	Trees/ hectare	Volume of wood harvested at 50 years (m³/hectare)	Diameter at 1.3 m above ground level (cm)
Natural untreated stands	±2,500	350	30
Planted stands with improved seeds	±1,500	560	38
Natural or planted stands with residual spacing (15 years)	±750	700	43
Fertilizer applied at 225 kg nitrogen at 25 years		840	46
Commercial thinning at 35 years	370	1,050	50

Sources: Modified from Travers (1993), 213; M'Gonigle and Parfitt (1994).

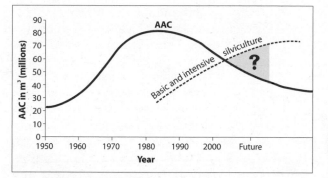

Figure 8.8 The fall-down effect

and intensive silviculture practices. The question mark represents the government's level of investment in these programs. With large investments the fall-down effect is substantially reduced. The annual allowable cut (AAC) increased by approximately 300 percent between 1950 and 1980. As discussed earlier in this chapter, this increase is related to the consolidation and integration of the forest industry, which saw new corporate structures, harvesting techniques, investments, and a major expansion into the interior to meet an ever growing demand for forest products. Maintaining a high AAC is in the provincial government's interest because it means increased revenues. As Halleran asks, however, can we maintain this level of cutting? The graph shows an abrupt decline, or fall-down, in the AAC for the 1990s and on.

Several conditions are responsible for the fall-down effect. In the coastal forest region, the base of old growth trees has largely been cut, and most harvesting in this region today is of second growth timber (Drushka 1999). These trees are considerably smaller and produce much less volume of wood per hectare. Another factor is the shrinking of the forest base. By the 1980s, many groups were demanding that forests be set aside for uses other than wood for fibre. Conflicts centred on old growth, wilderness, wildlife, water quality, recreation, grazing land, hydroelectric dams and the reservoirs they create, and Aboriginal title. Some of these issues have been addressed by creating parks in environmentally sensitive areas such as watersheds, but other issues have yet to be resolved. All these "other" uses of the forest require reduction of the AAC. Areas where harvesting has been the most intense are the first to experience the fall-down

effect. The north end of Vancouver Island, for example, had experienced a 38 percent reduction in its AAC, and the Sunshine Coast a 24 percent reduction, by the early 1990s.

There are ways to reduce the fall-down effect. Basic and intensive silviculture could increase the volume of wood considerably for the future. Herb Hammond (1991) and others suggest that there must be a fundamental change in philosophy toward the way we grow and harvest trees. Hammond advocates holistic forest practices that avoid clearcuts and use selective logging techniques; these could increase yields and reduce many environmental problems and conflicts. On the other hand, a recent report by the provincial government shows second growth in the interior at 35 percent to 55 percent higher than expected and suggests that the figures may be more impressive in the coastal zone, where growing conditions are better (Wilson 1998, A49). These figures mean that the supply of wood can become available much more quickly than forecast, reducing the time line of the fall-down effect.

MANAGING PRESENT AND FUTURE FORESTS IN A CLIMATE OF UNCERTAINTY

The forest industry has been the backbone of the economy for British Columbia for a long time; up until the 1970s it was widely proclaimed that fifty cents of every dollar spent in the province was generated by this industry (Farley 1972, 87). It still remains crucial to both the provincial economy and individual communities and regions. Figure 8.9 outlines the geographic distribution of dependence on forestry-related employment. The British Columbia Round Table found in 1993 that "94,000 British Columbians were directly employed in the forest sector. The livelihoods of as many as 140,000 more people depended largely on the forest sector" (p. 43). Unfortunately, the long history of exploitation, made easier by sophisticated technologies and rules that encourage overharvesting, along with globalization, has brought conflict and crisis to some regions and a climate of uncertainty to the whole industry (Marchak 1983; Clapp 1998; Fulton 1999).

The forests of British Columbia are a component of a dynamic global marketplace in which the multinational corporations that have effective control over the resource

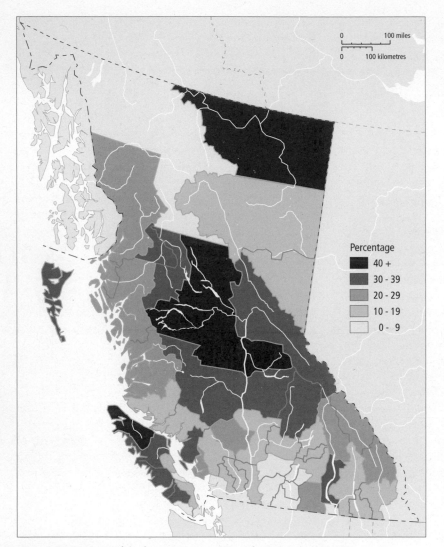

Figure 8.9 Importance of the forest sector to regions of British Columbia, 1996
Source: British Columbia, Ministry of Finance and Corporate Relations (1999), 28.

able and the cost of labour a fraction of that in British Columbia, put the older, less productive coastal mills, especially, in jeopardy. Price fluctuations can make this situation critical for any specific mill. The price per tonne of pulp in October 1995 was US$1,000, for example, and by March 1996 it had fallen to US$525 with little hope of an immediate recovery (Savings and Credit Unions of British Columbia 1996, 9). Skeena Cellulose, the old pulpmill in Prince Rupert, was one of the hardest hit and required provincial assistance to prevent closure. Gold River, on the west coast of Vancouver Island, closed in early 1999. In this climate of uncertainty, forest companies employ strategies of restructuring and flexible specialization.

Hayter and Barnes (1997) document well the economic viability of producing wood products from sawmills, pulpmills, and plywood plants, along with mills producing fibreboard, particle board, and shakes and shingles and show that competition and uncertainty have resulted in major restructuring. Flexible specialization has meant fewer employees, more contracting out, and lower pay. In short, the old, centralized, Fordist methods of production have been challenged.

make decisions based on world market prices for wood products – and both the prices and the demand for wood products fluctuate considerably. Moreover, British Columbia's competitive position in the global arena has been seriously challenged in some areas. The province used to be one of the cheapest places in the world to produce pulp, for example, but today it is one of the most expensive ("Coastal Mills" 1997, A10). Countries such as Indonesia, where fibre is inexpensive and avail-

Compounding the problems of uncertainty for forest workers and managers alike in this province are environmental conflicts over practices from clearcutting to logging in watersheds. As well, through the modern treaty process, First Nations are demanding a share of the forests and the management of them. At the local level, confrontations have resulted in blockades and even court action. On a global scale, organizations such as Greenpeace have successfully lobbied European nations and California to boycott lumber from MacMillan Bloedel because of its old growth content.

Escalation in the annual allowable cut has reduced much of the old growth on the coast and produced some huge clearcuts both on the coast and in the interior. For First Nations, the escalation in AAC is not only an environmental issue but also challenges their traditional territories. Some bands and tribal councils have sought redress in court action, and many have been successful in gaining injunctions. Logging was stopped on Meares Island, Stein Valley, and southern Moresby Island by this method. Willems-Braun (1997) gives a good overview of how the provincial government and forest companies have denied Aboriginal title and shut out First Nations. The new treaty process, discussed in Chapter 5, is designed to resolve many outstanding problems for First Nations and will no doubt have an impact on forest land and its management.

Environmental issues have brought the greatest public awareness and criticism of provincial management of the forest industry. The New Democratic Party, elected in 1991, established a number of policies to try to extinguish the environmental "brush fires." The Council on Resources and Environment (CORE) was created with a mandate to resolve environmental conflicts by increasing the number of parks throughout the province. In 1991, parks amounted to 6 percent of the BC land base, and the objective of CORE was to increase this to at least 12 percent of the province, evenly distributed among regions. The 12 percent objective within geographical regions such as Vancouver Island, the Kootenays, and Cariboo represented a significant challenge. Vancouver Island was the first region that CORE tackled, and their recommendations were to set aside a total of 13 percent for parks. The logging interests were not happy, and nor, as it turned out, were the environmentalists. Clayoquot Sound, where some of the last stands of old growth remain, was not included in the park allotment. Blockades in 1993, for which many protesters were jailed and fined, drew international attention to and condemnation of BC logging practices. The CORE process did not resolve the conflicts, and negotiations over Clayoquot continue.

Many environmental and land use issues are at a much more local level, and two other processes were implemented to deal with them. Local resource use plans (LRUPs) established a round table of groups and organizations having a vested interest in the use of the local forests – the stakeholders. A number of forest activities occur in the watersheds of communities, leading to conflicts over water quality. Here, integrated watershed management plans (IWMPs) were established to resolve the issues. The creation of more parks to preserve the wilderness has addressed some of the issues, but it is not yet known how well these local processes work. Some LRUPs have produced more park designation while others are not yet complete. IWMPs have certainly brought the quality and importance of water to public attention. Logging activities have been deferred in favour of watershed restoration in some IWMP regions. A resolution to watershed issues may lie in creating a Quality of Water Act under the Ministry of Environment, Lands, and Parks (rather than the Ministry of Forests). This implies that water quality, quantity, and timing of flows form the most important values in any watershed and all other activities, including cutting trees, must defer to them.

A Forestry Practices Code was enacted in 1994 to increase fines and penalties for environmental violations, set limits on clearcuts, and pay much more attention to stream setbacks and damage to biodiversity, wilderness, and other aspects of the environment (British Columbia, Ministry of Forests 1994). Now that the code is being implemented, forest companies are protesting the amount of paperwork. Other groups, such as the Sierra Legal Defence Fund, question the effectiveness of the legislation. Clark Binkley's specific criticism of the code is that it recommends the rapid development of road systems: "Since the very worst environmental problems associated with timber production in British Columbia come from the failure of roads, it is ironic that new legislation purporting to protect environmental values actually demands more road building" (1997, 25). In 1998, the provincial government promised a reduction in paperwork, which caused some to question whether environmental standards would be relaxed. There is a basic conflict in having the Ministry of Forests, whose mandate is to encourage the harvesting of trees, act as the gatekeeper of environmental standards.

As a result of reduced AAC and accompanying unemployment, a new crown corporation was created in 1995.

Forest Renewal BC was created with a mandate to retrain and employ forest workers for a variety of new forest-related activities. Stumpage rates were increased to finance these projects, which included reforestation, value-added initiatives, restoration of environmentally damaged areas, retraining of forest workers, and assistance to forest-based communities. This policy change is directly related to the impact of flexible specialization, the need to employ displaced forest workers, and encouragement of the more labour-intensive value-added sector.

A fundamental environmental concern is raised over the principles behind calculation of the annual allowable cut. Hammond (1991), Drushka (1999), Drushka, Nixon, and Travers (1993), M'Gonigle and Parfitt (1994), Binkley (1997), Fulton (1999), and others see it as a process that liquidates old growth and replaces it with uniform plantations, creating serious reductions in biodiversity and sustainability. This gets right to the heart of whether the provincial government, as manager of the forests, is more concerned with a high AAC and revenues than with an ecosystemic basis of sustainability. "Over 83 percent of the land base has been designated as provincial forests to meet the needs of forest harvesting" (Gunton 1997, 66). Since British Columbia comprises 95 million hectares, this amounts to approximately 79 million hectares. The Council of Forest Industries (1994, 38) states that only 26 million hectares is viable "commercial forest land," indicating a considerable discrepancy between the amount of productive forest land and the amount of land that the Ministry of Forests wants to influence and control. A partial solution may be to assess forest land in the same manner as agricultural land and for the best forest land to be put into reserves that minimize conflict among potential uses.

Harvesting techniques, or prescriptions, are equally important to the environmental condition. Kimmins (1997) suggests that the technique of clearcutting may be appropriate in fire-prone interior forests but completely inappropriate in many coastal zones. Not only does this point to the issue of clearcutting versus selective logging but it also questions road-based logging, methods of hauling logs, and other logging practices that affect the environment. The method of cutting must be sensitive to the individual ecosphere, and environmental conflicts must be reduced or eliminated.

The Clayoquot forest harvest agreement, completed in the summer of 1999, may point to yet another model for harvesting timber. In this environmentally sensitive region, a joint venture corporation has been created between the local First Nations (Nuu'chah'nulth), with 51 percent ownership, and MacMillan Bloedel (now Weyerhaeuser), with a 49 percent holding. The agreement includes a logging plan that has the consensus of environmental groups. Tim Aitkins stresses its importance: "Greenpeace and other environmental groups have agreed to endorse and promote this project as a global model of ecologically sustainable forestry. The groups will also help develop markets for the eco-friendly, processed timber products as well as non-timber economic ventures" (1999, 7). Certification, or a guarantee that wood has been harvested in a sustainable manner, is bringing pressure to bear on all corporations in the province to amend their clearcutting practices.

Forestry issues are not all environmental; many are economic. The threatened closure of the Skeena Cellulose pulpmill in Prince Rupert in September 1997, along with the potential closure of the company's sawmills in nearby communities, highlights the dependence of Prince Rupert on this mill for employment. At the same time, the economic viability of producing pulp in this relatively old pulpmill is dubious. The decision by the provincial government to provide loan guarantees saved many jobs, at least in the short run. A different fate befell the pulpmill in Gold River. This mill closed in early 1999, resulting in job loss and the sale of many houses at "firesale" prices in an attempt to save the town. Binkley points out the dilemma faced by this industry: "Productivity in the British Columbia forest sector is squeezed between a rising floor of raw material costs and a fixed ceiling for product prices set by international competitors in the forest product industry and by the cost of substitute products" (1997, 22). Figure 8.10 illustrates the problems in the lumber component of the industry.

Adding value to wood products is another component in the economic equation. The restructuring of the forest industry in the mid-1980s was in part a response to dependence on mass production of dimension lumber, mainly destined for the United States. Markets have

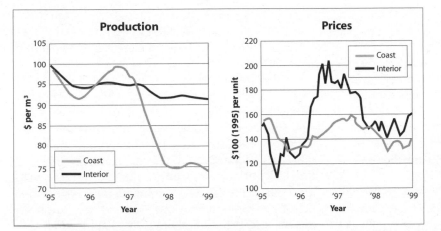

Figure 8.10 BC lumber production and prices, 1995-9
Source: Savings and Credit Unions of British Columbia (1999), 2. Reproduced with permission.

diversified somewhat, implying some diversification in lumber products (Figure 8.11). The Japanese market, for example, requires metric sizing of lumber products and different dimensions (and sometimes species of wood) than the United States. The production of dimension lumber, regardless of the market, adds value to the wood, and this forward linkage provides employment in British Columbia. Far more forward and final demand linkages are possible, however, and these can increase employment and revenues. M'Gonigle and Parfitt (1994, 90) point out the unfortunate fact that approximately

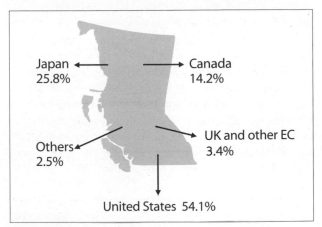

Figure 8.11 Destination of BC wood products by value, 1996
Source: Compiled from Internet, Statistics Canada.

50 percent of the lumber exported to the United States is used in remanufacturing plants. In other words, BC lumber is turned into other wood products, from wooden doors to guitars, that gain even greater value. The challenge is to capture the remanufacturing in British Columbia, and the provincial government has launched an initiative to encourage this activity (see British Columbia, Ministry of Forests 1996).

A thornier issue, because it is highly political and often viewed as the root cause of so many of the issues above, arises from obsolete forms of tenure and their control by large corporations. Forest licences and tree farm licences are the main forms of tenure, and Table 8.5 shows both the control of tenures and volume of wood harvested by the ten largest corporations over time. These tenures, which give large corporations a great deal of control over provincial forests, are the target of those concerned about issues such as declining employment and underproduction of high value-added products. M'Gonigle and Parfitt (1994) argue that the large multinationals have sewn up the wood supply and left insufficient wood available for small, innovative remanufacturing mills wanting to produce higher value-added products. Along the same lines, the Truck Loggers Association has expressed concern that the multinational corporations have considerable interest in turning saw logs into pulp chips. The stumpage is low and so are the labour and value-added components. The initiative by the provincial government to encourage higher value-added is admirable. Whether it can be achieved under the present tenures and their holders still has to be demonstrated.

The conditions of tenure have changed, and although tenures were intended to produce a sense of security for forest company investment (especially through the twenty-five-year renewable clause of TFLs), corporations recognize that they are still leases and thus very much subject to changes in government policy. The consequence, according to Drushka, Nixon, and Travers, is that "silviculture [has been] an obligation instead of an economic opportunity" (1993, 14). Changes to tenure conditions also affect local communities: "With the demise of the old appurtenance clauses that linked tenure to the maintenance of specific mills, companies can dispose of wood as they see fit" (M'Gonigle and Parfitt 1994, 96). Even though a region may still have timber, it is no longer guaranteed that the local mill will receive it. Corporations are at liberty to designate the wood to locations outside the region, thus jeopardizing the survival of local mills and local employment.

Tenure, and its corporate concentration, is also the target of many environmental concerns. Large corporations with tree farm licences such as MacMillan Bloedel have to defend their clearcut harvesting techniques, harvesting of old growth, cutting beside salmon-bearing streams, and all other practices under attack by environmental groups and a public sensitive to destruction of the landscape.

Are there new and different forms of tenure that address these many issues? As noted earlier in this chapter, Hammond (1991), who has practised holistic forestry over a lifetime, suggests turning the forest into a smaller, area-based tenure for individuals sensitive to principles of sustainability and the ecosystem. Others suggest the Swedish model, in which much of the forest is privately owned and managed. M'Gonigle and Parfitt (1994) argue for much more community control of the forests. Whatever the solution, if the provincial government is willing to entertain alternative tenure systems, it has the ability to impose them. The contract between TFL holders and the government allows the province to reallocate 5 percent of the volume of wood for any year. The provincial government can also change the form of tenure

when a licence expires. As M'Gonigle and Parfitt point out, "157 volume-based Forest Licenses come up for renewal [every fifteen years] and there is no obligation to renew" (1994, 96).

For the north end of Vancouver Island, the Queen Charlottes, and the central interior, where so many communities depend on this single industry, the issues are real. The forest base is declining, employment is unstable, large forest corporations are in control, environmentalists and First Nations are in confrontation with foresters, and government decisions to increase parks are seen as representative of southern urban values. Neither the perceptions nor the issues are easy to resolve but changes, even radical ones, are required.

SUMMARY

Forestry is the dominant industry in British Columbia from the perspective of generating revenues and employment and sustaining communities. It favoured the coastal region initially because of the advantages of using the ocean as a means of transportation. Technological developments affected all aspects of the industry and allowed wood to be harvested from more remote locations and processed for many end uses, the most important being lumber, pulp, and paper. The most significant changes occurred from the 1960s on, when the industry expanded into the interior and opened up whole new areas, and communities, in the province. These new technologies also consumed a huge volume of wood each year, putting into question the sustainability of this renewable resource.

As the manager of this resource, the provincial government retains the ownership of 96 percent of the forests and has used various forms of tenure to allow private companies to harvest and process timber from public lands. In the early days, some of the best forest land on the coast was given away through crown grants and other forms of tenure. These private forest lands may be small in comparison to the whole province, but they dominated the AAC until the 1940s and still generate a significant portion of it.

It was not until the Sloan Commission in 1945 that sustained yield and reforestation became important concepts. The two main forms of tenure today come from this commission – tree farm licences (area based) and

Table 8.5

Concentration by the ten largest corporations

	Harvest rights (%)	Volume of wood (m³)
1954	37	26,600,000
1975	59	59,996,000
1990	69	63,800,000

Source: Modified from M'Gonigle and Parfitt (1994), 44.

public sustained yield units (volume based) – although PSYUs were reorganized into timber supply areas in 1979 with a number of licences to harvest timber.

The large volumes of wood harvested until the 1980s provoked concerns about economic stability and the fall-down effect and a host of land use and environmental conflicts. New forums and policies – treaties, CORE, LRUPs, IWMPs, the Forestry Practices Code, and Forest Renewal BC – have addressed many issues and made adjustments to park land and harvesting methods but many concerns still exist. The geography of forestry shows the domination of large corporations, which control much of the timber. It also shows how many Vancouver Island and interior communities depend on the forest resource although they have little influence in how it is used and managed. Viewing the forests from the ecosystem perspective of sustainability challenges the basis of the present tenure system. Forests are a renewable resource, but renewable only if they are properly managed.

REFERENCES

Aitkins, T. 1999. "Clayoquot: A Model for the Future." *Greenpeace* 7 (3): 7.

Barker, M.L. 1977. *Natural Resources of British Columbia and the Yukon.* Vancouver: Douglas, David and Charles.

Binkley, C.S. 1997. "A Crossroad in the Forest: The Path to a Sustainable Forest Sector in British Columbia." In *Troubles in the Rainforest: British Columbia's Forest Economy in Transition*, ed. T.J. Barnes and R. Hayter, 15-35. Canadian Western Geographical Series vol. 33. Victoria: Western Geographical Press.

British Columbia, Ministry of Finance and Corporate Relations. 1998. *1998 Financial and Economic Review.* Victoria: Crown Publications.

–. 1999. *British Columbia Local Area Economic Dependencies and Impact Rations, 1996.* Victoria: The Ministry.

British Columbia, Ministry of Forests. 1988. "Background Information to the Government's Policy to Replace Forest Licenses with Tree Farm Licences." Victoria: The Ministry.

–. 1994. *British Columbia Forestry Practices Code.* Victoria: The Ministry.

–. 1996. *Five-Year Forest and Range Resource Program 1996-2001.* Victoria: The Ministry.

British Columbia Round Table on the Environment and Economy. 1993. *An Economic Framework for Sustainability.* Victoria: The Round Table.

Canadian Council of Forest Ministers. 1997. *Compendium of Canadian Forestry Statistics.* Ottawa: Natural Resources Canada, Canadian Forest Service.

Clapp, R.A. 1998. "The Resource Cycle in Forestry and Fishing." *Canadian Geographer* 42 (2): 129-44.

"Coastal Mills Are Now World's Highest-Cost Pulp Producers." 1997. *Vancouver Sun*, 21 November, A10.

Council of Forest Industries (COFI). 1994. *British Columbia Forest Industry Fact Book – 1994.* Vancouver: COFI.

Drushka, K. 1999. "British Columbia's Forests: A New Way to Grow." *Vancouver Sun*, 31 March, D1, 2, 19, 20.

Drushka, K., B. Nixon, and R. Travers, eds. 1993. *Touch Wood: B.C. Forests at the Crossroads.* Madeira Park, BC: Harbour.

Edgell, M.C.R. 1987. "Forestry." In *British Columbia: Its Resources and People*, ed. C.N. Forward, 109-37. Western Geographical Series vol. 22. Victoria: University of Victoria.

Farley, A.L. 1972. "The Forest Resource." In *Studies in Canadian Geography: British Columbia*, 87-118. Toronto: University of Toronto Press.

–. 1979. *Atlas of British Columbia: People, Environment, and Resource Use.* Vancouver: University of British Columbia Press.

Fulton, J. 1999. "British Columbia's Struggling Forest Industry." *Vancouver Sun*, 20 January, A12.

Gunton, T. 1997. "Forest Land Use and Public Policy in British Columbia: The Dynamics of Change." In *Troubles in the Rainforest: British Columbia's Forest Economy in Transition*, ed. T.J. Barnes and R. Hayter, 65-73. Canadian Western Geographical Series vol. 33. Victoria: Western Geographical Press.

Haddock, M. 1995. *Forests on the Line.* New York: Natural Resources Defence Council and Sierra Legal Defence Fund.

Haig-Brown, R. 1961. *The Living Land.* Toronto: Macmillan.

Hammond, H. 1991. *Seeing the Forest among the Trees.* Vancouver: Polestar.

Hardwick, W.G. 1960. "Changing Logging and Sawmill Sites in Coastal British Columbia." In *Occasional Papers in Geography* no. 2, 1-7. Vancouver: Tantalus.

Hayter, R., and T. Barnes. 1997. "The Restructuring of British Columbia's Coastal Forest Sector: Flexible Perspectives." *BC Studies* no. 113 (Spring): 6-34.

Hayter, R., and R. Galois. 1991. "The Wheel of Fortune: B.C. Lumber and Global Economics." In *British Columbia Geographical Essays: Geographical Essays in Honour of A. MacPherson*, ed. P. Koroscil, 169-201. Burnaby: Department of Geography, Simon Fraser University.

Husband, V. 1995. Interview by Peter Gzowski. Canadian Broadcasting Corporation, *Morningside*, 5 December.

Jones, T. 1983. *Wilderness or Logging? Case Studies of Two Conflicts in B.C.* Vancouver: Federation of Mountain Clubs of British Columbia.

Kimmins, H. 1997. *Balancing Act: Environmental Issues in Forestry*, 2nd ed. Vancouver: University of British Columbia Press.

M'Gonigle, M., and B. Parfitt. 1994. *Forestopia: A Practical Guide to the New Forest Economy*. Madeira Park, BC: Harbour.

Marchak, P. 1983. *Green Gold: The Forest Industry in British Columbia*. Vancouver: University of British Columbia Press.

Savings and Credit Unions of British Columbia. 1996. *Economic Analysis of British Columbia* 16 (9): n.p.

–. 1999. *Economic Analysis of British Columbia* 19 (3): n.p.

Travers, O.R. 1993. "Forest Policy: Rhetoric and Reality." In *Touch Wood: B.C. Forests at the Crossroads*, ed. K. Drushka, B. Nixon, and R. Travers, 171-224. Madeira Park, BC: Harbour.

Willems-Braun, B. "Colonial Vestiges: Representing Forest Landscapes on Canada's West Coast." In *Troubles in the Rainforest: British Columbia's Forest Economy in Transition*, ed. T.J. Barnes and R. Hayter, 99-127. Canadian Western Geographical Series vol. 33. Victoria: Western Geographical Press.

Wilson, M. 1998. "Munro: Trees Ready Early." *The Province* (Vancouver), 31 May, A49.

FILMS

Collins, G., and G. Cuthbert, directors. 1980. *The Fall-Down Effect: A Discussion on the British Columbia Timber Supply Analysis*. Selkirk College, Castlegar, BC.

INTERNET

BC Stats: Quick Facts – The Economy, www.bcstats.gov.bc.ca/data/QF_econo.HTM

British Columbia, Ministry of Energy and Mines, "Total Coal, Production and Value – 1836 to 1998," www.em.gov.bc.ca/mining/miningstats

Statistics Canada, www.statcan.ca

The Salmon Fishing Industry:
Managing a Mobile Resource

9

Fish, like trees, are a renewable resource, but one very difficult to manage because it is mobile and not easy to see. The five species of Pacific salmon have different physical characteristics and economic value, but they are all born in fresh water, migrate to the ocean, and return to fresh water, where they spawn and die. Understanding the fresh and saltwater environments is central to our ability to manage the salmon and ensure survival.

Like other resource extraction industries in British Columbia, the fishery has a long history of technological change. The invention of canning solved the problem of preserving and exporting the salmon, and canneries were once located at the mouths of salmon-bearing rivers up and down the coast of the province. The canneries went through many changes, moving from a labour-intensive process to the capital-intensive Fordist process relatively early in the twentieth century. Consolidation of canneries by large corporations also occurred at this time. New markets and methods of preserving salmon developed over time, reducing the need for canneries.

The process of salmon fishing also underwent major technological change. Low technology gillnetters, which used oars and sails, were eventually supplanted by the most modern fishing fleet in the world, with a huge capacity to catch fish. This transformation had a major impact on cannery location. Historically, boats had little ability to travel any distance and were therefore tied to canneries. As fishing vessels developed the ability to travel farther and farther from the canneries, and technologies were developed to preserve fish on board, canneries in remote coastal locations were no longer needed. Abandonment of canneries and communities became the norm.

Salmon is a food resource, and its markets therefore fluctuate because of world events such as war and depression and also according to perceptions of how a particular food affects health. These considerations are reflected in changing world market prices and demands, putting the economic viability of fishing into question.

Most salmon are caught in the ocean, making management a federal responsibility. Management includes preservation, enhancement of stocks, and protection of habitat. That salmon migrate into international waters, often return through American waters, and finally end their journey in freshwater streams, where the provincial government allows conflicting resource extraction activities, makes the management task extremely difficult. The balancing act for the federal government is in allowing interested parties to harvest a certain number of salmon while simultaneously allowing a certain number of fish to escape upstream to spawn and start the cycle again. The industry has intensely competing interests that want a share of this resource.

A seemingly endless array of commissions and hearings reflects the difficulties in management. Unfortunately, the conflicts and issues of the past are with us today, with even greater concerns about the sustainability of salmon. The loss and deterioration of freshwater habitat has prompted salmon enhancement programs that range from rebuilding spawning beds to establishing hatcheries. The use of hatcheries to raise and release salmon fry artificially is controversial in itself, but the development of the salmon farming industry has provoked even greater friction. The price and demand for salmon has increased, creating pressure for larger harvests. Fish wars with the United States have resulted in overfishing. First Nations are demanding traditional rights to a food fishery and want a share of the commercial catch. The sport fishery, which is the basis of a growing, lucrative tourism industry, wants more fish. Trollers, gillnetters, and seiners, shown in Figure 9.1, make up the commercial fleet, and they have a large investment in boats and gear that needs to be paid off; they too want more salmon.

Clearly there is not enough salmon to satisfy the demands of all the interests. The provincial government is demanding more of a role in management and so are local fishing communities and Aboriginal groups. A more fundamental question of management is whether this resource should be privatized.

SPECIES, CYCLES, RACES, AND HABITAT OF PACIFIC SALMON

There are five species of Pacific salmon, commonly known as sockeye, chinook (spring), coho, pink, and chum. Sockeye is the best known in terms of external demand and commands the highest price because of its

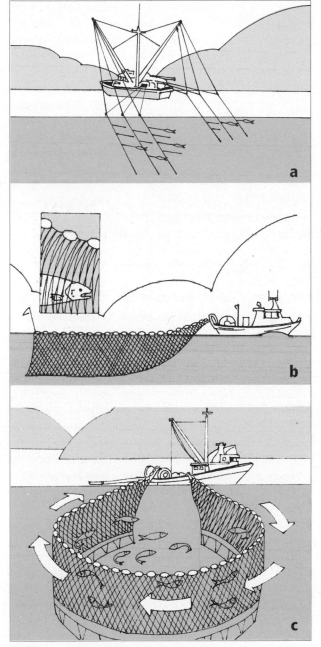

Figure 9.1 Commercial salmon fishing vessels:
a, troller; b, gillnetter; c, seiner
Source: Fisheries and Oceans Canada (1999), 53-4. Used with permission.

rich red meat. Chinook, also referred to as spring or king salmon, is the largest of the five species, weighing up to 30 kilograms. Coho from northern and southern British Columbia differ in size. Large coho in southern waters weigh approximately four kilograms, whereas in northern waters they can reach ten or more kilograms. Coho, along with spring, are referred to as the hook-and-line species because they are the most sought-after by anglers. Both species are essential to the sport fishery and tourism. Pinks and chum are referred to as the schooling varieties because they travel together at or near the surface in schools. This makes them accessible to the seiner fishers who encircle them with the nets.

Salmon are anadromous: they are born and spend part of their early life in freshwater lakes and streams, migrate to the ocean, and then return to the fresh water to spawn and die. Although the cycle is common to all five species, there are variations within it. Sockeye spawn in rivers with lakes in them, for example, and spend a year or more there before migrating to the ocean. Sockeye in the Fraser follow a four-year cycle while those in the Skeena and Nass River systems range from four to six years (Barker 1977, 115). Some of the chinooks in the Columbia River system before it was dammed migrated 1,400 kilometres or more and were on a seven-year cycle. Coho often have a three-year cycle, and pinks follow a two-year cycle. Chum and some pink salmon rarely migrate any distance upstream, preferring the gravel bed at the mouth of the river.

One way to view the relationship between the various salmon species and the rivers and streams where they are born is in terms of ownership. Because most salmon return and spawn in exactly the same stream in which they were born, that stream "owns" those salmon, and the salmon from it are referred to as a race of salmon bearing its name. Sockeye born in the Adams River, for example, is referred to as the Adams River race, which is the largest run of sockeye salmon in the world. The Quesnel River also has a large race of sockeye, which by coincidence is on an identical four-year cycle to the Adams run. Both the Adams and Quesnel Rivers are tributaries of the Fraser. Consequently, every four years there is an enormous run of sockeye salmon on the Fraser. Of course, a river or stream may have several races of different salmon species each year.

Since salmon return to their birth streams, the physical condition of those streams is essential to the continuation of salmon stocks. Figure 9.2 shows the main salmon-bearing rivers in British Columbia.

The federal government manages the salmon resource and through the Fisheries Act is responsible for both saltwater and freshwater habitats of salmon. Unfortunately, provincial jurisdiction over land-based resources such as forestry, mining, agriculture, and hydroelectric production and its encouragement of development has permitted many freshwater habitats to suffer serious damage, some of it irreversible. The gold rush on the Fraser and into the Cariboo turned streams upside down and inside out, destroying spawning beds. Later mining ventures and forestry increased siltation, added pollutants, and warmed the freshwater environment. Dams bring an end to salmon migration, which often means the extinction of races of salmon. Increased urban development not only adds wastes to our river systems but often results in streams being culverted and physically changed.

One specific catastrophe was the Hell's Gate slide in 1913. Hell's Gate is an extremely narrow restriction in the Fraser River just north of Yale that the salmon must navigate to reach the spawning beds of much of the Fraser River basin (Figure 9.2, insert). The construction of the Canadian Pacific Railway (CPR) in the 1880s on one side of the Fraser and then the Canadian National Railway (CNR) in 1912-13 on the other side added to the already difficult passage by dumping huge amounts of blasted rock into the Fraser. These problems were compounded when the rock loosened from blasting gave way. With the ensuing rockslide in the late summer of 1913, few salmon were able to overcome the barrier (Meggs 1992; Hume 1997). Tragically, 1913 was the fourth year of the sockeye cycle, and the rockslide occurred just as the salmon were about to return to their spawning grounds. Figure 9.3 shows the large numbers of salmon every four years prior to 1913 and the consequences of the slide. After the slide "eleven concrete fishways [were] designed in Seattle and built for $24 million cost shared equally by Canada and the United States" (Hume 1997, B3). Even though fish ladders allow salmon to get past Hell's Gate, rebuilding these stocks to historical levels is a difficult process and still in progress.

The saltwater environment presents other factors to consider in managing the salmon. Seals and sea lions take their share of the salmon, and in the past war was waged on both of these mammals. Forester and Forester (1975) give us some idea of how much of a threat these animals were perceived to be: "Notwithstanding the fact that 749 seals and 2,875 sea lions were destroyed during the season of 1915, it was only a drop in the bucket, especially as in the Fraser River, where the depredations from seals appear to be greatest, only forty-eight of these mammals were destroyed, and it is prophesied by certain fishermen that unless a greater destruction takes place the spring-salmon fishery is doomed" (p. 217). They go on to explain how bounties, patrol officers with machine guns, and homemade bombs were employed to exterminate these salmon predators, but with limited success.

Figure 9.2 Major salmon-bearing rivers and lakes

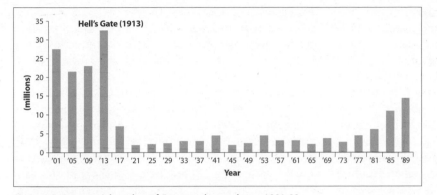

Figure 9.3 Commercial catches of Fraser sockeye salmon, 1901-89
Source: Roos (1991), 412, 413.

Whales, the natural predators of the seals and sea lions, were drastically reduced in numbers by a pervasive whaling industry. Herring, the main food supply for salmon, has been radically reduced by the herring fishery. Human interference has brought on some serious imbalances in the saltwater ecosystem. Today, the seal population appears to be very large but environmental constraints prevent its destruction.

Recent evidence suggests that the El Niño effect is responsible for bringing warmer water to the Pacific northwest, and with it, mackerel, which normally do not migrate this far north. The mackerel become another consumer of young salmon. Of course, the greatest consumers of salmon in the saltwater environment are the many fishing interests, both foreign and local.

With all these circumstances taken together, one wonders how any salmon make it back to their spawning grounds. Mary Barker points out the survival-of-the-fittest conditions for salmon: "3,000 eggs may be deposited by a sockeye, but only 100 will become fingerlings and travel to the sea. Of these, less than twenty will return as adults" (1977, 115).

TECHNOLOGICAL CHANGES TO PRESERVING AND CATCHING SALMON

Salmon have always been an important resource to First Nations, which preserved them by smoking and sun drying. The Hudson's Bay Company forts, such as Fort Langley on the Fraser, became interested in salmon for export and used the traditional European process of salting and barrelling. For this industry to develop commer-

cially required a means of preserving salmon without altering its taste the way salting did. The invention of canning was the key, and canneries began to appear at the mouth of the Fraser River in the 1870s.

Duncan Stacey (1982) defines two periods: the manual canning era from 1871 to 1903, and the mechanical canning era from 1903 to 1913. Technology was responsible for the transition from labour-intensive to capital-intensive methods of processing fish. The manual canning era, as the name suggests, required much of the canning process to be carried out by labourers. As Stacey explains, "In the earliest canneries each can was cut out of sheet tinplate, formed, and soldered, by hand. By 1890 a number of machines had been introduced to punch out body pieces, top, and bottoms and to apply solder, but these were still aids to the hand process" (1982, 4).

In 1876, 9,847 cases of salmon were exported, with twenty-four one-pound tins to a case (Stacey 1982, 4). In 1905, the figure was 837,489. These statistics may cause one to jump to the conclusion that a major change in technology was responsible for the changes in this twenty-nine-year period. Part of the answer lies in a market trend after 1888. Prior to 1888, only sockeye salmon were harvested. There was no market for the other, lighter coloured salmon species with less oil content. Commercial fishing was therefore only a five- to six-week occupation, and in 1876 there were just three canneries on the Fraser. From 1889 onward, however, the market was expanded to include all five species of salmon. The fishing season lasted five to six months and by 1901 there were forty-nine canneries.

The mechanized era brought a number of changes to canning, the two most significant being efficiency on the production line and corporate concentration. The introduction of electricity for lights and motors and Fordist assembly line techniques replaced the high demand for labour. Stacey tells us that "hand-butchering gangs gave way to butchering machines, manually-soldered cans began to be replaced by the mechanized, solderless, or sanitary, can, and all other sections of the canning line experienced varying forms of mechanization" (1982, 19). The Smith Butchering Machine, developed in 1907, was the most important mechanical invention of this era. With the assistance of three people it could clean and cut up sixty to seventy-five fish per minute, replacing butchering gangs of about thirty people (p. 21). Those replaced were frequently Chinese labourers; hence the Smith Butchering Machine became known by the racist term "the Iron Chink."

Consolidation occurred very early in this industry, when the British Columbia Packers Association acquired twenty-nine of the forty or so canneries on the Fraser in 1902 (Stacey 1982, 19). Consequently, this large corporation was able to invest in the capital equipment necessary to adopt Fordist technologies.

The assembly line process of canning fish still exists but few canneries remain. With the fresh and fresh-frozen markets and products such as smoked salmon becoming increasingly popular from the 1970s on, canneries have given way to fish-processing plants. Canneries were abandoned in isolated coastal locations such as Rivers Inlet, Namu, and Klemtu as changes to fish boat technology and refrigeration allowed vessels to travel the whole length of the BC coastline. The 185 fish processing plants in operation by 1996 were concentrated in the Vancouver, Vancouver Island, and Prince Rupert areas (Internet, BC Stats: Quick Facts).

In the nineteenth century, gillnetting was the most popular method of fishing in the salmon industry. Gillnets were set up in the mouth of the rivers to intercept the salmon as they went upstream to spawn. Gillnets would be set from Columbia River boats. Similar to the east coast dory, this boat is approximately six metres long, pointed at both ends, and operated by two people with oars and a sail. Typically, the gillnetter would row or sail upstream, set a net, drift downstream, pull the gillnet in by hand, and row upstream to repeat the process. When the boat was full it was time to travel to the cannery to unload the catch. This process changed little prior to 1903. The "big" invention was the placement of a roller at the stern of the vessel to facilitate hauling in the net.

Because they were propelled only by oar or sail, these low technology gillnetters were able to range only short distances from the canneries where they had to unload their catch. With the proliferation of canneries and fishers, waters became crowded and the competition intense. Some competition was relieved with the steam-powered tender boat, used by some of the larger canneries to tow the gillnetter farther out with the expectation of intercepting the salmon.

For the individual fisher, the invention of the gasoline engine in 1907 heralded a major change: "The mechanized gillnetter could make more sets since it could move more quickly upriver to start a new drift. The gasoline engine also enabled the fishery to increase the fishing area by working farther offshore. It increased fishing time since it took less time to travel to and from the grounds and vessels could fish in rougher weather" (Stacey 1982, 26). The gasoline motor was also responsible for a vast improvement in the purse seine fishery, whereby a motorboat was used to set out a very long net that could rapidly encircle schools of salmon (Figure 9.1).

It should be noted that a rather unusual set of rules over the use of motors divided the fishery into two regions. In the northern fishery, mainly in the Skeena River and Prince Rupert region, the canneries were much more in control of the whole fishing industry. They owned the vessels and licences and did not wish to invest in motors. Nor did they want competition from independent fishers, so they had motors legally banned in 1911 (Stacey 1982, 26).

The ability of motorized fishing boats to travel greater distances from the cannery posed another problem: perishability. Once salmon are caught and hauled in, the deterioration process begins. This problem was resolved by 1915 with a diesel-powered packer that used ice as a means of refrigeration. The motorized gillnetters could now unload their catch onto these relatively large vessels at sea. The diesel-powered packer was designed to stack the salmon in a hold without crushing it.

Fishing is a highly competitive industry in which new technologies are constantly sought to gain advantage. Many inventions arising from the First and Second World Wars to identify submarines, to communicate, and to lift war materials diffused down into the fishing industry. Vessels installed echo sounders, asdic, sonar, radar, radio telephones, hydraulics, and other devices. Along with these sophisticated technologies came synthetic nets, methods of freezing salt water, and a host of other inventions. Individual vessels got larger, their ability to travel great distances to catch salmon improved, and so did their capacity to harvest salmon. Yet increased capacity was not matched by increases in the resource. In fact, there have been great concerns that the salmon fishing industry may meet the same end as the cod fishing industry on the east coast.

MARKET FACTORS AFFECTING A FOOD RESOURCE

Salmon, like most resources, is subject to many factors that influence supply and demand, and these are reflected in the world market price. Differences in demand and value apply to the five species, with sockeye being the most lucrative. Global recessions and depressions have reduced the demand and price for salmon generally. The natural cycle of salmon causes fluctuations in supply, making for a lack of continuity of markets and income for fishers. Because it is a food resource, other unique influences affect the demand for salmon.

The Yukon gold rush of 1898 is an example of unpredictable changes in the demand for canned salmon. This gold rush was met with fear by the authorities, who thought that the stampede of miners entering the Yukon with little knowledge of this northern environment would starve to death. One of the requirements put in place for miners entering the territory was that they must be in possession of 1,100 pounds of provisions. This was a boom to canned BC salmon since would-be-miners frequently purchased their supplies in Victoria and Vancouver and tinned salmon was relatively easy to pack.

Wars have also had a positive influence on the demand for this resource. Both world wars saw an increase in salmon processed to feed the troops. It is a high source of protein and easily stored. Statistics for 1930 and 1939 show the impact of the Depression of the 1930s and the market adjustments with the start of the Second World War (Table 9.1). The end of the war had an unexpected impact also. As a result of the great expenses Britain incurred by the war, BC salmon became a "luxury" item. By 1948, the British market for canned salmon had collapsed.

Food resources are subject to rumours and misinformation that can have a negative impact on their market. The "canned beef scandal" of 1906 is a typical example. Contaminated beef, tinned in Chicago, found its way to

Table 9.1

BC salmon exports, 1930 and 1939

	1930 (cases)	1939 (cases)
To the British Empire		
United Kingdom	189,227	406,943
Australia	227,423	278,047
New Zealand	62,404	64,470
British South Africa	35,202	71,997
Fiji	11,015	15,020
British India	6,113	8,225
Other British countries	22,217	0
Total	553,601	844,702
To other countries		
France	219,854	108,283
Belgium	58,721	9,306
Italy	152,044	0
Colombia	5,011	9,252
Other South American countries	57,319	0
Portuguese East Africa	8,723	6,806
All other countries	73,526	43,363
Total	575,198	177,010

Source: Lyons (1969), 433.

the British market. The backlash to this disastrous affair was that consumers in Britain rejected imported canned food, including salmon from North America. As well, salmon producers have had to combat rumours that sockeye salmon has dye added to achieve its distinctive red colour and, worse yet, that it is linked to "salmonella." This deadly virus, associated with poultry, was discovered in the late 1930s by Dr Daniel Salmon. Although the virus had nothing to do with salmon, the word association was enough that many consumers were not going to take the risk. Several cases of a potentially lethal disease known as botulism, resulting from improper canning, did occur in the 1970s. Where food is involved in a perceived risk to health, the demand for that food will plummet.

Today, the fresh and fresh-frozen salmon market is marked by another controversy: wild versus farmed salmon. Commercial trollers and salmon farmers often compete for the same market, and those catching wild salmon claim that farmed salmon is an inferior product that should be labelled for consumer information. Bumper stickers stating "Real salmon don't eat pellets!" reflect, at least, the war of words in this battle.

MANAGING SALMON: THE POLITICS OF REGULATION

British Columbia entered confederation in 1871 with the condition that the federal government had jurisdiction over oceans and their resources. Coincidentally, the first canneries were being established in British Columbia at this time.

From a commercial perspective there was little concern over the resource until 1888. Only sockeye were being caught prior to that time, but by 1889 the market had expanded for all species of salmon, increasing the season to five or six months. Regulations were put in place for the 1889 season that included the following:

1 Introduction of Licenses with a restriction of 500. 350 for canneries and 150 to "outside" fishers.
2 Aboriginals prohibited from selling salmon.
3 Nets restricted to tidal waters and banned from fresh water.
4 Nets not permitted to obstruct more than one-third of a river.
5 Minimum size of mesh for nets – 6 inches.
6 Ensnarement, or fish traps, illegal.
7 Anchoring nets on shore, illegal.
8 No form of seining in the Fraser River permitted.
9 Closure: Saturday 6 p.m. to Monday 6 a.m. (Lyons 1969, 192).

Salmon have always been viewed as a **common property resource,** a resource for all to use that brings with it, unfortunately, the possibility for abuse. The introduction of regulations kept salmon as a common property resource but defined who was allowed to compete for it (licence holders) and the conditions under which they could compete. In this form of resource management, the federal government acts as the gatekeeper and operates under the assumption that it understands migrations of salmon and that the regulations are sufficient to allow salmon to escape upriver to spawn. The move from a completely unregulated industry to a fully regulated one, however, caused many in the industry to question both the assumption and the regulations. Whenever the public demonstrates sufficient concern over any rights or rules, the usual government response is to strike a commission to investigate and recommend changes. The first of many commissions on salmon occurred in 1890 as a reaction to the imposition of regulations in 1889.

The Wilmot Commission of 1890 was rather unsympathetic to the concerns of the fishing industry on the west coast and recommended two small changes to the regulations: a reduction in minimum mesh size from six inches to five and three-quarter inches, to enable a minor increase in catch; and a reduction in the closure to twenty-four hours, from Saturday 6 p.m. to Sunday 6 p.m. These changes were minimal, and the fishers demanded another commission to have the regulations relaxed. A new commission was struck two years later. It came out with new recommendations in 1892, the most significant being the removal of licence restrictions. Anyone could now purchase a licence to fish, with two long-term results: "The traditional fishing grounds became overcrowded and a change took place in the type of license ownership. The traditional ownership of the fishing license by the cannery gave way to non-cannery, or private, ownership of licenses" (Stacey 1982, 13). For the managers of this resource, more fishers meant more competition. Competition gave rise to new technologies, which increased the capacity for each vessel. So many fishers entered the industry that for the first time there was some anxiety about sustainability.

The federal government made some efforts to increase the number of salmon through hatcheries. The first one was located at Port Mann on the lower Fraser River in 1883, with others to follow. Meanwhile, streams were being damaged through mining and logging activities, but the federal government had little say about these provincial activities.

The questions and recommendations of the Duff Commission in 1922 showed other considerations in managing salmon. The three questions addressed by the commission were:

1 Should motor boats be allowed in the northern waters?
2 Do the Japanese have too many licences?

3 What should be done about the drastically reduced Fraser sockeye?

Recommendations brought down in 1924 included permission to use motor boats in northern waters and a 40 percent reduction in licences to the Japanese. The commission stated that it could do nothing about the Fraser River sockeye as it was an international problem (Lyons 1969, 352). The significance of the 1913 Hell's Gate disaster was not mentioned.

The need to manage salmon from an international perspective can be seen from Figure 9.4, which indicates the migration boundaries of the various species of salmon. Many of these runs of salmon migrate through international waters and the jurisdictional waters of Alaska and Washington State before returning to rivers in British Columbia to spawn. It should be noted, however, that national water jurisdiction has not remained static. Prior to 1964, jurisdiction only extended to three miles from the shoreline. It changed to nine miles in 1964, to twelve miles by 1971, and finally to 200 miles (322 km) in 1977. Canada contends that the salmon born in the fresh waters of British Columbia belong to Canada – this has led to many disagreements with the United States.

When the border between British Columbia and Washington State was established in 1846, it was agreed that Vancouver Island would be included as British territory although it dipped below the forty-ninth parallel. Unclear at that time was where the boundary went with respect to the San Juan Islands (Figure 9.5). The British claimed that the international boundary should be through the Rosario Strait, and the Americans countered that it should be the Haro Strait. The Middle Channel was proposed as a compromise, but American insistence on the Haro Strait saw the British accede the territory. The importance of this boundary, in terms of the salmon industry, was that many of the runs of salmon destined for the Fraser River come through American waters. The

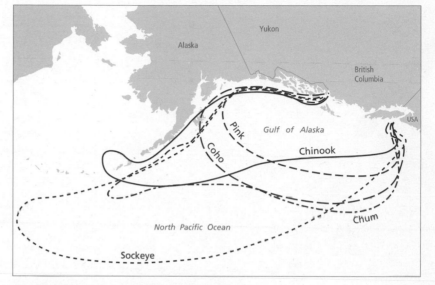

Figure 9.4 Migration routes for BC salmon species
Source: Modified with permission from Fisheries and Oceans Canada (n.d.), 2-6.

closures and restrictions set out in the Canadian regulations in 1889 did not apply in the United States and Americans were taking more Fraser-bound salmon than were Canadians.

Commission after commission attempted to negotiate treaties with the United States. All ended in failure and frustration, with the Americans taking the lion's share of the resource. The Report of the Commissioner of Fisheries for 1931, which ends with a threat, expresses these sentiments:

The failure of the Senate to ratify this treaty is a matter of grave concern in Canada. For years she has sought to maintain the sockeye-salmon fishery of the Fraser System by closely restricting the operations of her fishermen and by extensive hatchery operations. Notwithstanding her continuous efforts for the last thirty years, the fishery has steadily declined. She has failed to maintain it because of lack of support and assistance on the part of the United States. Her patience is close to exhaustion. Failing favourable action on the treaty now before the Senate, it is seriously proposed that Canada abandon further efforts to maintain the sockeye-salmon fisheries of the Fraser System and, in lieu thereof, collect all the sockeye-eggs obtainable in the

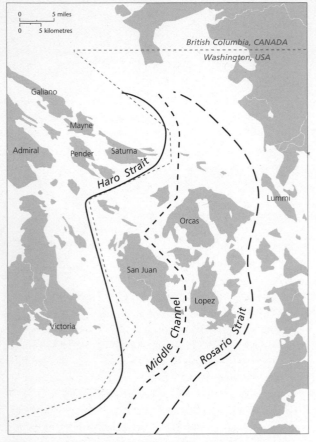

Figure 9.5 Salmon migration and international borders
Source: Modified from Lyons (1969), 214.

Table 9.2

Fraser River sockeye salmon catches for selected years

| | Canada | | United States | | Total |
	cases	%	cases	%	cases
1913	719,796	30.1	1,673,099	69.9	2,392,895
1917	148,164	26.5	411,538	73.5	559,702
1921	39,631	27.8	102,967	72.2	142,598
1925	35,385	24.0	112,023	76.0	147,408
1929	61,569	35.5	111,898	64.5	173,467
1933	52,465	29.0	128,518	71.0	180,983
1937	100,272	62.5	60,259	37.5	160,531
1941	171,290	60.8	110,605	39.2	281,895
1945	79,977	60.1	53,055	39.9	133,032
1949	96,159	54.4	80,547	45.6	176,706
1953	191,123	51.7	178,323	48.3	369,446

Source: Lyons (1969), 508.

A treaty was signed in 1957, only after a very sophisticated seine fleet was sent off Vancouver Island, threatening to intercept the pinks before they reached Washington State waters. It took until 1985 to negotiate an agreement on all species of salmon. This agreement had an eight-year review clause attached to it and the treaty was cancelled at the 1993 review, putting the Canada-US relationship back sixty years. The inevitable fish war serves the interests of no one, particularly with respect to future salmon stocks.

Managing salmon at the international level has been only intermittently successful. The 1997 round of treaty negotiations failed, marked by the blockade of an Alaskan ferry in Prince Rupert followed by threats and counterthreats of lawsuits. The 1998 negotiations fared little better. Federal fisheries minister David Anderson announced a total ban on fishing coho salmon in Canadian waters because of the fear that some important races of coho, notably the Skeena and Thompson River runs, were close to extinction. Alaska issued a statement denying that the coho was threatened and stating that they intended to catch their share (Internet, Coho Monitoring Bulletins). The Pacific salmon "fish war" between Canada and the United States ended with an agreement in June 1999 (Internet, Fisheries and Oceans Canada),

Fraser River basin, transfer them to her hatcheries on the Skeena and Rivers Inlet, and liberate the resulting fry in these waters, over which she has complete control. The suggestion is entirely practical, because it has been repeatedly demonstrated that adult sockeye return to spawn in the fresh waters in which they spent the first year of their life (Canada 1931, 12).

Tough words, but no ratification of a treaty occurred until 1937, and even then it only included sockeye. An examination of Table 9.2 indicates the positive results of these negotiations.

Many of the pink salmon return through the San Juan Islands, and this became the next target of negotiations.

and thus the managers of this resource could turn their attention to the many issues within their jurisdiction.

The BC fishery also experienced many internal conflicts. By the 1950s, the main issue was how to regulate the increasing capacity of the salmon fishing fleet. The Sinclair Report of 1958 "proposed a system of restricted vessel licenses and levies on catches to dampen incentives for over-investment" (Ross 1987, 189). It was the Davis Plan of 1968 that actually resulted in the reduction of the fleet, from 6,104 vessels in 1969 to 4,707 vessels in 1980. The unexpected consequence was new investment in the remaining vessels; some were upgraded and others were replaced. The seine fleet, for example, "actually increased from 286 to 316, because the new rules allowed combining several old gillnetter licenses into a seiner license" (Meggs 1992, 195). In the end, "the capacity of the fleet is estimated to have doubled or perhaps trebled" (Ross 1987, 189).

The Pearse Commission of 1981 summarized the problems of capacity by stating that there were "too many fishermen chasing too few fish" (Pearse 1982, 2). The report touched on the sophistication and overcapacity of fishing vessels, and observed that the size of salmon being caught was getting smaller. Pearse also expressed concerns about habitat destruction and its role in eliminating races of salmon, and the lack of a sufficient information base for managers to make decisions about run openings or escapement (allowing salmon to proceed up river to spawn):

> Salmon managers face four major difficulties in developing the pre-season plans. First, they seldom know with much confidence how many fish will enter a fishery. Second, they cannot reliably predict the time the stock will enter the fishery. Third, they do not know how many vessels will participate in a particular fishery. The highly mobile fleet in the salmon fishery responds quickly and often unpredictably to fishing opportunities along the coast. Sometimes managers refrain from planning openings for small runs because of the threat of excessive fishing effort being directed to the available stocks. Fourth, information about the stocks and their spawning requirements is so weak that the escapement targets are little better than guesses (p. 40).

Many of Pearse's "solutions" were controversial. The recommendation for more research on salmon was reasonable, but the reduction of 50 percent of the fleet through auctioning licences was extremely controversial. Equally controversial was the suggestion of tying salmon quota to vessel and gear type, which in effect would privatize the resource. In other words, commercial fishers would no longer be able to compete for as many salmon – or as much of the common property resource – as they could within an opening. Instead, a licence would entitle the fisher to a specific quota of salmon, in effect giving ownership of a portion of the resource and thus privatizing it. Pearse also recommended licences and restrictions for the sport fishery and, finally, he encouraged investment in hatcheries and fish farming. Few of his controversial recommendations were acted upon, though he had pointed out many of the problems for the federal managers.

FACING THE PROBLEMS TODAY

The freshwater habitat, which is crucial to both ends of the salmon's life cycle, requires much more attention than it has received. The expansion of urbanization and many forms of industry are destroying salmon spawning grounds, and this process must be reversed if the number of salmon is to increase (Glavin 1996; Clapp 1998). The Sierra Legal Defence Fund states that over 140 runs in British Columbia became extinct in the twentieth century and over 600 more are expected to become so. The Salmon Enhancement Program (SEP) is designed to reverse the trend, although this program is going in two fundamentally different directions. One strategy is to enhance existing streams by cleaning up obstructions to salmon migration, constructing artificial gravel beds, fertilizing lakes to increase food, and encouraging schools and community groups to establish incubation boxes in streams. The second strategy is to construct hatcheries and artificially raise salmon to be released into the ocean. Hatcheries are expensive to maintain, however, and they have the potential to spread diseases and reduce genetic diversity. Hatchery-raised salmon are also used to supply ocean ranches and fish farms, two threats to the commercial fishery.

Ocean ranching utilizes the freshwater habitat to raise a great number of fingerlings in hatcheries, allow them

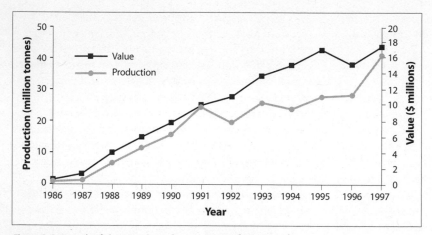

Figure 9.6 Growth of the BC salmon farm industry, 1986-97
Sources: ARA Consulting Group (1994), 2-2; Internet, BC Salmon Marketing Council

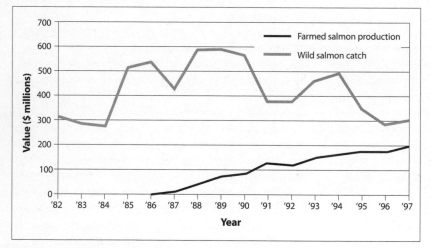

Figure 9.7 BC production of farmed and wild salmon, 1982-97
Source: British Columbia, Ministry of Finance and Corporate Relations (1998), 80.

proportions from the mid- to late 1980s, much of it focused on the Sunshine Coast. Figure 9.6 documents the very rapid growth in farmed salmon, both in tonnes produced and in dollars earned by the industry. Farmed salmon was the leading agricultural export in 1993 (ARA Consulting Group 1994, 2-2). Figure 9.7 shows the narrowing gap between farmed salmon production and commercial "wild" salmon production.

Because of inherent diseases related to the stress of being caged, not all species of salmon can be farmed. Sockeye, for example, cannot be farmed. The main species raised in netpens are chinook, coho, some rainbow trout, and Atlantic salmon. The preferred netpen locations are sheltered coves and inlets, and the nets are best anchored from shore. This requires a foreshore lease. The foreshore is the area between the high and low tidelines, and in British Columbia it is another common property resource, for which there is no private ownership. The proliferation of foreshore leases for fish farms has essentially privatized the foreshore and also excluded the public from many small bays and inlets. Objection was raised by recreationists and the tourism industry. In 1996, there were 149 foreshore leases or tenures in coastal British Columbia (Table 9.3).

Several environmental hazards are related to this industry, including the possibility of disease transfer, especially if the salmon in netpens escape and mix with wild stocks. Atlantic salmon, which are the preferred farmed species because they bring high value and are easy to raise in netpens, have different inherent diseases from those of Pacific salmon, and when these farmed

to migrate to the sea, and net them all several years later when they return to the hatchery. This eliminates the need for investment in commercial vessels, and damage to streams is of little consequence as the streams are not used for spawning. British Columbia has no ocean ranches, but Oregon and Alaska have experimented with the concept.

The raising of salmon in netpens – nets anchored to the ocean shore in bays and inlets – reached "gold rush"

Table 9.3

Location of salmon netpen tenures, 1996

Salmon farm location	Number of tenures
South coast: Georgia and Johnstone Straits	71
Northern Vancouver Island	38
North coast	5
West coast of Vancouver Island	35
Total	149

Source: Ellis and Associates (1996), 36.

species escape it intensifies the problem. Contamination of the aquatic environment beneath the netpens is also of concern and has led to further regulation (See Blore 1997; Ellis and Associates 1996; Western Canada Wilderness Committee 1998).

The commercial fishery views the fish farming industry as a threat on a number of fronts. Worldwide production of farmed salmon has increased the supply and is blamed for driving the price of all salmon down, while disease transfer to wild stocks looms to reduce the commercial catch. Andrew Nikiforuk remarks that "besides affecting price, the controversial raising of salmon in marine feedlots, complete with ocean-sterilizing pesticides and antibiotics, has devastated BC's salmon export business" (1996, 103).

The fish farming industry has struggled to produce a product of good quality. Initial locations on the Sunshine Coast in the mid-1980s proved to be disastrous, not only because of the problems mentioned above but because the water was too warm in summer and plankton bloom suffocated the fish. Millions of farmed salmon died and were disposed of at the landfill site near Sechelt, creating a distinct odour. Few salmon farms exist on the Sunshine Coast today. A five-year moratorium restricting any further increase has been in place since 1994. A 1997 review on the state of this industry suggested that further research on its environmental impact is required. Once new regulations are established, the industry has plans for a 600 percent increase in netpens, but this move is not without critics (Western Canada Wilderness Committee 1998, 1). The main solution envisioned for the fish farm industry is to lo-

cate it on land and in tanks, therefore treating the process as any other farming operation. This option, however, is expensive.

First Nations have a recognized food fishery right to salmon through the Indian Act. Early regulations prohibited First Nations from selling salmon except through the normal regulatory system of licensing. More recently, pursuit of Aboriginal rights through the courts has produced an interesting development in the fishing industry. The Supreme Court ruled that:

- aboriginal and treaty rights are capable of evolving over time and must be interpreted in a generous and liberal manner;
- governments may regulate existing aboriginal rights only for a compelling and substantial objective such as the conservation and management of resources; and,
- after conservation goals are met, aboriginal people must be given priority to fish for food over other user groups (British Columbia, Ministry of Aboriginal Affairs 1993, 2).

The Nisga'a agreement-in-principle, for example, has negotiated an allotment of salmon on the Nass River. This allotment is on the basis of a percentage of the run, to encourage future enhancement of salmon up the Nass. If the agreement is ratified it will be the first precedent in turning salmon from a common property resource into a **private property resource** in the sense that it will allocate salmon to a group and not individuals. Basing the entitlement on a percentage, though, may supply the incentive to enhance the stocks for everyone.

The sport fishery harvests perhaps only 4 to 10 percent of the overall provincial catch, but a major multiplier effect is attached to this industry, especially with growing tourism. The purchase of fishing lodge accommodation, charters, fishing licences, equipment, and even cases of beer accounts for the highest dollar value per fish caught. Many of these sport fishing interests and the communities that benefit are very concerned over the decline in salmon stocks and the outright ban on catching chinook and coho in some regions. Loss of employment because of the chinook ban in 1996 was

Table 9.4

Commercial boats and salmon catch, 1995

Boat type	Number of boats	Catch (millions of fish)
Troller	1,500	2.8
Gillnetter	2,300	5.0
Seiner	540	10.2

Source: Alden (1996), A19.

estimated at 2,175 jobs, and the loss of revenues has been estimated at $135.7 million (ARA Consulting Group 1996, S-4). Competition for salmon provokes frequent finger pointing at commercial interests, as the sport fishing industry asks how it can be guaranteed an allotment of the hook-and-line species. The 1998 announcement – "As of May 24, 1998, until further notice, there will be a non-retention of coho in all B.C. tidal and non-tidal waters" – has had a negative impact on the tourist and recreational side of the fishery (Internet, Coho Monitoring Bulletins).

The commercial fishery is not unified, as trollers, gillnetters, and seiners all compete for a share of the dwindling resource (Table 9.4). The threat of overcapacity has been handled largely through restricting openings and closings for various runs. The Mifflin Plan of 1995 has pursued the old idea of eliminating fishing vessels with a government buy-back program at the same time that it has introduced area licensing and the stacking of licences. The coast is divided into three areas, or zones, for gillnetters and trollers and two zones for seiners (Figure 9.8). A fisher's licence is tied to the area he or she has traditionally fished. To fish another area, a fisher must buy out a licence from that area, thereby acquiring multiple licences – or "stacking the licence" – and replacing another fishing vessel. By 1997, more than 800 vessels had been retired, with considerable discontent over the process.

The Mifflin Plan is not without controversy. Figure 9.7 shows that the plan is being implemented at a time of record low salmon catches. The ARA Consulting Group documents the combined commercial fishing sector job loss in 1996 to be 5,625, due to the poor catch and the Mifflin Plan. With the 2,175 jobs lost in the recreation sector, a total of 7,800 jobs have vanished (ARA Consulting Group 1996, S-9). As we can see from these statistics, forcing vessels out of the industry with a buy-back plan pays off the owner but increases the already high unemployment of deck hands. Fewer licences mean that the value of each one increases. The small fishers left to fish in one area only may find it difficult to make a living each year if they cannot afford to purchase another licence in another area. Concern has been expressed, particularly in the more remote commercial fishing communities, that the existing large corporate concentration of the industry will become even more dominant because corporations can afford to purchase licences. Nikiforuk's comment on the Fisheries Council of British Columbia puts corporate concentration in perspective: "Although it represents 240 processors, eight companies – including B.C. Packers Ltd., the Canadian Fishing Co., Ocean Fisheries Ltd. and J.S. McMillan Fisheries Ltd. – control the bulk of trade. Its members own most of the seine fleet, which has enough capacity to wipe out all the salmon stocks in one season, yet they still complain they can't get a regular supply of fish" (1996, 100). Some of these corporations have more control that others; BC Packers and Canadian Fishing have in excess of 100 seine licences and over 140 others (Seale 1996, 21).

The federal role of gatekeeper is not easy. The government attempts to regulate the number of salmon caught by designating run openings. Unfortunately, it does so based on inaccurate information of salmon stocks, primarily because it has little to no control over international and provincial activities that affect salmon: "We have 3,000 ecologically and genetically distinct populations of salmon on the West Coast, 1,000 spawning streams and about 150 DFO [Department of Fisheries and Oceans] types in the field" (Nikiforuk 1996, 101). It is a daunting task to gather full information about the various salmon species, their migrations and habitats, or the number of vessels that will participate in any opening. Under this system of management, harvesters are free to take as many fish as they are able to in any opening. For the federal government, which is responsible for both conserving salmon stocks and keeping the fishing industry economically viable, it is a "no-win" situation.

to market price and their own priorities. Certainly, it is a benefit to put those who use the resource more in control of conservation. Privatizing salmon would not be as easy as for halibut. Salmon are anadromous, have huge migration paths, and have far greater fluctuations in cycles and runs. As well, many more groups than commercial fishers want a share of the resource. Any quota designations would be extremely complex to allocate.

Even if the federal government were to establish an appropriate formula, privatization has its critics. The Cruickshank Commission's findings (1991) warned that the small fisher would lose to large corporate interests located in large communities. With no restrictions on corporate concentration under the Mifflin Plan, this fear may materialize even before there is serious discussion of the privatization question.

SUMMARY

All the competing interests for the salmon resource share the desire to increase stocks. To produce more salmon may well be possible, although it will take considerable effort to undo the environmental damage to the freshwater habitat in many areas. On the one hand, it is encouraging that salmon can be introduced to enhanced water systems and that their regeneration period is only a few years. On the other, all the activities that negatively influence the aquatic environment and the complexity of managing all the competing interests are discouraging. There have been some positive signs. Canada has reached an agreement with the United States, many First Nations are advanced in their treaty negotiations, and the federal government is planning to allocate salmon to the sport fishery. Though satisfying all interests may not be possible, the fear is that no interests will be satisfied if salmon stocks continue to be depleted at the current rate.

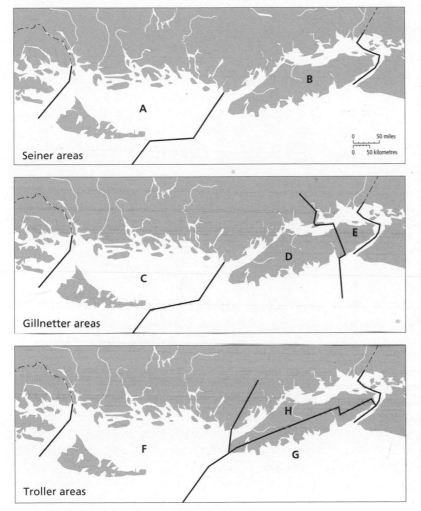

Figure 9.8 Salmon areas under the Mifflin Plan

With so many interests and so few fish to go around, Edward Alden (1996) and others wonder whether the principle of salmon as a common property resource should be abandoned in favour of privatization. For an alternative approach, Alden (1996) draws attention to the halibut fishery, in which a quota has been tied to vessel size and traditional catch volume, thus privatizing the resource. This has eliminated the intense competition of uncertain openings, caused less damage to the environment because cheaper longline fishing has replaced dragnets, and allowed fishers to work in relation

REFERENCES

Alden, E. 1996. "To Save Our Salmon, Privatize Its Future." *Vancouver Sun,* 26 July, A19.

ARA Consulting Group. 1994. *The British Columbia Farmed Salmon Industry: Regional Economic Impacts.* Vancouver: ARA Consulting Group.

–. 1996. *Fishing for Answers: Coastal Communities and the British Columbia Salmon Fishery.* Vancouver: ARA Consulting Group.

Barker, M.L. 1977. *Natural Resources of British Columbia and the Yukon.* Vancouver: Douglas, David and Charles.

Blore, S. 1997. "Betting the Farm." *The Georgia Straight* (Vancouver), 10 July, 15-18.

British Columbia, Ministry of Aboriginal Affairs. 1993. *Information about Landmark Court Cases.* Brochure. Victoria: Crown Publications.

British Columbia, Ministry of Finance and Corporate Relations. 1996. *1996 British Columbia Financial and Economic Review.* Victoria: Crown Publications.

–. 1997. *1997 British Columbia Financial and Economic Review.* Victoria: Crown Publications.

–. 1998. *1998 British Columbia Financial and Economic Review.* Victoria: Crown Publications.

Canada, Department of Fisheries. 1931. *Report of the Commission of Fisheries for the Year 1931.* Ottawa: Queen's Printer.

Clapp, R.A. 1998. "The Resource Cycle in Forestry and Fishing." *Canadian Geographer* 42 (2): 129-44.

Cruickshank, D. 1991. *A Commission of Inquiry into Licensing and Related Policies of the Department of Fisheries and Oceans – The Fisherman's Report.* Victoria: Mock.

Ellis, D.W., and Associates. 1996. *Net Loss: The Salmon Netcage Industry in British Columbia.* Vancouver: David Suzuki Foundation.

Fisheries and Oceans Canada. 1999. *1999/2000 British Columbia Tidal Waters Sports Fishing Guide.* Ottawa: Ministry of Supply and Services Canada.

–. n.d. *The Incredible Salmonids.* Brochure. Ottawa: Ministry of Supply and Services Canada.

Forester, E., and A. Forester. 1975. *British Columbia's Commercial Fishing History.* Saanichton, BC: Hancock.

Glavin, T. 1996. *Dead Reckoning: Confronting the Crisis in Pacific Fisheries.* Vancouver: Greystone.

Hume, S. 1997. "Carelessness, Greed Nearly Destroyed Salmon Run." *Vancouver Sun,* 7 July, B3.

Lyons, C.P. 1969. *Salmon: Our Heritage.* Vancouver: Mitchell Press.

Meggs, G. 1992. *Salmon: The Decline of the British Columbia Fishery.* Vancouver: Douglas and McIntyre.

Nikiforuk, A. 1996. "The Empty Net Syndrome." *Canadian Business* (October): 99-109.

Pearse, P. 1982. *Turning the Tide: A New Policy for Canada's Pacific Fisheries.* Vancouver: Commission on Pacific Fisheries Policy.

Roos, J.F. 1991. *Restoring Fraser River Salmon.* Vancouver: The Pacific Salmon Commission.

Ross, W.M. 1987. "Fisheries." In *British Columbia: Its Resources and People,* ed. C.N. Forward, 179-96. Western Geographical Series vol. 22. Victoria: University of Victoria.

Seale, R. 1996. "Fishing for Answers." *Nature Canada* (Autumn): 20-4.

Stacey, D. 1982. *Sockeye and Tinplate.* Victoria: Royal BC Museum.

Western Canada Wilderness Committee. 1998. *Wilderness Committee Educational Report* 17 (2): 1-4.

INTERNET

BC Salmon Marketing Council, "Go Wild," www.bcsalmon.ca

BC Stats: Quick Facts – The Economy, www.bcstats.gov.bc.ca/data/QF_econo.HTM

British Columbia, Ministry of Fisheries, Aquaculture and Commercial Fisheries Home Page, www.agf.gov.bc.ca/fish/acfb.htm

British Columbia, Ministry of Fisheries, BC Aquaculture Production 1994-1996, www.agf.gov.bc.ca/fish/aquastat.htm

Coho Monitoring Bulletins, www.pac.dfo-mpo.gc.ca/ops/fm/Salmon/Coho/Coho.htm

Fisheries and Aboriginals: The Enclosing Paradigm, www.indians.org/library/fish.html

Fisheries and Oceans Canada, Pacific Salmon Treaty, June 1999, www.dfo-mpo.ca/COMMUNIC/BACKGROU/1999/hq29e(a)_e.htm

Minister Announces Modifications to Pacific Salmon Revitalization Plan, 9 May 1996, www.pac.dfo.ca/pac/comm/pages/english/newscat/p-releas/hq9632e.htm

Sierra Legal Defence Fund, 1997. "B.C. Salmon on the Brink," www.sierralegal.org/m97_01.htm

Metal Mining: The Opening and Closing of Mines

10

M etal mining has the reputation of having opened up British Columbia to non-Native settlement and development. It began with the discovery of gold in the lower reaches of the Fraser River in 1858 and subsequent discoveries up the Fraser and into the Cariboo by the 1860s. The region had been held by the British mainly through Hudson's Bay Company fur trade posts with a few hundred Europeans, and it was to face a huge challenge when approximately 30,000 miners descended on the landscape (Chapter 4). The various mining booms have had a powerful and permanent impact on the geography of the area. Because minerals are a nonrenewable resource, ore eventually runs out and mines close. A community that has depended solely on a mine for its economic well-being often becomes a ghost town.

This chapter examines the metal mining industry of British Columbia. Mining of energy minerals such as coal, natural gas, and oil is discussed in the following chapter. The focus here is on gold, silver, copper, and other valuable minerals throughout the province. Understanding the physical landscape, and especially its geology, is important to the discovery of minerals, but a host of other factors influence the economic viability of any mineral going into production. Technology is very important to all aspects of mining, from discovery to the end use of the metal. Technological innovations can affect the production costs of and supply and demand for the final product. The economics of mining is also affected by access to often isolated mine sites, by political decisions, by the cost of energy and other inputs, and, more recently, by environmental considerations. The world market price reflects the overall supply and demand for any mineral.

This chapter begins with an historical overview of the production of metals in British Columbia and moves on to a more specific examination of mines that have been into and out of production in the 1990s. Examining the fundamental question of why mines open and

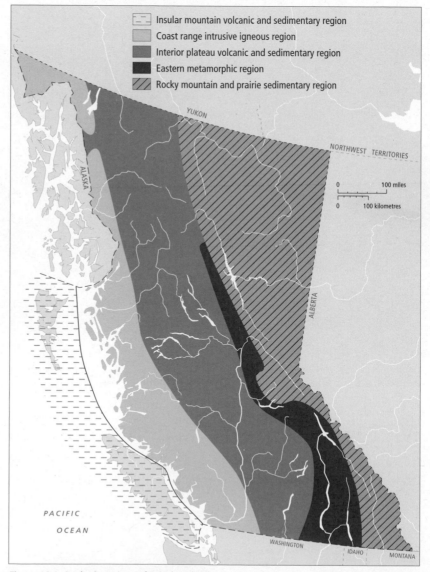

Figure 10.1 Geologic regions of British Columbia
Sources: Data from Rider (1978), 13; Geological Survey of Canada (1970), 30.

close reveals numerous interrelated factors. Finally, it is suggested that few new metal mines will open, putting the future of this industry into question.

FACTORS INFLUENCING METAL MINING

The complex geology of British Columbia, including the ages of different rock structures and the tectonic forces responsible for their formation, is explained in greater detail in Chapter 2. Figure 10.1 divides the province into five general geological regions according to their mineralization characteristics.

The zone farthest west is the Insular Mountain Volcanic and Sedimentary Region, encompassing Vancouver Island, the Queen Charlottes, and the northwestern tip of British Columbia. Geologically, this region is mainly a mix of intrusive and extrusive igneous rock with sedimentary layers where some gold has been discovered and considerable quantities of coal mined, but the most important metal has been copper. The Coast Mountains range is referred to as the Coast Range Intrusive Igneous Region. In this mainly granitic rock formation, deposits of gold, copper, and iron ore have been mined. The central portion of British Columbia, much of it referred to as the Interior Plateau, has been created largely by massive lava flows. This is the Interior Plateau Volcanic and Sedimentary Region and it has high mineralization, with major deposits of copper, molybdenum, gold, mercury, and asbestos. Next to it is the Eastern Metamorphic Region, made up of very old and very hard rock. It is also rich in minerals. In the Kootenay region, especially, silver, lead, zinc, copper, and even gold have been mined. The most easterly region, the Rocky Mountain and Prairie Sedimentary Region, is, as the name suggests, made up of sedimentary rock. Similar to Alberta, deposits of coal, oil, and natural gas have been developed there.

Some minerals were mined prior to the gold rush: coal on the north coast and Vancouver Island and some gold in the Queen Charlottes. Coal was mined for some time, but the gold finds were small, short-lived ventures. The history of metal mining and the relative value of each metal can be seen in Table 10.1. The table compares only the most important metals mined in the province. As can be seen, gold dominated the industry until a silver boom in the Kootenays late in the nineteenth century.

Table 10.1

Value of mineral production as a percentage of total metal mined

	Gold	Silver	Copper	Lead	Zinc	Molybdenum	Iron
1860	100						
1870	100						
1880	100						
1890	83	17					
1900	42	20	14	23			
1910	44	10	35	10	2		
1920	13	16	40	14	16		
1930	8	11	28	31	17		
1939	41	8	13	22	15		
1950	10	6	7	37	40		
1960	6	6	7	31	41		9
1970	2	3	42	12	15	19	7
1980	13	11	48	4	4	20	
1990	15	8	56	1	8	6	
1997	17	7	47	3	20	6	

Source: Internet, British Columbia, Ministry of Energy and Mines, "Time Series of BC Metal Values."

New mining techniques and new demands led to the development of copper, lead, and zinc mining. By the 1960s, some iron ore was being mined on Texada Island in the Gulf of Georgia. The big discovery, though, was molybdenum, which hardens steel when used as an alloy.

The Ministry of Energy and Mines often combines all mining activities in its statistics – metals, industrial minerals (e.g., sand and gravel), structural minerals (e.g., asbestos and gypsum), and energy minerals – as Figure 10.2 shows from 1995 to 1997. Revenues from mining have undergone considerable swings in these years, particularly from metal mining.

British Columbia has a long history in mining and a potentially productive future, as there is no lack of minerals. Being rich in minerals and making production of them economical, however, are two different things. Figure 10.3 categorizes the various stages of metal mining, from exploration to end use. All have to be considered in the industry. Technological innovation plays a role at each stage. The historical search for gold, for example, was initially accomplished with the simple technology of a shovel, a gold pan, and plenty of hard work. This

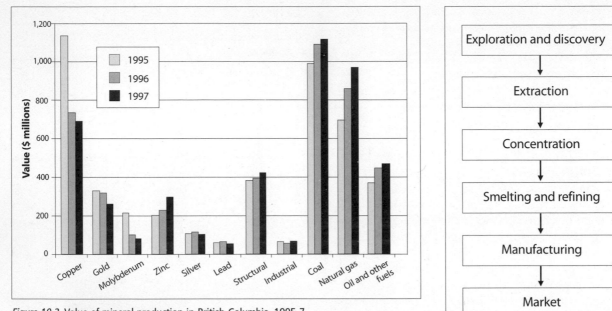

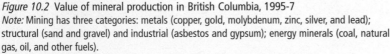

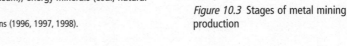

Figure 10.2 Value of mineral production in British Columbia, 1995-7
Note: Mining has three categories: metals (copper, gold, molybdenum, zinc, silver, and lead); structural (sand and gravel) and industrial (asbestos and gypsum); energy minerals (coal, natural gas, oil, and other fuels).
Sources: British Columbia, Ministry of Finance and Corporate Relations (1996, 1997, 1998).

Figure 10.3 Stages of metal mining production

was placer mining, where gold came in a pure form in existing or old stream beds. Next came sluices, dredges, and hydraulics, more costly technologies that allowed the exploitation of even greater gold deposits. Not all gold is in a pure form, though; most minerals or ores are "bound" to other minerals and require the accompanying rock to be smashed and crushed in order to extract the metal. This is **lode mining,** and it requires more sophisticated technologies with much higher costs than placer mining. Figure 10.4 puts into perspective the amount of gold produced by both methods.

Today's prospectors, searching for metal-bearing outcroppings and armed with a rock hammer and a knowledge of geology, supplement their search with techniques ranging from remote sensing to diamond hole drilling. Assessing the quality and quantity of any ore body is crucial to production. The development of processing technologies can change the quality/quantity equation considerably. Copper, for example, was only economically viable at 10 percent grade or higher before the twentieth century. By 1900, new techniques and new

demands – such as copper wire for the distribution of electricity – made 2 percent copper ore viable, and by the 1940s, 1 percent was considered viable. These percentages are very significant. A 10 percent copper ore body, for example, would result in 100 kilograms of pure copper being refined from 1,000 kilograms of copper ore. When the percentage drops to 1 percent, only 10 kilograms of copper will become available and 990 kilograms will remain as wastes, or tailings. The 1960s development of the open-pit method of mining, using huge earth-moving equipment, was able to operate with 0.5 percent copper, and 0.4 percent may be feasible today (Barker 1977, 28; Ross 1987, 163). As open-pit technologies replaced underground mining, low grade ores could be economically processed only if there were huge quantities.

Changes in technology reduced the labour required as mining made the transition from a labour-intensive to a capital-intensive industry. Hard rock mining with a pick and shovel and manual loading of a rail cart is an occupation of the past. Today's open-pit miners operate

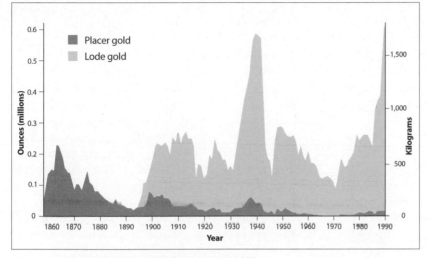

Figure 10.4 Placer and lode gold production, 1860-1990
Source: British Columbia, Geological Survey Branch (1990), 20.

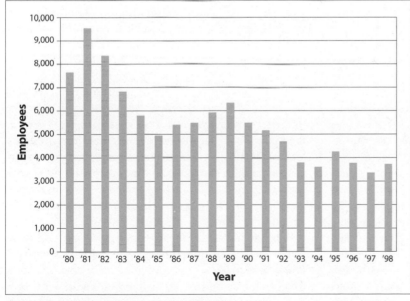

Figure 10.5 Employment in metal mining, 1980-98
Source: Internet, BC Stats – Metal Mining.

ers inhaled the dust into their lungs and died from silicosis.) There are, however, far fewer workers per tonne of minerals produced today. Figure 10.5 reveals the change in metal mining employment for British Columbia between 1980 and 1998.

Low grade ores are crushed and concentrated, and most are exported at this stage. Smelters at Trail and Kamloops refine or smelt these concentrates into a pure state. There is a third smelter at Kitimat, but it is used for aluminum and therefore operates on an entirely different basis. The key factor for the production of aluminum is inexpensive electrical energy. The raw material required is bauxite, which is not mined in British Columbia but must be imported from tropical regions such as the West Indies.

The last two stages of the diagram in Figure 10.3 – manufacturing and market – refer to the end uses of the mineral being mined. Demand for particular end uses influences the world market price, and price is central to economic viability. Both manufacturing and market are affected by changes in technology. One only needs to consider some of the traditional uses of copper, such as electrical wire and pipes for plumbing. Fibre optics from silica-derived glass threads and plastic pipe from petroleum compete with these traditional uses or markets today.

The technological development and market demand for molybdenum, or "moly," is a different story. Molybdenum was discovered and used to make armour-piercing shells during the Second World War. With the Korean War, Vietnam War, and Cold War, a high demand was

enormous trucks, drills, and shovels after assessing the lode by computer. With far fewer underground mines and no more old rock drills, the workplace is safer. (The drills were known as "widow makers" because most drill-

maintained for military uses and even for the "star wars" scenarios of the aerospace industry. This type of demand puts molybdenum in the category of a "war-based" mineral. British Columbia is one of the leading moly producers in the world. With the end of the Cold War and few prospects of any major war in the near future, the demand for this mineral has been dropping and so has the world market price.

Technological aspects of the processes of mineral production from discovery to market are not the only factors influencing costs in this industry. British Columbia has a vertical and rugged topography, often making access by rail, boat, or road a significant cost. Nelson, Trail, Greenwood, and a number of other places in the Kootenays had smelters in the 1890s and early 1900s, but several factors were necessary to make them economically viable: "A smelter needs three things: a satisfactory process of separation, an assured supply of ore, and reasonably priced fuel. In these early days neither the metallurgical process nor the quality of coke was satisfactory" (Taylor 1978, 132). Pioneer smelters appeared along the main line of the CPR, but not surprisingly, most of them failed. A widely distributed infrastructure of railways, mainly owned by the Canadian Pacific Railway (CPR), was constructed throughout the Kootenays to link the region east to southern Alberta, south to the United States, north to the CPR main line, and west to the Okanagan and Vancouver. With transportation in place, going from exploration and discovery to production was much easier.

Politics was very much a part of the decision to provide rail systems to the Kootenays, and politics is still a significant factor in this industry. The provincial government, as the manager of this resource, has played a mainly facilitative role in mining development by building infrastructure such as railways, roads, and hydroelectric developments, as well as by providing direct subsidies, crown grants, tax incentives, and other beneficial legislation. Nevertheless, the mining industry has also perceived politics as playing a negative role.

Bill 31, passed in 1973, illustrates the relationship between demand, world market price, competition, profits, and taxation. Copper is often referred to as a "war-based" mineral because one of the big demands for it is in the manufacture of shell casings for bullets. The ac-

celeration of the Vietnam War by the late 1960s and early '70s increased the demand for copper. On the supply side, in the United States, a major strike in the copper mines lasted for months, while in Chile, which was a major producer of copper, the government decided to nationalize foreign-run mines. Large exports from the United States and Chile were thus interrupted, driving the world market price for copper to over $1.40 per pound. For BC copper producers this was a windfall. The newly elected NDP government thought that the people of the province should also benefit from the windfall, and Bill 31 was introduced as a "super tax" on these "excess" profits. The industry screamed that the taxes were unfair. The real irony was that by the time the government got around to implementing the tax, world conditions had changed and the price of copper had plummeted to thirty-seven cents per pound. The strikes were over, Chile was producing copper, as were other countries such as Zaïre and Australia, and the Vietnam War was winding down. Oversupply and underdemand created bankrupt conditions almost overnight, and there was nothing left to tax. It was the government of British Columbia that took the blame, though, for the downturn in the mining economy.

Politics around the issues of taxation, decreased employment, and insufficient government support for the mining industry continue. The September 1993 issue of the *Mining in BC Newsletter* remarks that "government is largely responsible for the fact that exploration is declining, and for the fact that other provinces and countries are luring B.C. mining companies to invest and create jobs in their jurisdictions" (Mining Association of British Columbia 1993, 1).

World market price fluctuates, in some cases dramatically, to reflect the many changes in supply and demand. Precious metals such as gold, which have few practical uses, are not as influenced by the over- and undersupply and demand and changes in end uses that metals such as copper have. Gold has a very long history as a currency, and even once paper money was introduced, it was based on gold. A gold standard had been widely adopted by most countries by 1900 but had to be abandoned in the 1930s during the Depression; even so, the price of gold remained fixed until 1970. Table 10.2 shows the dramatic increase in the price when the market was

Table 10.2

World market prices for selected minerals and years, 1971-99

	Gold (US$/oz)	Silver (US$/oz)	Copper (US$/lb)	Lead (US$/lb)	Zinc (US$/lb)	Molybdenum (US$/lb)
1971	35	1.56	0.38[a]	0.12	0.13	1.39
1974	166[a]	4.98[a]	0.70	0.16[a]	0.29[a]	1.65[a]
1980	708	23.95	0.94	0.32	0.27	9.63
1985	434	8.40	0.72	0.13	0.38	3.55
1990	469	5.60	1.13	0.30	0.67	2.68
1991	362	4.04	1.06	0.25	0.53	2.35
1992	344	3.94	1.03	0.25	0.58	2.18
1993	360	4.30	0.87	0.18	0.46	2.28
1994	384	5.28	1.05	0.25	0.49	4.50
1995	384	5.21	1.33	0.28	0.53	7.42
1996	388	5.18	1.04	0.35	0.51	3.61
1997	331	4.89	1.03	0.28	0.65	4.18
1998	294	5.53	0.75	0.24	0.51	3.31
1999[b]	287	5.30	0.64	0.23	0.50	2.61

Notes:

a The average price has been used since 1970 for copper and 1974 for gold, silver, lead, zinc, and molybdenum.

b First quarter only.

Source: Internet, British Columbia, Ministry of Energy and Mines, "Time Series of BC Metal Values."

allowed to establish its own level. With the relatively high world market price today for a metal that is one of the heaviest of all the elements, it is no wonder that most new mining ventures in British Columbia over the past two decades have been gold mines.

Table 10.2 also shows the fluctuating world market prices for other metals mined in British Columbia. As stated earlier, copper is our most important metal and unfortunately it is one of the most volatile with respect to world market price. Table 10.2 provides the average yearly price but not "spot" prices such as the high US$1.47 per pound reached in 1995 that triggered expectations of continued high prices. In June 1996, however, a $2.6 billion trading scandal in Japan pushed copper prices down (Savings and Credit Unions of British Columbia 1996, 10). The continued slump and crisis in Asian markets has kept the price of copper low and several BC mines have closed, including, temporarily, the Highland Valley Copper, the largest copper mine in the province (Nutt 1998, D1). The price per pound

of many minerals has rarely kept pace with inflation, and in others the price has declined since the 1970s. Individual mines are therefore in the position of maintaining the same efforts for production, often with the same costs, but with smaller returns. This is evident in Table 10.3, which shows the quantity and value of copper and molybdenum from 1990 to 1998. Comparing the figures specifically for 1995 and 1998 for both metals, it is evident that while production has declined only slightly, value has dropped precipitously.

More recently, the environmental costs associated with mining have played an increasing role in the costs of operation. The mining of low grade ores results in waste materials, or tailings. Legislation now requires mines to deal with waste materials effectively, an added cost to factor into production. Tailings are often high in acid, and plans must be implemented to prevent the contamination of ground- and surface water sources (Internet, Environmental Mining Council of British Columbia). This issue was so controversial in the proposed copper-rich Windy Craggy mine of the Tatshenshini River area of northwestern British Columbia that the proposal was rejected and the region was declared a park in 1993.

The contaminated process water in tailing ponds is another source of pollution to the environment. The discharge of wastes from smelters to the air, water, and

Table 10.3

Production and value of copper and molybdenum, 1990-8

	Copper		Molybdenum	
	Quantity (kg millions)	Value ($ millions)	Quantity (kg millions)	Value ($ millions)
1990	325	985	12	88
1991	331	847	11	67
1992	316	892	9	55
1993	289	722	10	85
1994	246	755	10	91
1995	279	1,119	9	303
1996	235	725	8	100
1997	220	699	8	88
1998	277	681	8	82

Note: 1998 figures are estimates.

Source: Internet, BC Stats: Quick Facts – The Economy, "Mining."

land is another environmental hazard. Mining also provokes potential land use conflicts. In consideration of all of these environmental concerns, the politics of regulation come into play. Again, government is viewed by this industry as a stumbling block.

THE OPENING AND CLOSING OF MINES IN BRITISH COLUMBIA

To this point, the discussion has included an overview of the history and many influences on the development of the metal mining industry. Figure 10.6 shows prospected mineral deposits throughout British Columbia, demonstrating that the province is rich in minerals. Further, it may lead one to expect a great number of mines in the future. Few of these will materialize, however, as new mines face high capital costs of production, stringent environmental regulations, and transportation costs from remote sites. Table 10.4 compares operating mines

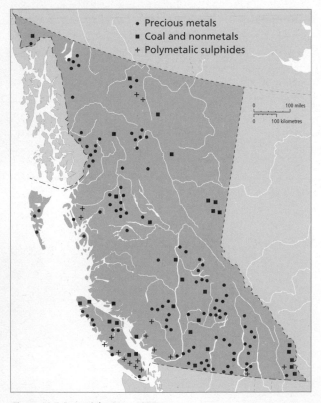

Figure 10.6 Potential mines, 1986
Source: British Columbia, Ministry of Mines (1987), 132.

in 1990 with those of 1999, showing where and when individual mines have opened and closed. Figure 10.7 shows their locations, an important factor in the analysis of the industry.

The list of producing metal mines is not numerous; a number of them have shut down for some portion of the ten-year period shown in Table 10.4, and many more have shut down permanently than have opened. There is a wide variation among mines in tonnes milled – the volume of minerals that have gone through the concentrator and been exported – one measure of the size of the mining operation. Only a few of the producing mines process just one mineral, even though mines such as Island Copper or Highland Valley are often referred to as copper mines. Gold and silver, as Table 10.2 shows, are measured in dollars per gram and are much more valuable than copper, which is measured in dollars per pound. The total average employment for each mine also varies considerably. The number of employees is related to the tonnes milled and also to mining processes. Notice, for example, the tonnes produced at Island Copper and at Myra Falls, both on Vancouver Island (Table 10.4). The higher number of employees in relation to output at Myra Falls is the difference between an underground operation at this mine versus an open-pit process at Island Copper.

Mines go out of production, either temporarily or permanently, for various reasons. In most cases permanent shut-downs have occurred because of exhaustion of the ore (e.g., Island Copper and Brenda Mines). These metals are non-renewable resources. Temporary closures are related to various factors, such as strikes and lockouts, environmental issues, a slump in the world market price of the metal, or decisions made by large mining corporations involving a combination of these reasons. The Similco mine at Princeton, for example, like many copper mines in British Columbia, is owned by a large multinational corporation. Head office made the decision to shut the mine down in 1993 because the price of copper was declining with little likelihood of higher returns. Prices did rebound and the mine reopened in 1994, but low prices in 1996 and 1997 resulted in another closure. As stated earlier, the recent collapse of the Asian markets for copper has depressed the price of that metal, and as a result the Gibraltar mine closed in February

Table 10.4

Production and employment for metal mines, 1990-9

Mine/owner or operator and principal mineral(s)	Tonnes milled	Average employment	1990	1991	1992	1993	1994	1995	1996	1997	1998	1999
Afton Copper, gold, silver	2,655,430	208	X	S/08			R/09	X	X	C/05		
Beaverdell/Tech Lead, silver, zinc	36,225	34	X	C/02								
Bell Copper, gold, silver	5,422,911	303	X		X	C/06						
Blackdome Gold, silver	72,912	102	X	C/01								
Brenda Mines Copper, gold, molybdenum, silver	4,281,870	190	C/06									
Endako/Placer Dome Molybdenum	9,772,865	217	X	X	X	X	X	X	X	X	X	X
Equity Silver Copper, gold, silver	3,145,900	205	X	X	X	X	C/01					
Eskay Creek Gold, silver	109,500	108					O/12	X	X	X	X	X
Gibraltar Copper, gold, molybdenum, silver	12,634,562	301	X	X	X	S/12	R/10	X	X	X	S/12	
Golden Bear Gold, silver	69,805	115	X	X	X	X	S/09			R/09	X	X
Highland Valley Copper/HVC Copper, gold, molybdenum, silver	46,263,361	1,230	X	X	X	X	X	X	X	X	X	S/05 R/09
Huckleberry Copper, gold, molybdenum, silver	6,570,000	n/a								O/10	X	X
Island Copper/BHP Copper, gold, molybdenum, silver	18,361,264	549	X	X	X	X	X	C/12				
Johnny Mountain Gold, silver	87,189	56	S/09			R/S						
Kemess South Copper, gold	n/a	n/a									O/06	X
Lawyers Gold, silver	184,587	145	X	X	S/08							
Mount Polley Copper, gold	6,570,000	n/a								O/09	X	X
Myra Falls/Westmin Copper, gold, silver, zinc	1,162,337	608	X	X	X	X	X	X	X	X	S/12	R/04

▶

◄ Table 10.4

Mine/owner or operator and principal mineral(s)	Tonnes milled	Average employment	1990	1991	1992	1993	1994	1995	1996	1997	1998	1999
Nickel Plate/Corona Gold, silver	1,141,255	173	X	X	X	X	X	X	C/07			
Premier/Westmin Gold, silver	735,598	166	X	X	X	X	X	X	S/04			
Quesnel River/Kinross Gold	438,000	80						O/04	X	X	S/04	
Samatosum Copper, gold, lead, silver, zinc	169,152	92	X	C/09								
Shasta Gold, silver	63,867	28	X	C/11								
Silvana Lead, silver, zinc	31,574	38	X	X	S/04							
Similkameen/Similco Copper, gold, silver	6,676,862	329	X	X	S/11	R/08	X	X	S/12			
Snip Gold, silver	155,000	134		O/01	X	X	X	X	X	X	X	X
Sullivan/Cominco Lead, silver, zinc	399,653	379	X	X	X	X	X	X	X	X	X	X
Table Mountain Gold	32,850	35					O/04	X	X	X	X	X

Note: C = Closed; O = Opened/new mine; R = Re-opened; S = Shut down; X = Operating; /01 = first month, etcetera (S/08 = shut down in August).
Source: Internet, Ministry of Energy and Mines, "Major Metal, Coal, and Asbestos Mines Operating Status."

1998, Myra Falls in December 1998, and Highland Valley in May 1999, not because they had exhausted their copper but because the price of copper was not high enough to make a profit (Savings and Credit Unions of British Columbia 1999). Afton, at Kamloops, has closed down for several periods as well, and again the closure was related to the world market price. Afton is unique in British Columbia and has other cost considerations because it is designed to smelt the particular type of copper ore in the Kamloops area.

One of the strengths of BC mines is that most produce more than one metal. British Columbia is a copper-producing region of Canada, but copper, as previously seen, has a very volatile world market price. Fortunately, these large copper mines also produce gold, silver, and in some cases, molybdenum. Three new copper mines opened in 1997-8 on the expectation of re-

bounding metal prices (which still had not materialized by the beginning of 2000), but all three have other valuable metals associated with their ores, such as gold. When a mine has only one metal, such as Endako with molybdenum, it is considerably more vulnerable to world market prices.

The story of Cassiar, just north of the Golden Bear Mine (Figure 10.7), is typical of what happens when a community relies on one resource and the demand for that resource declines. The mine at Cassiar produced asbestos, which is an industrial mineral rather than a metal. Since asbestos has become recognized as carcinogenic, there has been less and less demand for it. There is still plenty of asbestos left, but the low world market price does not make it economically viable. In 1993, the Cassiar mine closed and another ghost town was added to the list in British Columbia.

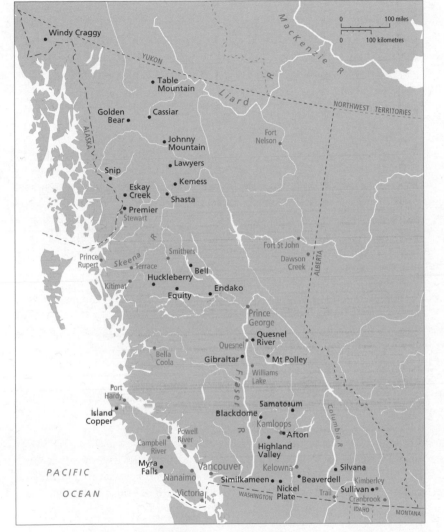

Figure 10.7 Operating metal mines, 1990-9
Source: Data from British Columbia, Ministry of Finance and Corporate Relations (1998).

natural hazards. In 1965, a landslide at the Granduc Mine near Stewart killed twenty-six workers and shut the mine down for a considerable length of time. The copper smelter community of Anyox, northeast of Prince Rupert, was reduced to ashes by a forest fire in 1939. Sandon, in the West Kootenays, was wiped out by a flood, and Britannia Beach, just north of Vancouver, was hit by both a landslide and a debris torrent. Many mines are at risk from natural hazards.

SUMMARY

British Columbia is rich in mineralization and has had a long history of metal mines opening and closing, adding to the pattern of boom and bust (and the numerous ghost towns) throughout the province. A host of economic and political factors, beginning with the economic recession of the 1980s, has caused the industry to decline. The primary metal mined in British Columbia is copper, although on the global scale it has little impact on the world market price with respect to supply. All the metals mined in the province are exported, usually as concentrates, and British Columbia therefore has little control over the demand side. The industry is very capital intensive, making any operation high in fixed costs even though revenues are volatile. The exception in metals is gold. Although the value fluctuates, it is nonetheless consistently so high that even small amounts of gold are often worth mining. Because of the high costs and high risks involved, metal mining in British Columbia is dominated by multinational corporations, which are frequently foreign owned.

When all factors are considered, the pattern of mine closures exceeding mine openings is not surprising. Three

A significant number of the mines listed in Table 10.4 produce gold, or gold and silver. They are relatively small and tend to start up, shut down, and start up again. These occurrences are primarily related to the high price of gold and the ability of a small mining company to offer shares and raise money for short runs of production.

The closure of a mine is usually related to economic considerations, but the province has a history of mines being closed and their communities dying because of

new copper mines opened in mid-1997 in expectation that the high prices of copper and other metals in 1995-6 would continue. These were the first copper mines to open in British Columbia in nearly twenty years. Unfortunately, the unpredictable global economic climate has caused most metal prices to decline from late 1996 to 1999. In the past decade, most mine openings have been in gold or gold and silver because of the high value of these minerals, and this trend will probably continue.

Metal mining peaked in the 1960s and '70s, with major investments in copper and molybdenum mining, especially. Many of these large mines had exhausted their deposits by the 1990s. The consequent mine closures, and overall decline of the industry, have brought tragic levels of unemployment to several communities. On the positive side, the vast experience gained in mining has proven to be an exportable commodity for British Columbia.

REFERENCES

Barker, M.L. 1977. *Natural Resources of British Columbia and the Yukon*. Vancouver: Douglas, David and Charles.

British Columbia, Geological Survey Branch. 1990. *Metal Mining in British Columbia*. Victoria: Ministry of Energy and Mines.

British Columbia, Ministry of Finance and Corporate Relations. 1996. *1996 British Columbia Financial and Economic Review*. Victoria: Crown Publications.

–. 1997. *1997 British Columbia Financial and Economic Review*. Victoria: Crown Publications.

–. 1998. *1998 British Columbia Financial and Economic Review*. Victoria: Crown Publications.

British Columbia, Ministry of Mines. 1987. *Mining Yearbook*. Victoria: The Ministry.

Geological Survey of Canada. 1970. *Geology and Canada*. Ottawa: Department of Energy, Mines, and Resources.

Mining Association of British Columbia. 1993. *Mining in BC Newsletter* 6, 8 (September): 1-4.

Nutt, R. 1998. "Gibraltar First Casualty of Weak Copper Prices." *Vancouver Sun*, 13 March, D1.

Rider, J.M. 1978. "Geology, Landforms, and Surficial Materials." In *The Soil Landscapes of British Columbia*, ed. K.W.G. Valentine, P.N. Sprout, T.E. Baker, and L.M. Lavkulich, 11-33. Victoria: Resource Branch, Ministry of the Environment.

Ross, W.M. 1987. "Mining." In *British Columbia: Its Resources and People*, ed. C.N. Forward, 162-76. Western Geographical Series vol. 22. Victoria: University of Victoria.

Savings and Credit Unions of British Columbia. 1996. *Economic Analysis of British Columbia* 16 (9): n.p.

–. 1999. *Economic Analysis of British Columbia* 19 (3): n.p.

Taylor, G.W. 1978. *Mining: A History of Mining in British Columbia*. Saanichton, BC: Hancock.

INTERNET

BC Stats: Quick Facts – The Economy, "Mining," www.bcstats.gov.bc.ca/data/bcfacts.htm

British Columbia, Ministry of Energy and Mines, "Time Series of BC Metal Values," 1999, www.em.gov.bc.ca/mining/miningstats

British Columbia, Ministry of Energy and Mines, "Major Metal, Coal, and Asbestos Mines Operating Status," www.em.gov.bc.ca/mining/miningstats/opstatus.htm

Environmental Mining Council of British Columbia, "Acid Mine Drainage," www.sunshine.net/bc_miningwatch

Energy: Supply and Demand

11

Energy resources in British Columbia have come in many forms, as animal power, wood, and water wheels gave way to coal, electricity, oil, and natural gas. Demand, both locally and for export, has changed considerably over time, and technology has played a lead role in the development of new energy sources for old and new uses. Within British Columbia are spatial patterns of supply and demand; not all energy resources are spread uniformly throughout the province, nor are the demands for energy (Chapman 1960).

This chapter describes the overall historical and spatial development of energy in British Columbia and then examines in turn coal, oil, natural gas, and electricity – the energy sources on which we rely most. Each of these energy sources is reviewed in terms of past to present development, and included in the review are some of the main issues arising from use of these forms of energy.

Amory Lovins (1977) challenges our dependence on traditional, centralized energy by rethinking energy strategies for the future. In Lovins's view, we must focus on decentralized, renewable forms of energy technologies and explore the demand side of energy. What role can individuals play in using and saving energy? How will British Columbia fulfill its demands for energy in the future?

HISTORICAL OVERVIEW OF ENERGY DEVELOPMENT
British Columbia's early energy needs were met by wind for sailing ships and wood for fuel. By the 1830s, coal was being mined on Vancouver Island and added to the energy options. The gold rush brought a great number of people to the region and signalled the beginning of many industries to follow, as well as a need for ever greater amounts of energy. Horses, oxen, donkeys, and people performed a fair amount of the work. Flumes moved logs and water wheels ran mills and mining operations, while wood and coal were often burned for heating, cooking, and running the steam engine. As railways became the main mode of transport and the steam engine became essential in industry, the demand for coal steadily increased.

Table 11.1 reflects the demand for energy in the province. As it shows, from the 1920s to the 1960s, oil, or petroleum, became increasingly important. It was used for transportation, for heating, creating electricity, for

Table 11.1

Energy use, 1925-96

	Coal (%)	Wood (%)	Petroleum (%)	Natural gas (%)	Electricity (%)
1925	52	16	24	–	8
1935	30	18	30	–	22
1945	31	12	29	–	28
1955	14	13	65	–	8
1965	2	8	60	15	15
1975	1	7	54	23	15
1985	1	24	37	19	20
1996	1	16[a]	36	29	18

Note: Totals do not always equal 100 percent, due to rounding.
a Hog fuel or wood wastes.
Sources: Barker (1977), 50; British Columbia, Ministry of Finance and Corporate Relations (1997), 88.

smelting, and, eventually, for myriad petrochemical products from paint to lipstick. A significant technological change occurred with the wide adoption of the diesel engine, which replaced the coal-driven steam engine and reduced the demand for coal. The growing dependence on oil in British Columbia, most of which has come from Alberta since the 1950s, is little different from the rest of Canada and the industrialized world. Nonetheless, the energy crisis of the 1970s caused a major reassessment of British Columbia's energy supply and demand, particularly with respect to dependence on oil. Natural gas, electricity, and hog fuel (wood wastes and black liquor from pulp operations) became increasingly important sources of energy and alternatives to oil.

The rising price of energy caused other developments not evident in the statistics. Solar energy and wood stoves became popular at the individual, residential level. BC Hydro, a crown corporation, investigated the potential of tidal power, geothermal production, nuclear power, coal-fired thermal power, and extensive hydroelectric megaprojects. The federal and provincial governments began to explore new ways to acquire oil in order to reduce dependence on Alberta's supplies. Attention was also given to conservation measures.

The demand for energy in British Columbia and supplying this demand with BC energy resources are two different components of the energy question. As electricity was required, dams were constructed to produce

Table 11.2

Energy consumption by sector, 1978-96

	1978 (%)	1986 (%)	1990 (%)	1996 (%)
Residential	15.5	15.0	14.0	14.0
Commercial	10.0	10.0	9.0	12.0
Transportation	26.0	25.0	24.0	35.0
Industrial	48.5	50.0	50.0	39.0
Total	100.0	100.0	100.0	100.0
Losses	–	–	3.0	–

Sources: Sewell (1987), 241; British Columbia, Ministry of Finance and Corporate Relations (1987), 95; (1991), 69; (1997), 88.

hydroelectricity. Supply of electricity met demand until the major development of both the Columbia and Peace River dams in the 1960s and '70s, at which time a surplus was produced and the province was able to export hydroelectricity. Oil was imported prior to its discovery in the Peace River region in the 1950s. Unfortunately, there has never been enough BC oil to meet the demand, and Alberta oil is still needed to supplement it. On the positive side, natural gas discoveries have produced a surplus of this energy source. Coal made a major production recovery in the late 1960s, but the demand (shown as only 1 percent in Table 11.1) was not from the domestic market. Japan, and later Korea, needed coking coal for the steel industry and thermal coal for electricity, thus reviving coal production in British Columbia during that decade.

Energy demand can be further broken down into various sectors (Table 11.2). Industry consumes the greatest amount of energy, followed by the transportation sector, and then residential and commercial use. These requirements can be fulfilled by many types of energy, although transportation demands are met almost exclusively by petroleum.

Another aspect of the industry is the spatial differences between where energy is produced and where it is consumed. The largest demand is in the Lower Mainland, which has the greatest concentration of population. Industrial, commercial, residential, and transportation demands are met with a range of energy sources, few of which come from this region. Pipelines transport oil and natural gas, transmission lines bring electricity, and

barges and trucks bring hog fuel. Until the 1990s, Vancouver Island did not have the option of natural gas, and the island does not have large river systems suitable for generating hydroelectricity. Consequently, the Cheekeye-Dunsmuir transmission line was built in the 1980s to channel electrical energy to the island, and a natural gas pipeline from the mainland was built in the 1990s to supply gas to the pulpmills and communities in Howe Sound and the Sunshine Coast (Figure 11.1).

A human and environmental price is paid by any region in which energy is generated. In regions where dams are constructed, for example, valleys are turned into reservoirs. Transportation of this energy via transmission lines and pipelines has a further environmental cost (Wilson 1978). The electrical demand in the Vancouver Island/Lower Mainland region is 70 percent of the provincial total, but the region produces only 12 percent of the electricity (BC Hydro 1994, 5). The region thus gains the benefit of electrical production and pays little of the environmental cost. For the development of energy

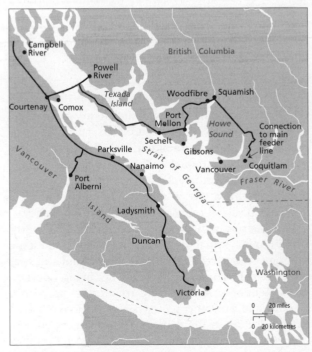

Figure 11.1 Vancouver Island natural gas line extension, 1995
Source: Modified from Pacific Coast Energy Corporation (1995).

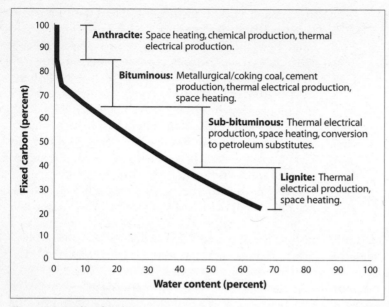

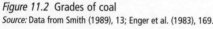

Figure 11.2 Grades of coal
Source: Data from Smith (1989), 13; Enger et al. (1983), 169.

sources in the future, it is important to assess all the costs and benefits for individual regions and attempt to equalize costs and benefits across the province.

COAL

Coal is a fossil fuel. It is graded according to the amount of carbon it contains because that determines its ability to burn and produce heat (Figure 11.2). Lignite coal has the least amount of carbon and gives off relatively low heat values. Bituminous coal has an increased carbon/heat ratio, and anthracite coal has the highest rating. British Columbia has all three grades of coal, and the known deposits of coal can be seen in Figure 11.3.

Coal was one of the first resources to be mined in British Columbia. Initial discoveries in the 1830s and some mining at the north end of Vancouver Island and near present-day Prince Rupert were the beginning. Substantial discoveries and developments in Nanaimo, Cumberland, and other central island locations followed (Seager 1996). As the demand for coal increased, discoveries were made in the East Kootenays and production got under way. Petroleum replaced coal by the 1950s and many coal mines closed, but coal production had

revived by the end of the 1960s, almost exclusively for export to Japan. The industry took off, with huge multinational corporate investments in capital-intensive, open-pit mining and concentrating of coal. Old mining communities such as Natal and Michel near the Crowsnest Pass were phased out and the new communities of Sparwood and Elkford were built to house the coal miners. With government help, dedicated rail lines were built from the Crowsnest Pass region to the new, large coal terminal of Roberts Bank. Coal became a major export commodity to Japan and later Korea. Figure 11.4 reveals these changes in coal production over the past century.

By the 1960s, demand for thermal coal and coking coal was responsible for new investment. The primary use of thermal coal is to create steam to turn turbines for electrical production. Coking coal is used almost exclusively in the iron and steel industry, which requires high grade bituminous coal. British Columbia's bituminous deposits in the East Kootenays, known as southeast coal, were targeted for both these end uses.

The story of northeast coal production is much more controversial, as decisions made for this region at national and provincial political levels were based on international events and influences. Figure 11.5 outlines the projected and real demands of the Japanese steel industry. The wildly exaggerated projected increase translated into a similarly exaggerated demand for BC coal. The energy crisis over oil began in 1973, causing global reconsideration of energy use. In Japan, most production was achieved by importing almost all raw materials, including oil as the energy base. As the price of oil and oil-based transportation increased, Japan's steel production became less competitive, and even though it projected major increases they did not materialize. The governments of British Columbia and Canada produced their own exaggerated projections, which were accompanied by foreign investment for plant construction, operation, and equipment, and contracts for coal. The two levels of government invested an enormous amount

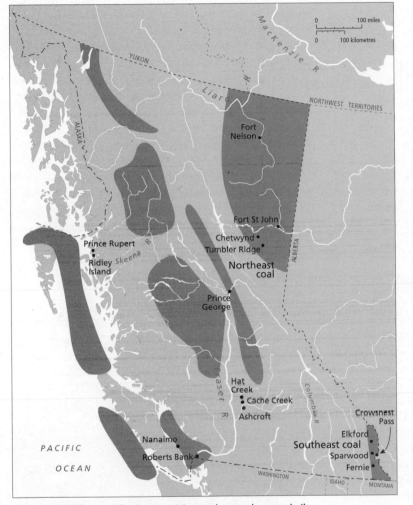

Figure 11.3 Sedimentary basins containing coal, natural gas, and oil
Source: Modified from Sewell (1987), 230.

was caught in global oversupply with a demand that had stabilized.

The energy crisis caused BC Hydro to consider the use of coal to produce electricity. Hat Creek Coal, a mine with huge reserves of low grade lignite coal, was acquired by BC Hydro. This deposit, along with coal from the East Kootenays and some from Vancouver Island, was proposed for thermal electrical production. None of these projects were initiated, for reasons to be reviewed below.

Burning coal to produce electricity damages the environment. As well as coming in various grades, coal contains varying amounts of sulphur, which is released into the atmosphere when burned and contributes to acid rain. The residents of the Hat Creek/Cache Creek/Ashcroft area expressed great concern over the impact of acid rain.

Coal remains British Columbia's leading energy export commodity. The province has plenty of it, but so do many other regions of the world, which keeps the world market price at relatively low levels. The recent economic slump of the Japanese and Korean markets is having a serious impact on northeast coal production, raising the question whether

of money in infrastructure for the northeast coal region: for the new community of Tumbler Ridge, the coal port facility at Ridley Island, the double tracking of the CN Rail line to Prince Rupert, and a new electric rail line between Tumbler Ridge and Chetwynd.

The Quintette and Bullmoose mines of northeast coal had just gone into operation when the recession of the 1980s began. Consequently, they experienced financial troubles from the beginning, particularly when the Japanese renegotiated the coal contract down from over $100 per tonne to less than $90 per tonne. British Columbia

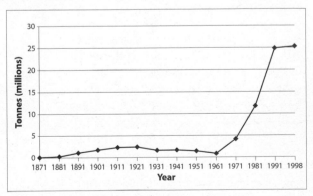

Figure 11.4 Coal production, 1871-1998
Source: Internet, British Columbia, Ministry of Energy and Mines.

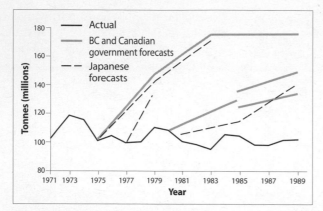

Figure 11.5 Forecasts of Japanese crude-steel production, 1971-89
Source: Modified from Halvorson (1990), 319.

Tumbler Ridge will become British Columbia's next ghost town. On a positive note, research and development is being conducted on more environmentally friendly ways of burning coal, thus reducing emissions that contribute to acid rain and global warming.

OIL

Of all the sedimentary basins shown in Figure 11.3, only the Peace River region in the northeastern part of the province has produced oil and natural gas. Both of these valuable energy sources are found at depths between 1,000 and 3,000 metres and both are the product of organic materials created through photosynthesis millions of years ago. This organic material was covered by sediments; pressure and heat converted it into petroleum or natural gas.

Prior to 1954, British Columbia's oil demands were met with imported oil shipped from California to the oil refineries of Burrard Inlet. This changed with the discovery of oil in the Peace region and the construction of the Transmountain pipeline, which brought both BC and Alberta oil to refineries in the Lower Mainland (Figure 11.6). As well, a new refinery was constructed

in Kamloops to serve the needs of the interior. The need for and dependence on oil continues today.

The energy crisis of the 1970s saw the price of oil climb from $3.01 per barrel in July 1973 to over $11.00 per barrel by January 1974. By the late 1970s, oil was close to $40.00 per barrel. This represented a significant increase in cost to everything that had an oil component, whether gasoline or any product made from crude petroleum such as plastics, fertilizers, paints, and so forth. An atmosphere of crisis was manufactured over cost and also because of fear of global oil shortages. Regions without oil, or with insufficient oil, worried about their balance

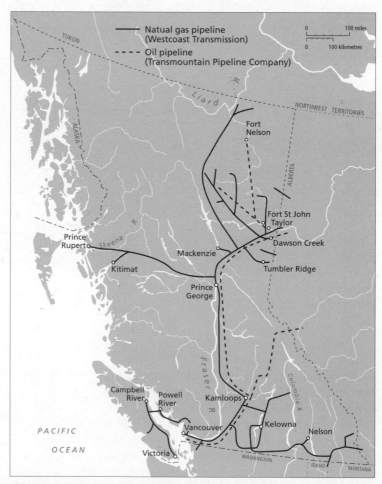

Figure 11.6 Natural gas and oil pipelines in British Columbia
Source: Modified from British Columbia, Ministry of Finance and Corporate Relations (1998), 87.

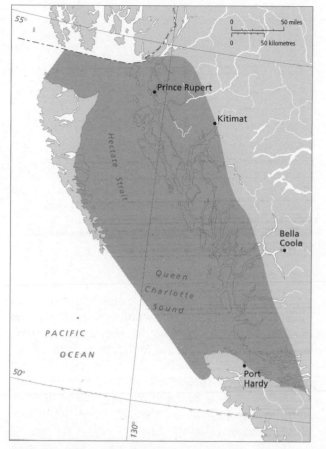

Figure 11.7 Offshore oil claim by BC government
Source: Modified from Whiteley (1982), E5.

of payments. For Alberta, the crisis produced a windfall in investment and development of the traditional oil patch and technologies to produce oil from tar sands.

British Columbia received some of the attention with investments in the Peace River/northeast region and prospects of offshore oil. Various companies staked out their territories, mainly between the Queen Charlotte Islands and the mainland, but offshore prospects met with two issues. First, the government of British Columbia claimed the inland waters and any oil revenues derived from them (Figure 11.7). This claim challenges the terms of Confederation, which gave the federal government sole jurisdiction over oceans. The second, and more serious issue, is the prospect of finding oil in the most

seismically sensitive area of Canada. An earthquake could rupture an oil well, with serious environmental consequences for the aquatic environment. A series of moratoria have prevented the drilling process. Meanwhile, the price of oil descended rapidly in the 1980s and has not reached such lofty prices since.

The energy crisis brought another response to prospective oil shortages that involved British Columbia. Alaska proposed to move its oil reserves by tanker either to the Cherry Point Refinery in Washington State's Puget Sound, which is already connected to the Burrard refineries in Vancouver by pipeline, or to the port of Kitimat, where a pipeline would then be built to the Lower Mainland. In either case, the risk of transporting oil in tankers along the west coast of British Columbia was met with a great deal of anxiety by its residents. Neither of these schemes materialized, and the oil spill of the *Exxon Valdez* in 1989 off the coast of Alaska serves as a real reminder of the catastrophic level of disaster that can be caused by oil spills.

Although the energy crisis resulted in some shifting to other energy sources, oil is still in the highest demand in the province. British Columbia produces approximately 25 percent of the provincial demand and imports 75 percent from Alberta. Its lack of self-sufficiency is another incentive to develop and use made-in-British Columbia energy sources.

NATURAL GAS

For a resource that was discovered by accident, natural gas has become a great success story. In the process of drilling for oil in the 1950s, energy companies discovered natural gas in considerable quantities in the northeast region. Sufficient gas was found to build the Westcoast Transmission pipeline in the 1950s, connecting the Fort Nelson/Fort St John area with the Lower Mainland and the Pacific northwestern region of the United States. From this main line, branches were built to Prince Rupert, and from Kamloops into the Okanagan and the Kootenays (Figure 11.6). Connections have also been made to Alberta's supply of natural gas. The most recent expansion of natural gas distribution, in the early 1990s, has been from the Vancouver area to Squamish and the Sunshine Coast and across to Vancouver Island (Figure 11.1). In this latest expansion, natural gas lines were connected to the coastal pulpmills in these areas and provided some communities and other industries with this energy option.

Having an abundance of natural gas and a shortage of oil, British Columbia has promoted the use of natural gas over oil products, including converting automobiles to this fuel. More recently, the substitution of natural gas for oil has been encouraged on environmental grounds since natural gas burns more cleanly and gives off fewer greenhouse emissions (British Columbia Energy Council 1994). Natural gas can also be used for a variety of other non-energy products, such as chemicals, paints, and plastics.

Natural gas co-generation plants – plants that produce electricity plus steam – are common in Alberta because of its surplus of natural gas and lack of major river systems to dam for hydroelectric power. British Columbia produces most of its 13,300 megawatts of electricity from damming river systems and only 2,500 megawatts through natural gas co-generation. The oldest co-generation station is on Burrard Inlet near Vancouver. A relatively new one is at Taylor on the Peace River. A few small plants are run by private corporations, and a 250 megawatt facility was built at Campbell River on Vancouver Island in 1999. Approximately 2,600 megawatts are planned for the future, as natural gas co-generation has fewer drawbacks than large hydroelectric dams (British Columbia Energy Council 1994).

ELECTRICITY

The most efficient way to produce electricity is by moving water through a turbine. The many large river systems in British Columbia, combined with high elevations, makes possible enormous hydroelectric generation. To develop hydroelectricity, though, dams for large reservoirs have to be constructed because of the major fluctuations in discharge rates throughout the year. Dams impound immense volumes of water, which can be released at controlled rates to maintain an even production of electricity.

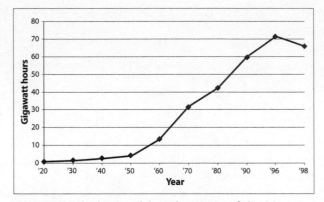

Figure 11.8 Hydroelectric and thermal generation of electricity, 1920-98
Sources: Barker (1977), 54; British Columbia, Ministry of Finance and Corporate Relations (1986), (1997), (1999).

The first electricity production in British Columbia was by steam at the Moodyville sawmill in North Vancouver in 1882 (BC Hydro 1979, 6), and the beginning of hydroelectricity production was in Nelson in 1896 (Sewell 1987, 234). From these humble beginnings, demand increased slowly throughout the province until the 1950s. The ensuing megaproject era involved the construction of large dams and a major increase in the supply of electricity (Figure 11.8).

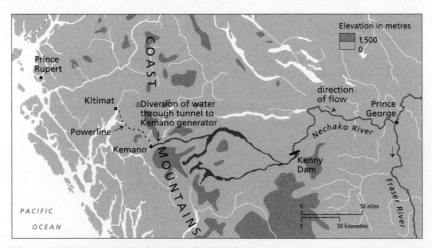

Figure 11.9 Kemano 1 Project: Damming the Nechako River
Source: Data from Sewell (1987), 231.

Escalation in electrical energy production was fuelled by the expansion of the mining and forest industries. The megaproject era for British Columbia began with the Kemano 1 project, built in the 1950s. The Kenny dam reversed a portion of the flow of the Nechako River system via a tunnel through the Coast Mountains to the Kemano generating station. This project was developed and built by Alcan to produce aluminum at Kitimat (Figure 11.9).

By the 1950s, the provincial government realized that electrical energy production was a crucial component to industrial expansion, especially for the interior of the province. In 1956, for example, the provincial government granted a vast area of north-central British Columbia to Swedish investor Axel Wenner-Gren for northern development. The region extended from Summit Lake, north of Prince George, to Lower Post on the Yukon border. Wenner-Gren's promises of pulpmills, mines, and a monorail were not to materialize, but both he and the provincial government recognized the large amount of hydroelectricity that could be produced by the Peace River.

The Columbia River was also recognized by both the BC and the US governments at that time as being able to provide the dual benefits of electricity and flood control by building dams. Dams hold back the peak flows, thus reducing floods, and the impounded water is then used to raise low flows in winter and thus to increase electrical production. Direct political involvement occurred in 1961, when the provincial government took over the private utility of BC Electric, created the crown corporation

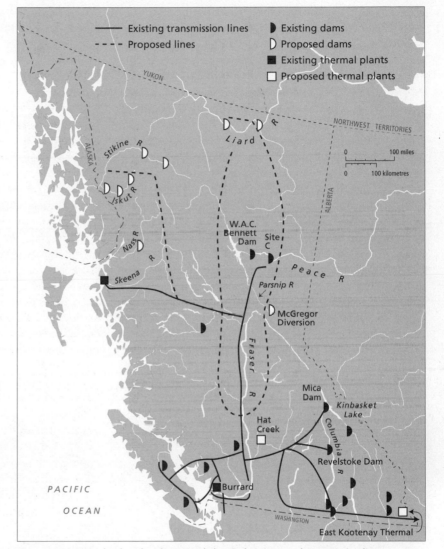

Figure 11.10 Main developed and proposed electrical projects and transmission lines
Sources: Data from Sewell (1987), 231; BC Hydro (1994), 3.

of BC Hydro, and launched the "two rivers policy." Major dams were designed and built for the Peace and Columbia Rivers. Through the Columbia River Treaty, Americans paid for three dams to be constructed in British Columbia and received the benefit of hydroelectricity and flood control from the Columbia River, while BC Hydro supplied British Columbians with electricity from the Peace.

The flooding of the Columbia River resulted in massive lakes, loss of plant and animal species, and the relocation of residents (Wilson 1978). Recent research suggests with the aid of hindsight that the people in the Kootenays paid the price for the Columbia River dams. According to Toller and Nemetz, "hydro dams have adversely affected the ecology, economy, and social fabric of the Kootenays" (1997, 27).

Another controversial hydroelectric proposal was the Moran Dam, approximately twenty-five kilometres north of Lillooet on the Fraser River. The benefits would be hydroelectric power and flood control (Hardwick 1962, 63). The costs would be mainly to the races of salmon no longer able to spawn in the upper half of the Fraser River basin. In this instance, fishing interests outweighed energy interests and the dam was not constructed.

Electrical energy increased greatly as dams were constructed. Through BC Hydro, the provincial government used electrical energy as an industrial strategy. Projections, or forecasts of future demands, are essential to creating electricity. Figure 11.10 shows the main dams and transmissions lines now in the province compared to the projected energy schemes of the 1970s.

There seemed to be no end to the investment, nor to the need for electrical energy, with the rapid expansion of the forest and mining industries in the 1960s. Large dams require a minimum of ten years lead time before they produce electricity, however, and it is therefore essential to anticipate electrical demand accurately (Marmorek 1981). In other words, huge amounts of money need to be invested at the start to build a dam, with the realization that it will take at least ten years to move from planning, through construction, to power generation. By the mid-1970s, the accuracy of BC Hydro projections was a matter of debate, and particularly after the energy crisis of 1973. The crown corporation did not expect the forest industry's reaction to increased energy costs, namely, to use hog fuel to generate large amounts of electricity for their own purposes. Similarly, it did not account for conservation measures requiring less energy use. Figure 11.11 illustrates the differences in projected demand according to BC Hydro, the BC Energy Commission, and environmental groups over a twenty-year period. Like compound interest, it took only a few years before a significant difference appeared. One can observe the large gap in projections by 1996, particularly between the sharp increase expected by BC Hydro and the 1 to 2 percent per year increase foreseen by environmental groups. The implications of these projections can be seen in the list of projects planned by BC Hydro in 1976 (Table 11.3).

To fulfill the exaggerated demand by BC Hydro, CANDU nuclear reactors, coal-fired thermal generators, and a great number of dams were on the drawing board: all to produce electricity for an industrial demand based on a past performance that had little relationship to the future. As the price of energy went up in the 1970s, the largest industrial user of electricity, the forest industry, began to produce its own electrical energy from hog fuel. The recession of the 1980s saw the shut-down of mines and mills and a further reduction of electrical energy demand. Only a few of the projects on the list were ever built, the largest being the Revelstoke Dam.

BC Hydro projections also raised the question of the environmental cost of progress and the impact of megaprojects. Many of the proposed projects were vehemently opposed by the public. Concerns over nuclear wastes, low level radiation, and possible contamination saw few citizens wanting a nuclear power plant in their backyard. Coal-fired thermal plants raised fears of acid

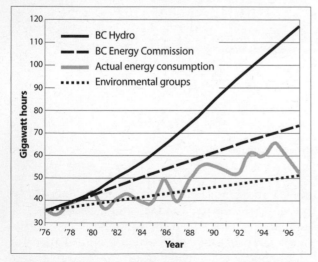

Figure 11.11 Electrical energy projections and consumption, 1976-97
Sources: Data from British Columbia Energy Commission (1976), 25; Farrow (1977), 36; British Columbia, Ministry of Economic Development (1987), 64; British Columbia, Ministry of Finance and Corporate Relations (1993), 252; (1998), 283.

Table 11.3

Major electrical energy proposed projects, 1976

Projects	At site capacity (megawatts)	Average annual energy (millions of Kwh)	Earliest in-service date	Completion date
Hydroelectric				
Site One	700	3,150	1979	1980
Seven Mile	700	3,150	1980	1979
Revelstoke	2,700	7,970	1981	1984
McGregor Diversion	0	3,110	1985	n.c.
Kootenay Diversion	0	875	1984	n.c.
Kemano 2 (Completion)	1,200	6,480	1983	n.c.
Peace Site C	900	5,010	1984	n.c.
Peace Site E	750	3,920	1984	n.c.
Murphy Creek	400	2,100	1983	n.c.
Duncan Bay (Vancouver Island)	150	270	1980	n.c.
Kokish (Vancouver Island)	160	255	1980	n.c.
Elaho	500	1,710	1983	n.c.
Homathko	1,450	7,550	1984	n.c.
Skeena (Cutoff Mountain)	1,080	6,370	1987	n.c.
Iskut-Stikine	3,025	19,500	1986	n.c.
Liard River	3,800	23,600	1987	n.c.
Yukon-Taku	3,700	22,800	1990	n.c.
Upstream Fraser	5,800	29,850	1987	n.c.
Downstream Fraser	2,420	15,510	1985	n.c.
Thermal projects				
Hat Creek (proven)	2,500	16,500	1982	n.c.
Hat Creek (possible)	2,500	16,500	1985	n.c.
East Kootenay	1,800	12,000	1983	n.c.
Comox (Vancouver Island)	550	3,500	1981	n.c.
CANDU nuclear plants (two)	1,200	8,400	1985	n.c.

Note: n.c. = not completed
Sources: British Columbia Energy Commission (1976), 72; British Columbia, Ministry of Finance and Corporate Relations (1998), 87.

rain. Damming the Fraser River and destroying the salmon runs was considered a sacrilege by many. Impounding water to create huge reservoirs behind dams causes enormous environmental problems and affects many people and their way of life. The proposed Site C dam, next to Fort St John on the Peace River, would have flooded some of the finest agricultural land in the province.

The plan to divert the McGregor River into the Parsnip River would have joined two entirely different ecosystems – one that flows to the Pacific and one that flows to the Arctic. The Parsnip/Peace Rivers have different parasites and organisms from the McGregor/Fraser system, and it was feared that the two might find ways to invade one another, with devastating consequences to fish stocks. Alcan, at Kitimat, had plans to expand its aluminum production that required increased hydroelectricity. Kemano 2 was a plan to dam the Nanika River, a tributary of the Skeena, and to reverse its flow into the Oatsa reservoir, thus increasing the capacity of the existing hydroelectric system. This controversial project was cancelled in 1987 in favour of the Kemano Completion Project. That equally controversial proposal included drilling a second tunnel through the Coast Mountains and draining far more water from the Nechako (Christensen 1995). This project was cancelled in 1995, although Alcan is pursuing compensation through the courts.

Transmission lines, which would have been built from northern dams to distribute power to southern consumers, raised other environmental concerns (Figure 11.10). If these dams were to be constructed, then a great swath of forests would have to be cleared for the transmission lines to run from one end of the province to the other. Uninsulated lines have a corona discharge, from escaping ions, and research suggests that many health risks are related to this hazard (see Brodeur 1993; Young 1974).

The environmental concerns are many and need to be kept in mind when considering where future kilowatt hours of electricity will come from. This province is growing in population, with a corresponding need for electrical energy. Natural gas co-generation is one alternative; small-scale in-stream hydroelectric generators are another. These units require no dams and have few negative impacts on the environment. The Columbia River Treaty made provision for the United States to transmit to British Columbia half of the energy it generated from the project, thirty years after the construction of the three dams on the BC side of the Columbia. These returns were due in 1998, 1999, and 2003 and will probably fulfill immediate demands, including those of a new aluminum plant. Maybe we should be asking how energy consumption can be decreased in the future.

RETHINKING ENERGY OPTIONS

Most of the energy used in British Columbia is supplied by large oil and gas corporations or crown corporations such as BC Hydro. There is often little thought given to types of energy or where it comes from, the most important concern being the cost.

Amory Lovins (1977) observed the energy directions being promoted in the 1970s by governments and industries as they reacted to the energy crisis. He used the term **hard energy path** to describe the typical forecasting that BC Hydro, and many other utilities in the world, were engaged in at that time. He saw that the perceived energy crisis of the period created strategies that centralized energy sources and focused on how to find more energy. From this supply side perspective, he also saw great increases in costs for energy because the rivers most easy to dam had already been dammed and the most accessible oil and gas had already been exploited. Future energy supplies, according to Lovins, would rely on nuclear and coal-fired generators, and costs would increase because of the inherent environmental problems attached to both of these non-renewable energy sources. The hard energy path is implied in Table 11.3, with BC Hydro projections of a need for CANDU reactors, coal-fired thermal plants, and many more hydroelectric dams.

An alternative to the hard energy path is the **soft energy path,** which focuses on decentralized forms of energy, soft technologies, conservation of energy, and recognizing the link between life style and energy use. The term "decentralized" implies that individuals – citizens, local government, corporations – begin to make decisions about the types of energy they use, including alternative forms of energy. Lovins (1977, 38-9) uses five principles to define the characteristics of the technologies along this path:

1 *Use renewable technologies.* Make use of the sun, wind, wood, wastes, and water.
2 *Use diverse technologies.* Avoid thinking in terms of "all or nothing" regarding energy use. If solar energy can be utilized to produce 10 percent of heating, then this represents a 10 percent reduction in conventional sources. Use many different energy sources, and renewable ones where possible.
3 *Use simple technologies.* Build and maintain renewable systems, such as a solar-heated hot water system that makes use of understandable technology and is relatively simple to construct and operate. A solar-heated system is not necessarily unsophisticated, but it does not depend on highly specialized, costly technical expertise.
4 *Match energy quality to end use needs.* Recognize how various energy sources are utilized to perform jobs, or end uses. A simplified classification, using heat, divides end uses or tasks into three categories:

   - High grade tasks: above 148°C (300°F)
   - Medium grade tasks: approximately 148°C (300°F)
   - Low grade tasks: below 148°C (300°F)

   Combustion in automobiles, smelters, and many electrical motors, for example, are high grade tasks requiring high grade energy sources such as oil and natural gas. Conversely, heating water, space heating, and air conditioning require energy below 148°C. Many renewable energy sources such as solar and wood can be employed for these demands, reserving high grade energy for high grade tasks.
5 *Match energy in scale and geographic distribution to end use needs.* The energy demands of the individual household, the neighbourhood, the community, and the region are matched with the renewable energy sources that exist in the region to fulfill the demands. Windy areas can utilize windmills, for example, while those having considerable amounts of sunshine use solar technologies.

The soft energy path shifts attention to the demand side of energy, or how to make energy go further. The need for conservation of energy recognizes that British Columbians and Canadians in general are the highest per capita consumers of energy in the world (Marmorek 1981, 77). Lovins refers to many of the conservation measures employed as "technical fixes." BC Hydro's Power Smart program, for example, has promoted the need to save energy and describes ways to do so. It suggests saving energy by using insulation and energy-efficient machinery, redesigning buildings, co-generating energy, and many other techniques to make resources go further but with no change to life style. A change in habits – by riding a bike, walking, car pooling, using public transit, turning down the thermostat, and turning off the air conditioning – may result in far greater savings of energy.

SUMMARY

Energy resources have been abundant and important to the development of British Columbia. Prompted by technological change and new demands for energy, the province has provided not only for domestic energy needs but also for an export market. The exception is oil, which is imported from Alberta.

Along with the spatial distribution of the many energy resources, world events and politics have also shaped the geography of energy in British Columbia. Unprecedented industrial expansion of the 1950s and '60s required a major increase in energy production and created an expectation of never-ending growth. The provincial government moved from a policy of facilitating and encouraging energy projects to one of becoming directly involved in megaprojects to increase electrical production.

With the energy crisis of the early 1970s, there was a move away from oil and toward natural gas and electricity. Increased environmental awareness provoked a sceptical reaction to proposed major developments of electrical energy generation, and the recession of the 1980s brought a new reality to energy development. The forest industry implemented a decentralized source of energy (hog fuel), improvements in insulation standards were reflected in the building code, and energy effi-

ciency was rewarded. The David Suzuki Foundation, however, is concerned that "renewable energy receives fewer incentives than other energy sources, while subsidies to fossil fuels and nuclear power continue" (1999, 5). BC Hydro's Power Smart program reflects an overall emphasis on conservation.

The Vancouver-based Ballard Industries may be pointing the way to the future, with a hydrogen cell designed to power vehicles. Other technologies are experimenting with electric batteries for the same purpose. In both cases, the message is clear, that the use of fossil fuels with emissions that contribute to greenhouse gases and acid rain has to be diminished. The environmental consequences of all uses of energy can no longer be ignored.

The lessons of Amory Lovins remind us that individuals can make decisions about supplying their own forms of alternative energy, and that energy use is a life style.

REFERENCES

Barker, M.L. 1977. *Natural Resources of British Columbia and the Yukon.* Vancouver: Douglas, David and Charles.

BC Hydro. 1979. *Power Perspectives '79.* Vancouver: BC Hydro.

–. 1994. *Making the Connection: The BC Hydro Electric System and How It Is Operated.* Vancouver: BC Hydro.

Bridges, G.E. 1993. *Decision-Making Processes and the Energy Sector.* Victoria: British Columbia Round Table on the Environment and Economy.

British Columbia, Ministry of Economic Development. 1987. *British Columbia Facts and Statistics.* Victoria: Crown Publications.

British Columbia, Ministry of Finance and Corporate Relations. 1986. *1986 British Columbia Economic and Statistical Review.* Victoria: Crown Publications.

–. 1987. *1987 British Columbia Economic and Statistical Review.* Victoria: Crown Publications.

–. 1991. *1991 British Columbia Economic and Statistical Review.* Victoria: Crown Publications.

–. 1993. *1993 British Columbia Economic and Statistical Review.* Victoria: Crown Publications.

–. 1997. *1997 British Columbia Economic and Statistical Review.* Victoria: Crown Publications.

–. 1998. *1998 British Columbia Financial and Economic Review.* Victoria: Crown Publications.

–. 1999. *1999 British Columbia Financial and Economic Review.* Victoria: Crown Publications.

British Columbia, Ministry of Industry and Small Business. 1981. *British Columbia Economic Activity 1980: Review and Outlook.* Victoria: The Ministry.

British Columbia Energy Commission. 1976. *British Columbia's Energy Outlook 1976-1991.* Victoria: The Commission.

British Columbia Energy Council. 1994. *Planning Today for Tomorrow's Energy.* Vancouver: The Council.

Brodeur, P. 1993. *The Great Power-Line Coverup: How the Utilities and the Government Are Trying to Hide Cancer Hazards Posed by Electromagnetic Fields.* Boston: Little, Brown.

Chapman, J.D. 1960, "The Geography of Energy – an Emerging Field." *Occasional Papers in Geography,* 31-40. Canadian Association of Geographers, BC Division, no. 1. Vancouver: Tantalus.

Christensen, B. 1995. *Too Good to Be True: Alcan's Kemano Completion Project.* Vancouver: Talon.

David Suzuki Foundation. 1999. "Energy Revolution." *Finding Solutions* nos. 4 and 5 (April): n.p.

Enger, E.D., J.R. Kormelink, B.F. Smith, and R.J. Smith. 1983. *Environmental Science: The Study of Relationships.* Dubuque, IA: Wm. C. Brown.

Farrow, M. 1977. "Hydro Energy Growth 'Suspect.'" *Vancouver Sun,* 3 March, 36.

Halvorson, H. 1990. "The British Columbia Coal Industry." In *The Pacific Rim: Investment, Development and Trade,* 2nd ed., ed. P.N. Nemetz, 310-22. Vancouver: UBC Press.

Hardwick, W.G. 1962. "The Moran Dam and Availability of Cultivated Land in Interior British Columbia." *Occasional Papers in Geography,* 62-8. Canadian Association of Geographers, BC Division, no. 3. Vancouver: Tantalus.

Lovins, A.B. 1977. *Soft Energy Paths.* New York: Harper.

Marmorek, J. 1981. *Over a Barrel: A Guide to the Canadian Energy Crisis.* Toronto: Doubleday.

Pacific Coast Energy Corporation. 1995. *Vancouver Island Natural Gas Line Extension – 1995.* Brochure. Vancouver: The Corporation.

Seager, A. 1996. "The Resource Economy, 1871-1921." In *The Pacific Province: A History of British Columbia,* ed. H.J.M. Johnston, 205-52. Vancouver: Douglas and McIntyre.

Sewell, W.R.D. 1987. "Energy." In *British Columbia: Its Resources and People,* ed. C.N. Forward, 226-56. Western Geographical Series vol. 22. Victoria: University of Victoria.

Smith, G.G. 1989. *Coal Resources of Canada.* Geological Survey of Canada Paper 89-4. Hull, ON: Ministry of Supply and Services Canada.

Toller, S., and P.N. Nemetz. 1997. "Assessing the Impact of Hydro Development: A Case Study of the Columbia River Basin in British Columbia." *BC Studies* no. 114 (Summer): 5-30.

Whiteley, D. 1982. "Offshore Claim Negotiations Less Friendly." *Vancouver Sun,* 10 September, E5.

Wilson, J.W. 1978. "Electric Power Development in British Columbia: A Case of Metropolitan Dominance?" In *Vancouver: Western Metropolis,* ed. L.J. Evenden, 79-93. Western Geographical Series vol. 16. Victoria: University of Victoria.

Young, L.B. 1974. *Power over People.* New York: Oxford.

INTERNET

BC Hydro, "The Power Is Yours," www.bchydro.bc.ca

BC Stats: Quick Facts – The Economy, www.bcstats.gov.bc.ca/data/QF_econo.HTM

British Columbia, Ministry of Energy and Mines, www.em.gov.bc.ca/mining/miningstats

Statistics Canada, www.statcan.ca

Agriculture: The Land and What Is Produced

12

British Columbia has a mountainous landscape and therefore lacks agricultural land for commercial production. Pockets of agricultural land have been developed, and this resource has been important to the settlement and development of many parts of the province. A host of physical conditions influence the range and types of agricultural commodities that the land is capable of producing. Many agricultural decisions are based not solely on the physical parameters but on human considerations such as the economic viability of producing a commodity. The range of viable agricultural products has changed as development of new seeds and products, farming methods and technologies, transportation routes and modes, political decision making, and consumer tastes have shaped and reshaped the agricultural landscape of British Columbia.

The most important ingredient in farming is land, and in many ways it is a non-renewable resource because it is so difficult to create "good" soil. Over time in some locations of the province agricultural land has been converted to other uses. An already limited agricultural land resource has shrunk, and along with it the capability to produce food for the future. To protect agricultural land, the BC government in 1973 established agricultural land reserves with a strict set of rules on its use. This tactic is unique in Canada and much of the world, but it is not without its critics.

From its beginning, farming in British Columbia has been in transition. As conditions change so do the products and methods of farming; today, approximately 200 agricultural commodities are produced. The combination of traditional and new, exotic products has created regional specialization and considerable hope for future growth and prosperity. The 1996 census shows that British Columbia is one of the few provinces in Canada to increase its number of farms, leading all the other provinces in the establishment of new farms between 1991 and 1996 (Table 12.1).

PHYSICAL CONDITIONS

There is very little good agricultural land in British Columbia. Figure 12.1 shows the pockets of primary farmland, along with larger areas of range lands and the few areas of potential farmland in the north. These lands make up the total of British Columbia's arable land: approximately 4 percent of the land base and another 30 percent with some agricultural capability (Internet, BC Stats: Quick Facts).

The soils of British Columbia are influenced by a combination of factors such as climate, vegetation, elevation, slope, and surficial geology, resulting in a variety of soils and agricultural capabilities (Valentine et al. 1978; and see Chapter 2). The Canada Land Inventory has a seven-point scale to grade soils in terms of agricultural capabilities, modified by descriptive subclasses giving details of crop limitations (Environment Canada 1972; see Table 12.2). Typically, a farm includes sloping land, land beside a stream, boggy land, and so forth, and will therefore have a number of designations, as can be seen in the following example:

$$5_w^6 - 4_w^4 \underset{\text{I}}{(4_w^6 - 3_w^4)}$$

The farmland thus described is in a wet area and has two designations. Both class 5 and class 4 lands are in an unimproved state. The small 6 above the W indicates that 60 percent of this land is class 5 and the land has excess water, which is the subclass W. The other 40

Table 12.1

Census farms in Canada, 1991 and 1996

	1991 farms	1996 farms	Percentage of change
Newfoundland	725	731	0.8
Prince Edward Island	2,361	2,200	-6.8
Nova Scotia	3,980	4,021	1.0
New Brunswick	3,252	3,206	-1.4
Quebec	38,076	35,716	-6.2
Ontario	68,633	67,118	-2.2
Manitoba	25,706	24,341	-5.3
Saskatchewan	60,840	56,979	-6.3
Alberta	57,245	58,990	3.0
British Columbia	19,225	21,653	12.6
Canada	280,043	274,955	-1.8

Note: The term census farm refers to a farm, ranch, or other agricultural operation that produces agricultural products for sale.
Source: Statistics Canada (1997), 4.

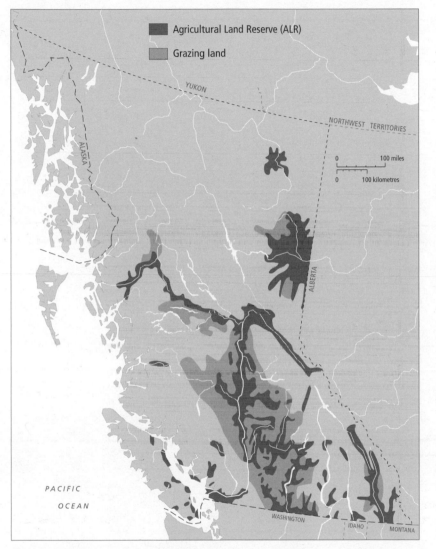

Agricultural Land Reserve (ALR)

Grazing land

Figure 12.1 Farming areas in British Columbia
Source: Modified from Dalichow (1972), 40; Farley (1979), 81.

flooding might still occur. The classification system is dynamic in that it recognizes the potential for land improvement. Perhaps the most important value of the Canada Land Inventory is as a scientific and objective means of assessing agricultural land.

Another major constraint to agriculture is climate. The combination of prevailing westerly winds bringing relatively mild moist conditions in winter, significant latitudinal changes influencing incoming solar radiation, and high ranges of mountains affecting air masses and moisture means that both temperature and precipitation vary greatly throughout the province (see Chapter 2). Figure 12.2 maps the average number of frost-free days per year throughout the province, revealing significant differences from region to region. High elevations and northerly latitudes subject much of the province to frosts even in summer, with fewer than fifty frost-free days per annum. The relatively warm Pacific Ocean combined with westerly flows of winds gives a distinct advantage to the coast. The southern interior river valleys of the province also have a considerable number of frost-free days. The number of frost-free days is crucial to planting and harvesting crops.

percent of the land (hence the small 4) is class 4; however, it has excess water and is subject to flooding or inundation (the I subclass). The bracketed classification, or second designation, is interesting because it suggests that this land could be improved if the moisture problems were taken care of through by installing a drainage system or some other technique. The improved land would then be 60 percent class 4 land and 40 percent class 3 land, even though excess moisture and some

Another frequently used statistic for calculating quality of agricultural land is growing degree days. When the temperature is above 6°C plant growth is active. (Some research suggests 5°C.) Each day that the temperature is above 6°C is recorded as a growing degree day, and a yearly total of degrees above that benchmark is calculated (Matthews and Morrow 1985, 38).

Table 12.3 provides a regional summary of physical characteristics, including potential agricultural limitation.

Table 12.2

Soil classification, area, and subclasses

Soil class	Canada Land Inventory characteristics	British Columbia		Agricultural Land Reserve	
		Area (ha)	% of total land	Area (ha)	% of total ALR
1	Widest range of crops	69,948	0.2	56,655	1.2
2	Moderate limits	397,688	1.3	302,163	6.4
3	Moderately severe limits	999,778	3.3	717,637	15.2
4	Severe limits	2,131,867	7.1	1,454,159	30.8
5	Only permanent pasture/forage	6,138,294	10.4	1,482,487	31.4
6	Natural grazing	5,357,900	17.9	448,523	9.5
7	No capability	14,900,513	49.7	259,671	5.5
Total		29,995,988	100.0	4,721,295	100.0

Subclass characteristics

C	Adverse climate	N	Salts
D	Undesirable soil structure	P	Stoniness
E	Erosion	R	Bedrock near the surface
F	Low fertility	T	Topography (slope)
I	Inundation (flooding)	W	Excess water
M	Moisture deficiency (droughtiness)	X	Combination of soil factors

Note: Totals do not always equal 100 percent, due to rounding.
Sources: British Columbia, Environment and Land Use Committee Secretariat (1976); Berry (1988), 25.

Table 12.3

Physical characteristics and potential limitations for agriculture

Agricultural area	Recording station	Altitude (m)	Average annual precipitation (mm)	Average summer precipitation (mm)	Frost-free days	Growing degree days (above 6°C)[a]	Potential limitations
Vancouver Island	Victoria	20	965	81	159	3,109	Climate, soil
Lower Mainland	Vancouver	3	1,524	150	206	2,018	Hydrography, vegetation
Okanagan	Kelowna	369	305	71	146	1,864	Climate, hydrography
Kootenay	Cranbrook	927	356	102	82	1,641	Climate, soil
South central interior	Kamloops	382	254	84	176	1,259	Climate, hydrography
North central interior	Prince George	573	533	137	81	1,238	Climate, soil
North coast	Terrace	31	1,194	157	132	1,440	Vegetation
Peace River	Fort St John	700	432	180	123	1,275	Climate

a Growing degree days are estimated by adding together the number of degrees above 6°C the temperature reaches each day over the course of the year.
Source: Modified from Dalichow (1972), 18, 44.

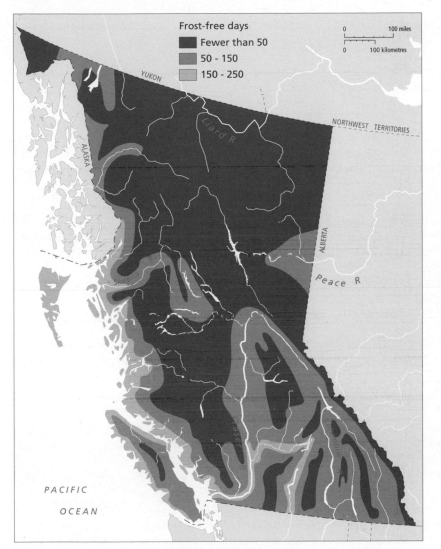

Figure 12.2 Frost-free days in British Columbia
Source: Modified from Internet, Ministry of Agriculture and Food.

in the important summer growing season, but because of its northern latitude it has fewer growing degree days.

All these physical characteristics are the foundations on which an agricultural industry is built. Human factors are also complex because farmers must attempt to make a living. The physical constraints of the land are just one factor with which a farmer contends. Others are the economic and political climate for agricultural production and the ability to adapt to it.

THE DEVELOPMENT OF AGRICULTURE IN BRITISH COLUMBIA

The above discussion was based on the development of a commercial agricultural industry. Nevertheless, the land has sustained people throughout the ages, and food has been abundant ever since the end of the last ice age. First Nations practised fishing, hunting, and gathering to support themselves, but agriculture was unknown in British Columbia (Dalichow 1972, 23; also see Chapter 5). The First Nations lived off the resources of the land without ever planting crops or domesticating animals.

The first Europeans in what is now British Columbia resided in fur trade forts and practised agriculture initially for subsistence, growing vegetables, planting a few fruit trees, and sometimes caring for cows and chickens. Horses were raised in the southern interior for transporting furs (see Chapter 4). By the 1840s, the Hudson's Bay Company attempted some commercial agriculture as well as promoting agricultural settlement. Neither was particularly successful. Perhaps the most important aspect of these limited and isolated agricultural endeavours is that

Kelowna, Cranbrook, and Kamloops, for example, are all in southern interior river valleys with little precipitation. The hot, dry, desert-like conditions of summer make irrigation systems a requirement for producing agricultural products. Cranbrook's higher elevation further limits agricultural conditions by engendering considerably fewer frost-free days, or growing degree days, than either Kelowna or Kamloops. Fort St John has more frost-free days than does Cranbrook and more precipitation

they built a base of information for the wave of settlers who came shortly after. Where there was a Hudson's Bay Company fort, there was information about the soil, which crops would grow and which would not, the limitations of water and climate, and the basics for a commercial agricultural industry.

The gold rush of 1858 brought a wave of gold seekers along with many who sought to earn a living providing food for the miners. Farms spread throughout southeastern Vancouver Island, the Fraser Valley, and, when gold was discovered there, north into the Cariboo. Ranchers followed gold settlement as well. Cattle drives, often originating in the United States, used the Okanagan-to-Kamloops route and the Thompson River system to bring meat to the Cariboo communities. Ranches also became popular through these areas. There were neither rules nor much concern about the number of cattle grazing on the natural grasses – mainly bunch grass – in these dry interior valleys. Serious overgrazing resulted. The newly built Cariboo Road, though, was important in facilitating the expansion of the agricultural industry. More wheat was grown, flour mills were built, the number of dairy farms increased, and more vegetables and other agricultural produce were grown to supply the remote interior.

The early colonial governments of Vancouver Island and British Columbia had a land policy that encouraged settlement on 160-acre parcels through a process of pre-emption, especially after 1861 (Harris 1997; Little 1996). Pre-emption was a system whereby "an occupant who improved a claim to the value of ten shillings ($2) per acre would receive a certificate of improvement, making the land eligible for sale or mortgaging" (Little 1996, 78). By this policy it was hoped that Europeans would be encouraged to settle; however, both Harris and Little note that many pre-emptions were for land speculation rather than agricultural development.

The decline of the gold rush by the late 1860s brought a decline in agriculture. Many farmers abandoned their land and others became subsistence farmers. Ranchers fared somewhat better because ranching was connected to distant markets by cattle drives (Dalichow 1972, 33).

Transportation became a major influence in commercial agriculture. The industry increased marginally until the building of the Canadian Pacific Railway (CPR) in the 1880s. The thousands of railway workers represented a market for beef and other agricultural produce, and the completed line provided connections to the Prairies and central Canada, while the port provided a connection to the world. The mining boom of the late 1880s and 1890s in the Kootenays resulted in more railway expansion, more people, and more demand for agricultural products. As transportation provided access to British Columbia's agricultural land and to markets, demand for agricultural land led to more land speculation and claims of the free 160-acre homestead available after 1875, just as it had across the Prairies (Demeritt 1995-6).

New farms and ranches became widespread in the southern portion of the province. The Kootenays, Okanagan, South Thompson River, Lower Fraser Valley, and southern Vancouver Island were popular areas. The 1891 total population was 98,173, of which some 22,000 were farmers on 6,500 farms (Internet, British Columbia, Ministry of Agriculture and Food). Agriculture was becoming an important industry in terms of employment, commodity export, and settlement. The mild winter climate, combined with an abundance of inexpensive fuel wood for energy, saw the development of a greenhouse industry in the Lower Mainland and southern Vancouver Island. Greenhouse vegetables could be transported easily to a Prairie market. The rapid growth of the Vancouver region provided its own market for agricultural products, and the Fraser Valley developed specializations in dairy products and market gardening.

Further railway construction, and even the speculation of railways to come, opened up other agricultural areas. Rumours that the Grand Trunk Pacific (absorbed into the Canadian National Railway, or CNR, in 1922) would terminate on the north end of Vancouver Island enticed a number of settlers, mostly Scandinavian, into Quatsino Sound, Cape Scott, Sointula, and farther up the coast to Bella Coola. The ultimate destination for the CNR was Prince Rupert, however, thus opening up the more northerly regions such as the Bulkley Valley for farming. The Peace River block of agricultural land became accessible on the Alberta side with the expansion of a railway from Edmonton by 1918. Extension of this railway to Dawson Creek by 1930 made the BC Peace River region somewhat accessible, but it was the extension of the Pacific Great Eastern (today the British

Table 12.4

Farms, farm size, population, and percentage urban for selected years

	Farms	Area (hectares)	Population	% Urban
1911	16,958	1,144,237[a]	392,480	50.9
1931	26,079	1,416,616	694,263	62.3
1951	26,406	1,880,909	1,165,210	68.6
1981	20,012	2,467,330	2,744,462	77.8
1991	19,225	2,392,300	3,282,061	80.4
1996	21,653	2,392,341[b]	3,724,500	82.0

Notes:
a 1921 census figure for area as 1911 figures are not available.
b 1991 census figure for area.
Sources: Census Canada, 1911, 1931, 1951, 1981, 1991, 1996.

Columbia Railway, or BCR) by 1958 that connected the region to Vancouver and provided much greater accessibility. In the Okanagan, Lower Mainland, and southern Vancouver Island, local railways were also built to facilitate agricultural production.

Over the years, many economic conditions changed. Small family farms and ranches became less competitive, giving way to large producers and corporate farms and ranches. New, capital-intensive technology – agribusiness – was taking over by the 1960s. Table 12.4 reflects early increases in the number of farms and in the overall population of the province, then decline in the number of farms from the 1950s to the early 1990s as farms grew larger and the population of the province mushroomed. More and more people in this period were living and working in urban environments. The surprising data from the 1996 census is that the number of farms (21,653) increased by 12.6 percent over 1991. When one classifies farms by income, however, it becomes apparent that the increase has been primarily in small farms. Some 11,400 farms show gross sales of less than $10,000 per farm. Most farmers in British Columbia today work part time.

Far fewer British Columbians are employed in agriculture today than historically. Just under 2 percent of the workforce is employed on farms, in greenhouses, nurseries, and other horticultural operations, and in veterinary offices, hatcheries, grooming, and other agriculture-related services (Internet, BC Stats: Quick Facts). Even so, the revenues from agriculture have

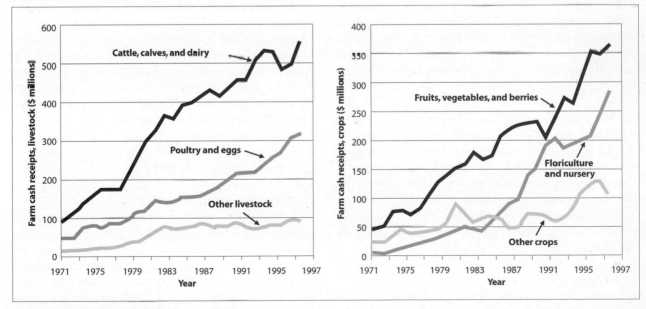

Figure 12.3 Value of agricultural products, 1971-96
Source: Data from Internet, BC Stats: Quick Facts – The Economy.

steadily increased (Figure 12.3). The variety of crops has also increased considerably since the 1970s.

As Chapter 7 described, time-space convergence occurs when changing transportation technologies shrink the time it takes for people, goods, or information to travel between places. After the CPR was built across Canada, for example, the trip from the east coast to the west took seven days, as opposed to weeks via the river systems and trails or sailing around South America. In a sense, the railway shrank the distance, or brought Halifax and Vancouver considerably closer together. The initial time-space convergence accomplished by wagon roads and railways was important to the expansion of agricultural settlements and agricultural production in British Columbia. The more recent time-space convergence of highway systems and refrigerated trucks, container ships, and jet air cargo carriers, however, have not always been so beneficial to agriculture in this province. These transportation technologies brought global competition. Vegetables and fruit grown in California, Mexico, New Zealand, and other countries are now common commodities at BC supermarkets.

Global competition has also been spurred by genetic engineering, large corporate investment and control of produce, changing consumer expectations and demands, and free-trade politics – a combination that has made agriculture in British Columbia a highly competitive industry. Both commodities and markets have changed drastically, and the challenge is often to develop new agricultural products and find new places to sell them.

The global movement of agricultural products and outside competition was initially met with provincial legislation in 1937. The Natural Products Marketing Act and subsequent acts created marketing boards that controlled the supply of agricultural produce coming into Canada and production both within and between the provinces. Marketing boards were intended to prevent overproduction leading to product price decline and farm bankruptcy. They are still in effect today. A milk quota, for example, was established in 1956, setting a limit on the number of gallons of fluid milk a farmer could legally produce. Because it is perishable, fluid milk is destined mainly for the local market. Industrial milk includes dairy products such as butter, cheese, yoghourt, ice cream, and powdered milk. These have a much longer

Table 12.5

Number of cows and farms in British Columbia for selected years

	Dairy cows (two years old +)	Dairy farms
1941	62,402	6,109
1951	51,819	6,526
1961	62,402	4,075
1962	57,000	n/a
1964	58,000	2,000
1996	n/a	890

Sources: Dalichow (1972), 53; Census Canada 1996.

shelf life, can be transported greater distances, and are subject to a federal quota system. British Columbia has a small part in the national quota of industrial milk. Quebec has claim to 47 percent of the market, Ontario has 31 percent, and British Columbia holds a paltry 7.6 percent (Forbes, Hughes, and Worley 1982, 109; Canada, Ministry of Supply and Services 1992, 22). The system has resulted in fewer cows and dairy farms, as can be seen in Table 12.5.

Marketing boards exist for producers of milk and dairy products, vegetables, fruit, wheat, and poultry. The last category includes chickens, turkeys, and eggs. Control of the supply side of agriculture, along with transportation subsidies for products such as wheat, have produced a rather artificial agricultural economy. The Free Trade Agreement (FTA) of 1989, the North American Free Trade Agreement (NAFTA) of 1994, and the 1995 global meetings of the General Agreement on Tariffs and Trade (GATT) have produced global economic pressures that seriously challenge our controlled system of agriculture production (Wilson 1990).

AGRICULTURAL LAND RESERVES
Agricultural land can be improved through drainage, adding nutrients to the soil, and a number of other measures designed to fit local conditions. The process is often slow and expensive, and in this sense agricultural land is a non-renewable resource. In other words, it is difficult to make more of it. Agricultural land is the base for producing agricultural products now and for the future, and it is therefore important to assess the value of

being able to feed the regional population versus depending on imported food. Once land is lost to other uses, options to produce more food are also lost (Rees 1993).

Population in British Columbia has steadily increased, albeit not evenly throughout the province. The Lower Fraser Valley, southern Vancouver Island, and the Okanagan have experienced some of the greatest population growth in the province since the Second World War. The trend has been accompanied by land development for transportation, industry, commercial functions, and housing. It has also meant the loss of agricultural land to construction. By the late 1960s and early 1970s, an estimated average of over 4,000 hectares of arable land per year were being converted to other uses (Farley 1979, 80). These losses were felt in the high growth regions such as the Lower Fraser Valley and Okanagan. The same land use conversion process was occurring in other regions, but more slowly. The newly elected NDP government in 1972 took the political action of "freezing" all agricultural land, and by 1973 Bill 42, the Land Commission Act, was passed, designed to preserve all lands assessed or zoned as farmland from class 1 to class 4 under the Canada Land Inventory. Initially, some 4.56 million hectares (11.4 million acres) of land came under the designation of Agricultural Land Reserve (ALR).

There was considerable opposition to the creation of the Agricultural Land Reserve. Developers, land speculators, local governments, and even farmers resisted reserving land for agricultural practices alone and appealed to have their land excluded. The Agricultural Land Commission (ALC) was the body responsible for appeals, but their mandate was to base decisions on the quality of the soil. As noted earlier, good soil can be scientifically and objectively defined through the Canada Land Inventory assessment.

The NDP were voted out in 1975, and the Social Credit were back in. By this time, the farmers and the public were largely in favour of the land reserve system, so Bill 42, which had created the ALRs, was kept on the books. Changes were made to the appeal process in 1977, making it easier but much more political to exclude land from the agricultural reserve. The new appeal process allowed appeals to go to the Environment and Land Use Committee (ELUC), and their mandate put priority on

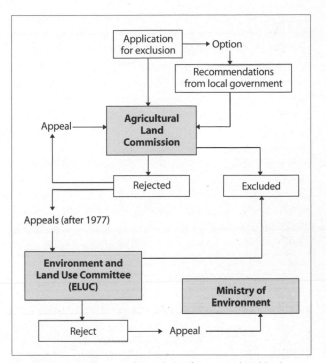

Figure 12.4 Appeal process for exclusion from agricultural land reserves
Source: Modified from Canada Works Project (1978), 5-6.

"best use" for the land in question. If an appeal to ELUC was unsuccessful, an applicant could appeal directly to the minister of environment (Figure 12.4). Many appeals resulted in ugly confrontations and headlines in the major newspapers. The 1989 decision by the provincial government to allow golf courses on agricultural land set off another round of charges of political manipulation and favouritism and the recognition that some of the best agricultural land was being lost. By 1991, the NDP was back in government and the original appeal process was re-instituted. The negative publicity and confrontations over individuals attempting to exclude their land from the reserve between 1975 and 1991 resulted in minimal exclusion of agricultural land.

Table 12.6 outlines the regional distribution of ALRs, including the changes each region has experienced since 1974. All areas have had more land excluded than included, with the exception of northern British Columbia. It is interesting to note that the northern region

Table 12.6

Changes in area of Agricultural Land Reserve by region, 1974-98

	ALR area		% net change (1998)	% of ALR
Region	1974 (ha)	1998 (ha)		
Vancouver Island	130,161.9	113,413.2	-12.9	2.0
Lower Mainland	181,821.8	168,043.1	-7.6	4.0
Okanagan-Shuswap	257,246.9	231,600.5	-10.0	5.0
Kootenay	399,109.3	385,686.9	-3.4	8.0
Central interior	1,494,210.6	1,489,178.0	-0.3	32.0
Northern British Columbia	2,258,744.8	2,321,753.9	2.7	49.0
Total	4,721,295.3	4,709,675.6	-0.2	100.0

Source: Internet, British Columbia, Ministry of Agriculture and Food, Tables A4 and A5.

contains nearly one-half of the agricultural land in the province. It is also interesting that the Vancouver Island, Lower Mainland, and Okanagan-Shuswap regions, which are extremely important agriculturally, have the highest exclusions but contain a very small land base.

British Columbia has taken one of the strongest stands of any province in Canada, and indeed most countries of the world, in protecting farmland. This bodes well for agricultural land and its production possibilities now and in the future. Some farmers have expressed concern, however, that agricultural land reserves save the land but not necessarily the farmer. Several programs have been implemented to address farm income, including crop insurance, a joint provincial-federal net income stabilization account program, and, in 1998, the whole farm insurance pilot program. Still, the farmer must contend with all the vagaries of the physical elements and the markets, along with marketing boards and global demands to remove agricultural subsidies. Perhaps it is no wonder that there are so many part-time farmers in British Columbia.

AGRICULTURAL COMMODITIES AND REGIONAL SPECIALIZATION

The geography of agriculture in British Columbia has been subject to many influences. The rugged, mountainous landscape, limited amounts of soil in classes 1 to 5,

and extensive variations in climatic conditions have meant that less than 5 percent of the land base is available for agriculture. On the human side of the scale, historical patterns of settlement, transportation and technology developments, and a host of economic and political changes have resulted in general agricultural specializations with regional associations.

Table 12.7 gives a breakdown, by value, of the main categories of the 199 agricultural commodities produced in 1996 and 1997. Livestock and livestock products are dominated by dairying (southern Vancouver Island, Lower Mainland, and Okanagan/Shuswap regions), cattle and calves (Cariboo/central region), and hens and chickens (Lower Mainland). Under crops, fruits (Okan-

Table 12.7

Agricultural production, 1996 and 1997

Agricultural product	1996		1997	
	Value ($ millions)	Percentage	Value ($ millions)	Percentage
Livestock and livestock products				
Dairy products	306	19.3	323	17.2
Hens and chickens	189	11.9	204	10.8
Cattle and calves	183	11.6	213	11.3
Eggs	91	5.7	86	4.6
Hogs	52	3.3	51	2.7
Turkeys	28	1.8	28	1.5
Other livestock	41	2.6	44	2.3
Subtotal	890	56.2	949	50.5
Crops				
Floriculture and nursery	213	13.4	265	14.1
Fruit	163	10.3	337	17.9
Vegetables	133	8.4	155	8.2
Potatoes	24	1.5	16	0.9
Forest products	32	2.0	32	1.7
Ginseng	26	1.6	41	2.2
Other crops	71	4.6	54	2.9
Subtotal	662	41.8	900	47.8
Other receipts	32	2.0	32	1.7
Total	1,584	100.0	1,881	100.0

Source: Internet, BC Stats: Quick Facts.

Through the census, British Columbia is divided into eight agricultural regions (Figure 12.5). Each has advantages and disadvantages in producing agricultural commodities.

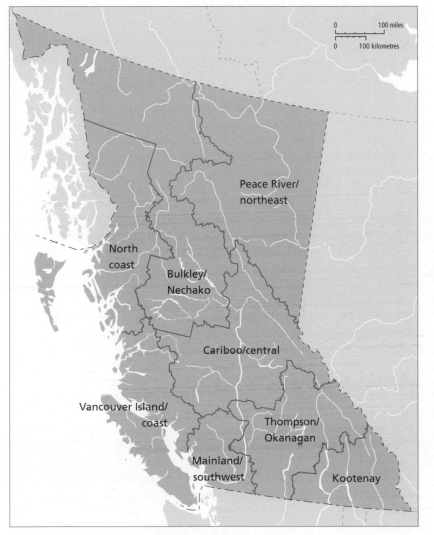

Figure 12.5 Agricultural regions of British Columbia
Source: Modified from Internet, Agricultural Profiles.

Vancouver Island/Coast Region

The Vancouver Island/coast region is accessible mainly by ferry, making transportation of agricultural products onto and off the island expensive. With a population in excess of one-half million, much of it concentrated on the southeastern end of the island, there is a substantial market as well as sufficient agricultural land for dairying and for growing vegetables, fruit, and a variety of speciality crops and livestock such as kiwifruit, organic vegetables, ostrich, and fallow deer. The southern end of the region has the mildest winter temperatures and the greatest number of growing days, which allows the 2,300, mostly small, farms to produce a wide range of crops. Greenhouse production of flowers and vegetables has become an important part of the industry. Nevertheless, there is still plenty of opportunity to increase local production, as most foods are imported.

The Mainland/Southwest Region

The greatest number of British Columbians live in this region, which includes the Lower Fraser Valley north to Lillooet. A wide variety of farming activities take place here: dairying, raising poultry and hogs, greenhouse production of flowers and vegetables (with a significant increase in organic vegetables), and nut and berry growing, especially cranberries. Many other speciality vegetable crops serve an ethnically diverse population. The combination of a relatively large market, ease of access, and a favourable growing season have made

agan) and floriculture and nursery production (Lower Mainland and southern Vancouver Island) made the most significant gains in 1997. Vegetables (Lower Mainland and southern Vancouver Island) have also been a valuable crop. Other agricultural concentrations have occurred, such as grain and canola production in the Peace River region, ginseng in the southern interior, grapes in the Okanagan, and berries in the Lower Mainland.

this the most important agricultural region in British Columbia in terms of value of goods produced and employment: 5,700 farms and over 15,000 people employed (Internet, BC Stats: Quick Facts). The price of farmland in this region is the highest in the province, and the conflicts between farmland and urbanization are great. The Right to Farm legislation passed in 1996 is intended to reduce these conflicts (British Columbia, Ministry of Agriculture, Fisheries, and Food 1996).

The Thompson/Okanagan Region

The Thompson/Okanagan region, with 5,800 farms, is arid and needs irrigation in order to produce crops. This has traditionally been cattle country and still is, primarily with cow and calf operations. The finishing-off stage of beef cattle ranching usually involves grain feeding cattle in feed lots, most of which are located in Alberta. Recently, cattle raising has been challenged by health concerns over the consumption of red meat, resulting in a decline in the demand for red meat. The Free Trade Agreement has offset this downward trend to a certain extent by giving Canadian ranchers greater access to the American market. A general increase in public awareness of healthy eating has provided opportunities for production of white meats with lower fat content, including chicken and pork, as well as of fish products and organic vegetables.

The Okanagan

The Okanagan is famous for its orchards, producing apples, pears, and soft fruits such as peaches, apricots, and cherries. It continues to produce more than 96 percent of the tree-fruit crops in the province (Internet, British Columbia, Ministry of Agriculture and Food). There has always been stiff competition in the fruit industry from American producers, who have access to cheaper labour. Recent strategies have seen the disappearance of the old, labour-intensive McIntosh and Delicious apple trees and the appearance of the newly developed Gala and Fuji apples. These varieties are planted thirty to sixty centimetres apart, 8,000 to 12,000 stems to the hectare, receive water and fertilizer by drip irrigation, and grow only two metres tall. With no need for ladders to pick the fruit, the labour hours are reduced. These new apple varieties are in high demand by both farmers and consumers. High technology has reached most forms of agriculture.

The Okanagan region produces dairy products and vegetables, and now two newer speciality crops have appeared on the market: wine grapes and ginseng. Ginseng, destined mainly for the Asian market, has taken off since the mid-1980s, with over 1,300 hectares of this high value crop in production in 1998. It takes four years to mature and is easy to recognize in the field by the black shade cloth necessary for protection from the intense summer sun.

Grape producers were one of the groups hardest hit by the Free Trade Agreement, as protectionism for locally grown grapes was dropped. The bad news is that approximately 50 percent of the Okanagan grape growers went out of business within several years following the signing of this agreement. The good news is that the provincial government allowed the importation of grape vines from around the world and introduced new regulations for production of wine from locally grown grapes. Two new categories of licence were introduced in the early 1990s – the Farm Gate Winery licence and the Estate Winery licence – allowing individual farmers to sell directly from their farms. The success of this industry is based on the new varieties of vines and a marketing system, the Vintner's Quality Association (VQA) labelling, which guarantees a quality wine. Many award-winning wines have been produced, building the reputation of the Okanagan region. With the new winery licences and a growing reputation for producing quality wines, this agricultural industry is firmly tied to the tourism industry. The Okanagan produces almost all the premium grapes in the province, making it the centre for this expanding industry (Internet, British Columbia, Ministry of Agriculture and Food).

The Kootenay Region

Agriculture is carried out mainly in the lowlands of the Columbia and Kootenay Rivers of the Kootenay region. This is a mountainous region with only 1,150 farms. Livestock and forage crops are the main commodities, along with some apples, poultry, honey, and Christmas tree production. With relatively few people in the region and fairly high transportation costs, commercial agriculture faces a number of economic constraints.

The Cariboo/Central Region

The Cariboo/central region includes much of the grassland and open forests of the Interior Plateau, which is ideal for grazing cattle. This region has been the centre for cattle ranching since the Cariboo gold rush of the early 1860s. The whole Cariboo/central region has been overgrazed, and in response range laws have been enacted and ranchers have turned to forage crops to overwinter their stock (Wood 1987, 150). Today this region is still devoted to ranching, producing one-third of the province's 800,000 beef cattle and calves (Internet, British Columbia, Ministry of Agriculture and Food). The 1,600 farms in this region have also diversified into some dairy and lamb production, along with potatoes, cabbages, turnips, cauliflower, and carrots. The exotic commodities produced here are fallow deer, bison, and ginseng.

The North Coast Region

The north coast region is wet and heavily treed, with the exception of inland areas such as Terrace. The forests must be cleared to practise farming, which makes it costly. The number of growing days is restricted this far north, and transportation to distant markets adds to the disadvantages of farming in this region. Few people live in this area and there are only 140 farms. The main agricultural product is cattle, although forage crops, eggs, and hogs are also produced.

The Bulkley/Nechako Region

Even fewer people live in the Bulkley/Nechako region, but it has some 800 farms. Again, ranching is the primary agricultural occupation, with a little production of cereal crops. On the one hand, the region has a short growing season, which puts a high cost on the winter feeding of cattle. Transportation is another high cost, along with the need to clear forest land. On the other hand, farmland is considerably less expensive than in the Lower Mainland region.

The Peace River/Northeast Region

The Peace River/northeast region is a prairie and its agriculture is dominated by grain. It has 1,700 farms and produces 85 percent of the provincial grain crop, as well as canola and cattle. Other products in this region are forage crops, some vegetables, bison, and honey. The short growing season limits the range of crops, but the rail link to Prince Rupert gives farmers some advantages in transportation costs. For new farmers starting up, the cost of land is less than in the rest of British Columbia. This region also has the largest block of Class 1, 2, and 3 land – the best land in the province.

SUMMARY

Farming is a balancing act between the many physical and human factors that affect the economic viability of producing agricultural commodities. Good agricultural land with few limitations is relatively rare in this province and confined mainly to valley bottoms. Some of this land is in the southern and western portions of the province, which have the physical advantage of a long growing season and mild winters, allowing for a greater range of agricultural commodities. The area also has the majority of the population, providing a ready market for products.

Transportation and time-space convergence have always played a key role in this industry, bringing to it the advantage of ready markets and the disadvantage of competition. Competition from outside British Columbia and overproduction from inside resulted in political decisions to control imports of agricultural products and in marketing boards to control the internal supply. Political decisions at the federal level – FTA, NAFTA, and GATT rulings – have introduced international and global competition and broken many of the safeguards of protectionism. In this process of change, old farming methods have given way to new ones and to new market opportunities, both local and international.

Political decisions in British Columbia have given serious consideration to the preservation of agricultural land, creating agricultural land reserves. Farming and non-farming uses for land still compete, particularly in the more heavily populated areas, but little agricultural land has been lost and Right to Farm legislation has reduced the conflicts.

As the "rules" of farming have changed, the number of agricultural commodities has increased and marketing has kept pace. The development of the wine industry and VQA labelling is a good example of these changes. Another development is direct marketing, which is popular in the Okanagan. Many farmers are establishing roadside outlets to sell their produce, along with value-added products such as preserves, dried fruits and vegetables, and even pastries.

Tastes have changed. Many people are concerned about their diet and about how the food they eat is produced. Production of white meat, fish, and organic vegetables has expanded to meet this demand, and consumption of red meat has declined, although production has not because the US market is available to Canadian producers.

A number of the 199 commodities produced in this province are exotic plants and animals. Llamas, snubnosed pigs, and herbs such as echinacea are being produced commercially, exemplifying the possibilities in farming today. Overall, this industry has developed with some remarkable regional strengths, and the number of farms is on the increase. With British Columbia still importing over half its food, there is plenty of opportunity for the agricultural industry to expand.

REFERENCES

Barker, M.L. 1977. *Natural Resources of British Columbia and the Yukon.* Vancouver: Douglas, David and Charles.

Berry, J. 1988. *Agriculture and Food in British Columbia.* Ottawa: Agriculture Canada.

British Columbia, Environment and Land Use Committee Secretariat. 1976. *Agricultural Land Capability in British Columbia.* Victoria: The Secretariat.

British Columbia, Ministry of Agriculture, Fisheries, and Food. 1996. *Good Neighbour Farming: B.C.'s New Farm Practices Protection (Right to Farm) Act.* Victoria: The Ministry.

Canada, Ministry of Supply and Services. 1992. *An Inquiry into the Allocation of Import Quotas.* Ottawa: The Ministry.

Canada Works Project. 1978. *Preserving Agricultural Land.* New Westminster, BC: Douglas College.

Dalichow, F. 1972. *Agricultural Geography of British Columbia.* Vancouver: Versatile.

Demeritt, D. 1995-6. "Visions of Agriculture in British Columbia." *BC Studies* no. 108 (Winter): 29-60.

Environment Canada. 1972. *The Canada Land Inventory: Soil Capability Classification for Agriculture,* Report no. 2. Ottawa: Ministry of the Environment.

Farley, A.L. 1979. *Atlas of British Columbia: People, Environment and Resource Use.* Vancouver: University of British Columbia Press.

Forbes, J.D., R.D. Hughes, and T.K. Worley. 1982. *Economic Intervention and Regulation in Canadian Agriculture.* Ottawa: Ministry of Supply and Services.

Greater Vancouver Regional District (GVRD). 1996. *Creating Our Future ... Steps to a More Livable Region.* Vancouver: Strategic Planning Department.

Harris, C. 1997. *The Resettlement of British Columbia: Essays on Colonialism and Geographical Change.* Vancouver: UBC Press.

Little, J.I. 1996. "The Foundations of Government." In *The Pacific Province: A History of British Columbia,* ed. H.J.M. Johnston, 68-96. Vancouver: Douglas and McIntyre.

Matthews, G.J., and R. Morrow, Jr. 1985. *Canada and the World: An Atlas Resource.* Toronto: Prentice-Hall.

Rees, W.E. 1993. *Why Preserve Agricultural Land?* Vancouver: UBC School of Community and Regional Planning.

Statistics Canada. 1997. *The Daily.* Catalogue 11-001E, 14 May, Ottawa.

Valentine, K.W.G., P.N. Sprout, T.E. Baker, and L.M. Lavkulich, eds. 1978. *The Soil Landscapes of British Columbia.* Victoria: Resource Branch, Ministry of Environment.

Wilson, B.K. 1990. *Farming the System.* Saskatoon: Western Producer Prairie Books.

Wood, C.J.B. 1987. "Agriculture." In *British Columbia: Its Resources and People,* ed. C.N. Forward, 138-161. Western Geographical Series vol. 22. Victoria: University of Victoria.

INTERNET

Agricultural Profiles, www.agf.gov.bc.ca/aboutind/fastfact/fastfact.htm

British Columbia, Federation of Agriculture, www.bcfa.bc.ca/bcfa

British Columbia, Ministry of Agriculture and Food, www.agf.gov.bc.ca

BC Stats: Quick Facts – The Economy, www.bcstats.gov.bc.ca/data/QF_econo.HTM

History of BC Agriculture, www.agf.gov.bc.ca/educate/history.htm

Water: An Essential Resource

13

Amap of British Columbia quickly reveals a province with many river and lakes. British Columbia has "access to more than one-third of Canada's runoff, which itself amounts to almost one-tenth of the world total" (Sewell 1987, 199). The province is fortunate to have such an abundance of freshwater supplies. The basic physical conditions of ground- and surface water are seen in the hydrologic cycle. Seasonal variation also plays a role, especially on the discharge rates of the river systems. These physical parameters determine water availability, including surpluses and deficits in different regions.

Water is both a renewable and a mobile resource that has many uses in society: drinking, transportation, irrigation, recreation, hydroelectricity, fish habitat, commercial and industrial uses, as a coolant, to dilute effluents, and as a place to discharge wastes. These can be categorized into uses that are compatible and those that are not. The various levels of government that manage this resource consequently often have difficult decisions to make. Over time more population, new industries, and new demands for water have placed some of our sources in jeopardy.

PHYSICAL PROPERTIES OF WATER

With one-third of Canada's supply of fresh water it appears that British Columbia has an abundance of this resource. Yet the water is not equally distributed, its quantity fluctuates seasonally, and its quality can vary.

The hydrologic cycle is the first step in understanding the physical parameters of water (Figure 13.1). Extreme climate conditions throughout British Columbia are responsible for a good deal of the spatial variation of water (Chapter 2). Westerlies bring moist, relatively warm air masses from the Pacific, which are then forced to rise over the mountain chains of Vancouver Island and the Coast Mountains. As the air rises, it cools, condenses, and releases moisture on the western slopes of the mountains. This process, as mentioned earlier, is referred to as the orographic effect. The farther inland and the greater the number of intervening mountain chains, the less influence is sustained from these Pacific air masses. Other

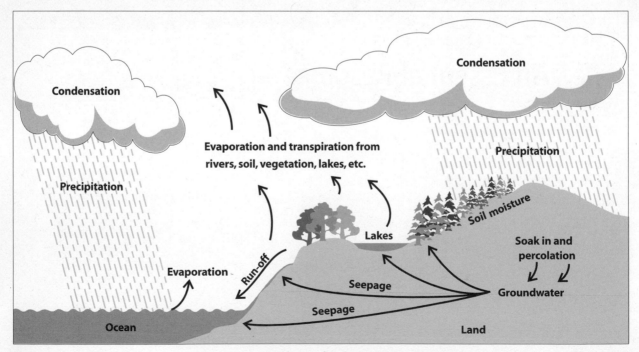

Figure 13.1 Hydrologic cycle

Table 13.1

Drainage area, length, and discharge of major rivers

River	Drainage area (km²)	Length (km)	Average annual discharge (m³/sec)
Fraser	231,312	1,368.5	3,600
Liard	143,418	483.0	1,300
Peace	128,451	281.7	1,400
Columbia	103,004	750.3	2,800
Skeena	54,488	579.6	1,000
Stikine	50,362	539.3	1,000
Nass	20,839	378.3	900

Source: Farley (1979), 38.

climatic influences vary the pattern of precipitation and account for record high rainfall in regions such as the west coast of Vancouver Island, but this contrasts with the extremely arid South Thompson and Okanagan regions.

Evaporation and transpiration – moisture given off to the air by plants – are important variables in the hydrologic cycle and act as the basis of the orographic effect. Once precipitation occurs over the land, surface water flows with gravity, meeting and merging with other streams in the watershed and often forming a large river. The watershed is the total geographic area where surface water is collected for any stream or river system. It is important to recognize that the Fraser River watershed is the largest in the province, with over 230,000 square kilometres – approximately one-quarter of British Columbia (Table 13.1). Other large river systems include the Columbia, Skeena, Nass, and Stikine, all of which flow to the Pacific. The Peace and Liard River systems flow into the Mackenzie and to the Arctic. Vancouver Island,

the Queen Charlottes, and much of coastal British Columbia have numerous rivers and streams but much smaller watersheds.

Figure 13.1 also shows groundwater as part of the hydrologic cycle and often as an important source of fresh water. Approximately 25 percent of British Columbians rely on groundwater as a source of water (Internet, British Columbia, Ministry of Environment, Lands, and Parks, "Groundwater"). Groundwater results from surface water percolating through the permeable soil of the earth to find its way to aquifers and eventually back to the ocean, completing the cycle.

Seasonal cycles influence this model in many ways. Transpiration is considerably higher in summer, when more vegetation is growing and incoming solar radiation, or insolation, is much more intense. The main seasonal influence on surface run-off is the amount of

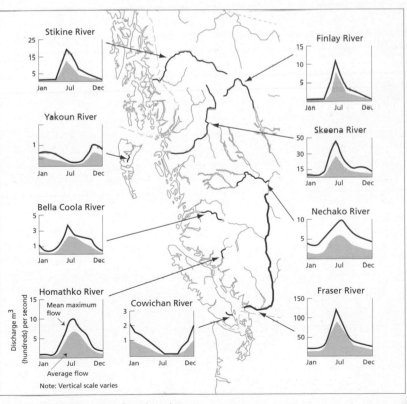

Figure 13.2 Discharge rates for selected rivers
Source: Modified from Farley (1979), 39.

precipitation, the accumulation of snowpack, and spring weather conditions that affect the rate of thaw. Seasonal variability of river systems is measured as the rate of discharge (Table 13.1, Figure 13.2). Essentially, two patterns emerge for British Columbia's rivers. First, the large rivers drain the interior, running either to the Pacific or to the Arctic, build up water in the form of snow, and discharge these stored volumes in the spring. Second, in coastal areas smaller river systems have considerably less water stored as snow, and their discharge peaks when the greatest amount of rainfall exists. The coastal and interior rivers peak at different times of the year, yet both types of system raise the concern of flooding (see Chapter 3).

One way to reduce the risk of flooding is by changing the seasonal pattern of discharge rates, which can be accomplished by building a dam. A dam impounds water, creating a reservoir of stored water and thus creating the ability to reduce the run-off at peak times and increase the flow at low water times. Despite these benefits, there are disadvantages to damming and impounding water.

Evapotranspiration ratios, with all their lines and labels, appear rather complex at first glance, although each line in itself is fairly straightforward (Figure 13.3). The precipitation line for a coastal region such as Vancouver is relatively high in winter and much lower in summer (Figure 13.3a). Conversely, the line of potential evapotranspiration, which graphs the theoretical relationship between the growing season, incoming solar radiation, and the ability for moisture to be given off to the atmosphere, has the reverse curve (Figure 13.3b). The line shows the volume of water that could be evaporated and transpired if water were available. The centre of Figure 13.3c shows, below the line of potential evapotranspiration, the line of actual evapotranspiration. Typically this is a considerable amount in the late spring and

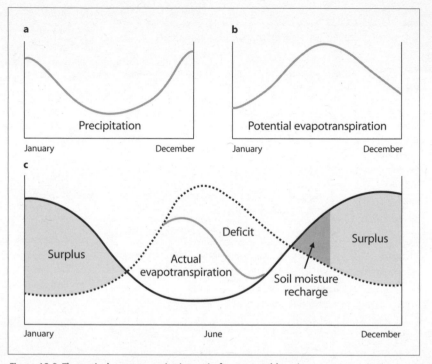

Figure 13.3 Theoretical evapotranspiration ratio for a coastal location

early summer, but by August there is very little moisture left and the gap between actual and potential evapotranspiration widens. By combining the graphs of annual precipitation, potential evapotranspiration, and actual evaporation, the periods of water surplus and deficit can be assessed (Figure 13.3c).

Following the model of Figure 13.3c for each month of the year, it is evident that the January/February rainy season has a surplus of water, and one can infer the need for storm drains at that time. As precipitation decreases plants grow, and as solar energy increases, greater evaporation and transpiration occur. By June, the potential for evapotranspiration outstrips available moisture and a deficit occurs. This deficit increases throughout the summer, requiring the irrigation of farms and the watering of gardens. By autumn, the leaves are falling, evapotranspiration has subsided with the decrease in incoming solar radiation, and precipitation is increasing. Soil moisture recharge is the term to describe the state of the soil. Using the analogy of the soil as a sponge, the summer months cause the sponge to dry out, whereas

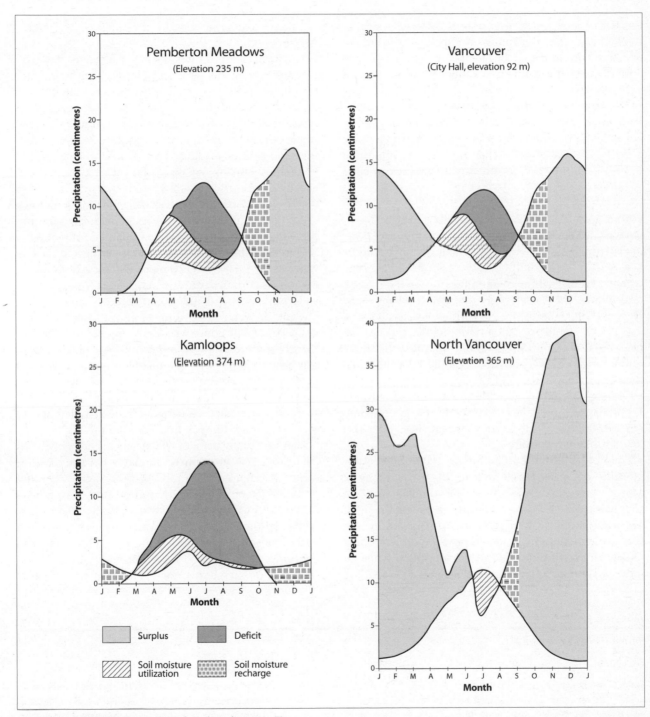

Figure 13.4 Evapotranspiration ratios for selected communities
Source: Canada, Department of Environment (1972).

the increase in precipitation and decrease in evapo-transpiration in the fall allows the sponge to absorb, or recharge, with moisture. By late fall, the soil is saturated and any further precipitation cannot be absorbed; it is surplus.

The evapotranspiration ratios for Pemberton, Kamloops, Vancouver, and North Vancouver illustrate very different water budgets and concerns about water supply (Figure 13.4). These physical characteristics are important to keep in mind when considering how water is used as a resource.

WATER AS A RESOURCE: USING IT AND ABUSING IT

Understanding the quality, quantity, and timing of flows for any water system is crucial. Water quality has to be assessed in terms of the many uses of the resource, whether for fish habitat, swimming, industrial purposes, or drinking. *The Guidelines for Canadian Drinking Water Quality* (Canada, Ministry of National Health and Welfare 1993) establish a standard for potable water, or water that is safe to drink. It should be noted that these standards can change, usually because of new information about toxicity levels of various elements in the water system. Arsenic, for example, often from groundwater supplies, was once believed to be safe at 0.05 milligram per litre, but that figure has now been reduced to half, at 0.025 milligram per litre.

Quantity is also important for any water source. Having sufficient water for all users during the year may be difficult, particularly when the source is limited. If a water supply is required for domestic uses including fire fighting, for example, as well as for industry, irrigation, and fish habitat, some uses may have to be curbed. Some regions of British Columbia are growing rapidly, and an increased population means more demand for water. The question of quantity may limit residential and industrial growth.

The third factor, timing of flows, recognizes that river and stream systems have variable discharge rates. Unfortunately, water use also tends to vary seasonally, and the highest use periods are opposite to highest supply. A typical example is the need in most communities to put watering restrictions in place for gardens and lawns during the summer months. This is precisely the period when most stream systems, as Figure 13.2 showed, are at their lowest discharge rate. Human activities such as logging in watersheds can also affect the timing of flows. Clearcut logging practices may result in much faster runoff, with even less water being available during the dry season. By impounding water and releasing it in the low water season, dams also change the timing of flows.

Table 13.2 categorizes the major uses of water, all of which affect water quality, quantity, and timing of flows. The withdrawal of water is increasing to meet the multiple needs of a province with a growing population. Domestic, commercial, and industrial uses, ranging from carwashes to the enormous quantities of water required by pulpmills, require prodigious daily withdrawals of water. This water is used and then returned in an altered state. In some cases, effluents have been added and in others the water temperature has been raised.

There are many uses for which water is not only withdrawn but also consumed. The largest consuming industry of water is agriculture. Irrigation removes water from surface or groundwater sources and does not return it to the same source. Of course, it still remains part of the hydrologic cycle as this water then evaporates and transpirates into the atmosphere. The resource is also consumed as potable water for drinking, in commercial products such as beverages, and in many industrial

Table 13.2

Categories of water use

Withdrawal	Consumed	In situ	In stream	Inadvertent
Water is withdrawn, altered, and returned. e.g., pulpmills	Water is withdrawn and not put back directly e.g., agriculture	Water is used in its natural state e.g., recreation	Water is used while remaining in river systems e.g., hydro dams	Water quality is affected negatively and accidentally e.g., leachate

Sources: Foster and Sewell (1981); Sewell (1987).

processes. For some municipalities and rural areas, limited potable water can restrict population growth.

In situ uses of water include all the values we have for water systems in their natural state. Recreational activities such as swimming, boating, and white-water rafting, to name a few, are important to local populations and to the tourism industry. The river systems leading to the Pacific are essential to the salmon fishing industry because freshwater lakes and streams provide the habitat for the beginning and the end of the salmon life cycle. A number of freshwater species of fish and animals, which are important to the recreation and tourism industry, require water in its natural, unaltered state.

The major in-stream use of water in British Columbia is the generation of hydroelectricity. The many rivers and streams permit a great deal of production, although most of our present hydroelectricity comes from dams on the Peace and Columbia River systems. Once a dam is built, the environment is considerably changed. The discharge rate of the river is modified below the dam, reducing high peak flows of the spring and increasing low discharge volumes of summer and fall. This may dry up downstream wetlands and deltas and thus affect the habitat of fish and wildlife. The W.A.C. Bennett Dam had just this impact on the Peace-Athabasca delta. Impounding water behind a dam creates a reservoir, in some cases so large that it affects the climate of the region. Williston Lake, the result of the W.A.C. Bennett Dam, is the largest lake in British Columbia and modifies the climate of the surrounding area. Furthermore, dams and lakes/reservoirs can conflict with other resource uses such as fishing, forestry, hunting, trapping, and farming.

Figure 13.5 North American Water and Power Alliance Plan (NAWAPA)
Source: Modified from Foster and Sewell (1981), 33.

Dams do not change the overall amount of water discharged in a river system unless one river system is diverted into another. The Kenny Dam on the Nechako River, as discussed earlier, diverts the flow of water from the Fraser River system and redirects it through a tunnel in the Coast Mountains to the Kemano hydroelectric plant, which supplies the aluminum plant at Kitimat (see Figure 11.10, p. 159). This dam, referred to as the Kemano 1 Project, was not without its controversy. It reduced salmon stocks on the Nechako, displaced people (the Cheslatta Band, particularly) without adequate compensation, and reduced the amount of water discharged to the rest of the Fraser, again affecting salmon habitat.

Other water diversion schemes have been proposed, such as the diversion of the McGregor River into the Parsnip River/Williston Lake/Peace River system and of the North Thompson River into the Columbia River system. Both these plans would increase the volume of water available to existing hydroelectric generators and thereby increase the production of electricity at a more economical rate. Again, these plans raise some serious environmental concerns.

By far the most ambitious and controversial proposal has been the North American Water and Power Alliance (NAWAPA), wherein diversions were proposed from Alaska, the Yukon, and British Columbia to quench the thirst of California, the American southwest, and Mexico (Figure 13.5). This massive scheme would have required the construction of 240 reservoirs, 112 irrigation systems, and 17 navigation channels (Foster and Sewell 1981, 31). The economic, physical, and environmental impacts would have been enormous for this megaproject of the early 1960s. This grand plan was put on the shelf, but smaller diversion projects involving part of the NAWAPA plan were carried out. Most notable was the Columbia River Treaty (1961), by which a number of dams were constructed on the BC side of the border to provide Americans with hydroelectric energy and flood control. More recently, the Free Trade Agreement of 1989 has raised concerns over the export of fresh water and further water diversion schemes.

The last category shown in Table 13.2, inadvertent, includes all the changes to water systems that alter quality. Historically, river systems were a powerful influence for settlement because they provided a ready source of drinking water, transportation, irrigation, and waste disposal. As communities grew and their agricultural and industrial base expanded, liquid wastes, solid wastes, industrial effluent, toxic wastes, farming chemicals, and so forth were increasingly discharged into the ground- and surface water systems. Road building, clearcut forestry practices, mining activities, overgrazing, and other activities within the watersheds have also had a negative impact on water systems, often changing the timing of flows. The most damaging effect of all has been an increase in the amount of sediment in stream systems. Sediment can be devastating to fish habitat, as it settles on the stream bottom and smothers the fish eggs there.

Sediment is also the main cause of turbidity, one of the major problems of potable water. Turbidity is the result of the fine sediments in water systems that are almost impossible to filter. Turbid water then flows through the water mains of communities, settling and coating the pipes. This environment can then become home for undesirable bacterial micro-organisms. To combat these bacteria, most communities add chlorine to their water systems. Ever increasing bottled water sales are evidence of water quality concerns.

Surface water systems have gone through an inordinate number of changes as a result of combined human and natural activity, reducing the options for water use along those systems from source to mouth. Groundwater is somewhat more protected, but even here leachates from agricultural activities – fertilizers especially – have rendered some supplies unusable without prior boiling. The level of nitrates is known to be increasing in areas such as the Lower Mainland, Osoyoos, and Grand Forks, a state of affairs normally caused by inadequate or badly managed septic systems, animal feed lots, manure storage facilities, and overuse of nitrogen-rich fertilizers (British Columbia Round Table 1990, 29).

Sewage, liquid wastes that are flushed into our water systems, is another controversial issue for most communities. Processes to purify sewage include primary, secondary, and tertiary treatment. Each stage removes greater amounts of materials and chemicals before re-entering the water system, but each is costly to implement. Few communities opt for tertiary treatment.

Several reports have been issued on the state of sewage discharge for major Canadian cities, outlining specific concerns. *The National Sewage Report Card* (Bonner and McDade 1994) included Victoria and Vancouver in its study. Victoria received a failing grade because the city has almost no sewage treatment prior to discharge into the Juan de Fuca Strait; Vancouver fared little better, with a D minus. More recently, waste management plans to improve sewage treatment have been completed and construction for new facilities is under way in both cities. The Fraser Basin Management Program (1995), as the name suggests, reviewed the many variables affecting the whole Fraser system. Some sectors, such as the pulpmill industry, received a B grade for reducing their effluent, while most municipalities along the Fraser

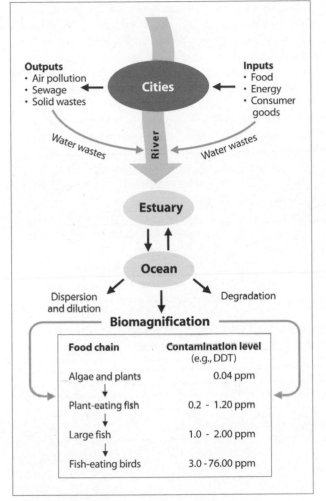

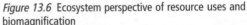

Figure 13.6 Ecosystem perspective of resource uses and biomagnification
Source: Modified from Tyler Miller (1975), 16-17; Raven, Berg, and Johnson (1993), 504.

and its tributaries received an F grade for lack of sewage discharge treatment. At least the process of understanding the problems and planning for them has begun.

Figure 13.6 gives an ecosystem perspective of water use and abuse. Each community and resource activity affects the next in line downstream, often resulting in huge costs for filtration and purification. The concept of **biomagnification** suggests that very small parts per million of toxic wastes are ingested at the lowest level of the food chain; as the next level up consumes the lowest level organisms, the toxicity builds. With people at the top of the food chain, great care has to be taken over the food we eat. One can also question whether the Canadian Drinking Water Quality Standards should tolerate any level of toxic chemicals in our drinking water.

SUMMARY

The historical view of water resources regarded them as inexhaustible, as having little or no value in economic decisions, and often as a means of disposing of wastes. Our overuse and abuse of water has brought home the reality that it not only has a value but is essential to our economic and physical well-being. With many uses for many interest groups, this mobile, renewable resource is not easy to manage.

The federal government has the responsibility for oceans and for any changes to freshwater systems crossing international boundaries. It also has interests and obligations with respect to water systems: flood control, salmon habitat, hatcheries, Aboriginal issues, and so forth. The provincial government manages many aspects of the freshwater systems as well and is responsible for regulating withdrawals, consumption, and quality of water, mainly through the Ministry of Environment, Lands, and Parks. Ministry responsibilities overlap with those of other managers, such as the Ministry of Forests and its Forestry Practices Code and local governments as purveyors of water for their communities. Water issues are complex, as they involve the quantity, quality, and timing of flows of surface water and quantity and quality of groundwater.

The multiministerial approach to water management – often with conflicting interests – should be phased out in favour of a Quality of Water Act that puts the interests of water quantity, quality, and timing of flows as the top priority at the provincial level. As population increases and as all the uses for water increase, the need for planning and action becomes more urgent.

REFERENCES

Bonner, M., and G. McDade. 1994. *The National Sewage Report Card.* Vancouver: Sierra Legal Defence Fund.

British Columbia Round Table. 1990. *Towards Sustainable Water Planning and Management in British Columbia.* Victoria: British Columbia Round Table on the Environment and Economy.

Canada, Department of Environment. 1972. *Canada Water Yearbook.* Ottawa: The Ministry.

Canada, Ministry of National Health and Welfare. 1993. *Guidelines for Canadian Drinking Water Quality,* 5th ed. Ottawa: The Ministry.

Farley, A.L. 1979. *Atlas of British Columbia: People, Environment, and Resource Use.* Vancouver: University of British Columbia Press.

Foster, H.D., and W.R.D. Sewell. 1981. *Water: The Emerging Crisis in Canada.* Toronto: James Lorimer.

Fraser Basin Management Program. 1995. *Board Report Card: Assessing Progress towards Sustainability in the Fraser Basin.* Vancouver: Fraser River Management Board.

Raven, P.H., L.H. Berg, and G.B. Johnson. 1993. *Environment.* Orlando, FL: Saunders College Publishing.

Sewell, W.R.D. 1987. "Water Resources." In *British Columbia: Its Resources and People,* ed. C.N. Forward, 198-225. Western Geographical Series vol. 22. Victoria: University of Victoria.

Tyler Miller Jr., G. 1975. *Living in the Environment: Concepts, Problems, and Alternatives.* Belmont, CA: Wadsworth.

INTERNET

British Columbia, Ministry of Environment, Lands, and Parks, "Lands and Parks," www.env.gov.bc.ca

British Columbia, Ministry of Environment, Lands, and Parks, "Groundwater," www.env.gov.bc.ca:80/sppl/soerpt/water/gndwater.html

British Columbia, Ministry of Environment, Lands, and Parks, "Water Quality," www.elp.gov.bc.ca:80/wat/wq/wqhome.html

Tourism: A New and Dynamic Industry

14

Tourism throughout the world has escalated at a remarkable pace over the past forty years and brought with it a great deal of revenue and opportunity (Table 14.1). A major recession in Canada from 1981 to 1986 seriously affected traditional resource-based industries but tourism not only maintained its share of economic activity, it increased it. Canada designated 1984 the year of tourism, a declaration that symbolized the growing importance of the industry.

Expo 86 became the catalyst for a whole new awareness of tourism and its potential in British Columbia. Tourists came from all over the world and from the rest of Canada to participate at the fair and to experience the BC landscape. The exposition achieved two major objectives: spreading the word on the many tourist attractions in the province, and having tourists return with their friends and relatives.

Defining tourists and tourism is difficult; the industry is complex, as it involves both private business and all levels of government. Like other industries in British Columbia, tourism is affected by events ranging from the global (e.g., the Asian recession) to the local (e.g., the closure of salmon fishing areas). Adding to the complexity is the conflict over land use between tourism and the traditional resource extraction industries. The need for planning is essential, as this new and expanding industry requires a great deal of co-ordination between sectors of government and business.

Tourism has been important to the economic growth and diversification of British Columbia, especially as traditional resources have been depleted. This dynamic serv-

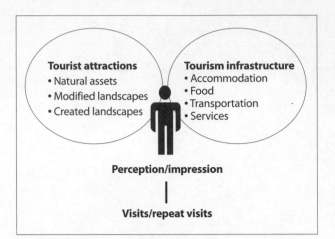

Figure 14.1 Relation of tourism industry to resources

ice industry is now the second largest resource industry in the province. In 1997, $8.5 billion was spent by 21 million overnight visitors (Internet, BC Stats: Quick Facts). Moreover, in 1996 tourism and tourism-related jobs employed 235,300, and over 15,700 businesses were involved in the sector (Internet, British Columbia, Ministry of Small Business, Tourism, and Culture). Nevertheless, the benefits of tourism have not been spread evenly over the nine tourist regions of the province.

TOURISM AS A RESOURCE INDUSTRY

Tourism as a resource industry is considerably different from the traditional extractive resources of forestry, mining, and fishing. Wood, minerals, and fish are harvested, often undergo some form of processing, and then for the most part are exported from British Columbia to gain income for the industry and the province. Tourism reverses this process by enticing tourists to come into the regions of the province.

British Columbia has a great variety of tourist attractions. The rugged topography and wilderness of both the coastal and the interior landscapes and the many recreational activities associated with them throughout the year are strong drawing cards. The combination of topography and climatic diversity enhances recreational opportunities and gives rise to a mix of unique flora and fauna species that attracts tourists. As well as the natural features, which are frequently promoted as "supernatural BC," many arts and cultural resources, heritage sites,

Table 14.1

International tourism estimates, globally, 1950-94

	Total international visits (millions)	Receipts from international tourism (US$ billions)
1950	25.3	2.1
1960	69.3	6.9
1970	165.8	17.9
1980	287.8	103.5
1990	455.9	261.0
1994	531.4	335.8

Source: British Columbia, Ministry of Tourism (1994), 2.

and sports events attract visitors. The industry also includes facilities and hospitality – not just how tourists are treated but the whole tourist infrastructure of accommodations and eating establishments. Figure 14.1 outlines many of the factors that attract tourists to a location.

The model categorizes tourist attractions into natural assets, modified landscapes, and created landscapes, each of which can attract tourists. The many natural attractions include scenic ocean landscapes, wildlife, and the rugged Cordilleran alpine landscape, to name a few. The modified landscape describes natural settings that have been modified for human use, such as swimming beaches and ski slopes. Created landscapes, or as Philip Dearden (1983, 78) calls them, "anthropocentric attractions," are created or built environments such as Butchart Gardens, Provincial Museum, the Royal British Columbia Museum, rodeos, First Nations centres, and industrial parks. Complementing these attractions is the essential tourism infrastructure: where tourists spend the night, where they eat, how they travel, and how they acquire services. The combination of all these experiences creates an impression on the tourist, and this impression, or perception, is essential to return visits, which in turn generate income and economic well-being for the region and the province as a whole.

To define a tourist one has to decide the distance a person must travel to qualify as one. How long does a person have to stay away from home, and at what point in an extensive stay does one cease to be a tourist? Is there a link between tourism and travelling because of work or business? Many individuals and government bodies have used definitions in an attempt to quantify the number of tourists and understand their impact on any region (see Mathieson and Wall 1982; Cooper et al. 1993; Internet, British Columbia, Ministry of Small Business, Tourism, and Culture). Peter Murphy uses precise distances and makes the definition specific to British Columbia, describing as tourism, "the travel, the activities, and services used by any British Columbia resident beyond a 40 kilometre (25 mile) radius from home for the purposes of personal enjoyment and the travel, the activities, and services used by non-residents who enter the province for any reason other than work. It is recognized that elements of tourism are often involved when people travel for business purposes" (1987, 402).

Although it is difficult and somewhat arbitrary to keep statistics about distances travelled, the process succeeds in identifying the range of places from where tourists come. The recognition that work and tourism may be related is especially relevant as more and more business is done in convention centres at locations such as Whistler, where tourist activities are part of the business package. Putting a time limit of one year for international tourists and six months for internal tourists, as the Ministry of Small Business, Tourism, and Culture does, may also be arbitrary, but it does recognize the number of tourists who see British Columbia as a safe, relaxing, and enjoyable destination for extended periods (Internet, British Columbia, Ministry of Small Business, Tourism, and Culture).

Tourism and recreation are closely related (as is the concept of leisure, since people require free time to engage in either recreation or tourism). Assessing tourism is rather complex, as it involves collecting information on a host of interrelated components, among them travel from communities, destination(s), activities, facilities used, and the creation of a positive impression on tourists in order to attract greater numbers in the future. Studying recreation, on the other hand, involves a somewhat narrower focus, namely, recording people engaged in recreational activities or events in particular locations. In British Columbia, recreational activities are often categorized in terms of either geographic location or of infrastructure requirements such as indoor versus outdoor settings. Marine-based recreation, for example, encompasses the many activities, from scuba diving to swimming or wind surfing, that take place in the diverse marine environment.

Tourism is a multifaceted industry that relies on all levels of government and involves a multitude of private interests. The federal and provincial governments are fundamental to the industry. They develop and maintain transportation systems (highways, airports, ferry systems, and ports) and provincial and national parks, and establish the policies and laws that affect the private sector. In 1996, for example, 518 provincial parks (totalling 8 million hectares) attracted 25 million park visitors, who used 11,000 campsites, 400 day use areas, and 114 boat ramps. As well, 139 ecological reserves covered another 159,500 hectares of provincial land

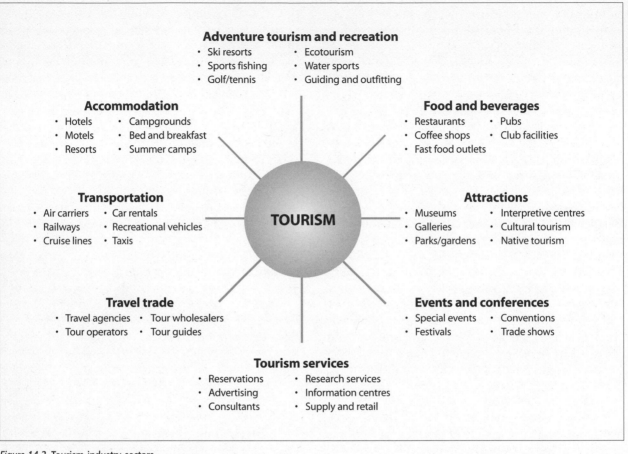

Figure 14.2 Tourism industry sectors
Source: Modified from Canadian Regional Forecasting (1991), 4.

(Internet, BC Stats: Quick Facts). The provincial government created a crown corporation called Tourism British Columbia to manage tourism in 1997, raising revenues through a hotel room tax and spending the money to promote the development and growth of the industry (British Columbia, Ministry of Finance and Corporate Relations 1998, 229).

Local governments play a crucial role in the tourism industry because they are responsible for zoning much of the tourist infrastructure and using the local tax base to build facilities for both local people and tourists. Municipalities and regional districts sponsor the development of a great number of facilities and activities that attract tourists: historical sites, museums, theatres, entertainment centres, parks, recreation complexes, and

so on. Moreover, tourism has added a degree of much needed economic diversity to communities that depend largely on one resource, such as forests or minerals, providing another way for people to make a living. For some towns, such as Chemainus and Tofino, the conversion from a forest-based economy to a tourist-based one has been profound.

The private sector contains the full range of tourist businesses, from large multinational corporations to many small, family-oriented businesses. Figure 14.2 gives an overview of the many components of the private sector, which generates the most tourism employment and investment. BC ski resorts, for example, had received more than $2 billion in investment by 1999. Sixty ski areas throughout the province employed more than

8,500 people. Also impressive, ecotourism generated $892 million in 1997 (Internet, Ministry of Small Business, Tourism, and Culture).

Co-ordination and planning are essential among the many levels of governments and thousands of private tourist-related businesses. A major area of conflict has been the traditional resource harvesting of fish, minerals, and forests. Overharvesting of salmon has resulted in sport fishing closures and restrictions on fishing for certain salmon species. The loss of employment and revenues to fishing lodges, fishing charters, restaurants, and coastal communities generally are obvious. Less obvious though still serious are conflicts over forest clearcutting and mining operations. The provincial government's move to increase parks and protected areas is one sign that it is attempting to preserve the natural setting and to enhance tourist values.

The issue of capacity also generates conflict and demonstrates the need for planning. Tourism is often referred to as a "clean" industry in the sense that tourists come to look or to enjoy recreation, spending money in the process and then leaving. This view of tourism – that it does not alter the environment – is not entirely accurate. Tourism and tourist-related facilities and activities do change the landscape, and in some regions it may be better to establish thresholds for tourism. When a community begins to erect many tourist attractions – such as billboards, amusement parks, water slides, and restaurants – local residents may not be content with the noise, excess traffic, or the aesthetics of the billboard landscape. Similarly, as a region attracts too many tourists, causing delays on roads and ferries, the regular users may become annoyed. Tourist overuse of provincial parks, marine parks, scuba-diving sites, and the many other destinations in the public domain can have a negative impact on the environment, making tourism less than a clean industry.

TOFINO: THE EVOLUTION OF A TOURIST LANDSCAPE

The transition from a resource extraction-based landscape to a tourist-based landscape is not without conflicts. Tofino and the adjacent Clayoquot Sound represent many of the struggles and changes to a specific landscape (Figure 14.3).

Table 14.2 summarizes the many changes to the area from pre-European times to the 1990s. The trajectory shows first the transition from traditional First Nations use of the land to the imposition of new landscape values based on resource extraction. The Nuu'chah'nulth were deterritorialized and their land was taken over by fur traders, missionaries, prospectors, fishers, foresters, and settlers. Throughout the fairly long period from exploration to corporate hegemony, the Tofino/Clayoquot region was portrayed as a rather exotic wilderness, appealing to the adventurous traveller. Both travellers and tourists became considerably more numerous once forest roads, built by the 1970s, gave access to Long Beach. The real struggle over use and development of this landscape evolved in the 1980s and 1990s with improved forest technology of clearcut logging.

The institutional structure and history of logging interests, including the establishment of long-term forest tenure for large corporate entities such as MacMillan Bloedel and Fletcher Challenge, was to come into direct

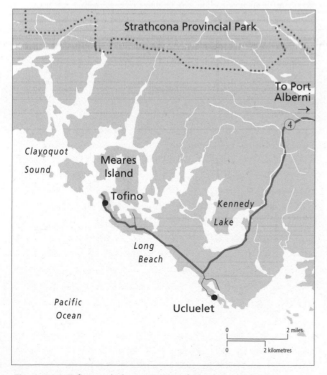

Figure 14.3 Tofino and Clayoquot Sound, Vancouver Island

Table 14.2

Overview of factors influencing the Clayoquot landscape

	To 1770s	1770s-1820s	1820s-1880s	1880s-1920s	1920s-1980s	1980s-1990s
Phase	Traditional precontact	Exploration, fur trade	Strategic control of trade	Settlement of frontier	Corporate hegemony	Environmental amenity
Authors	Elders, shamans chiefs	Explorers, naval officers, furtraders	Traders, missionaries, government agents, travellers	Land developers, government agents, travellers	Corporations, government public relations, travel writers	Environmentalists, tourism operators, Natives
Landscape images	Mythic, totemic images	Native life, landscape resources	Frontier posts, wild coast, scenic beauty, mining	Land clearance, settlements, mining, canneries	Canneries, "working forests,"[a] fish farms, scenic beauty	Scenic beauty, clearcuts, wildlife
Landscape impacts	Use of forest resources, shoreline fish traps, seasonal settlements	Trading posts, cutting of spars, extirpation of sea otters	Fortified villages, missions, trading posts, destruction of fish traps	Land clearance, settlements, mines, canneries	Clearcuts, logging roads, sawmills, mines	Viewscape protection, new forest practices, parks
Institutional arrangements	Defence of hereditary ownership	British defence of contested trade hegemony	Timber grants, Indian reserves, pre-emptions, mining claims	Pre-emptions, resource sales, transportation infrastructure	Timber licences, tree farms, integrated resource management	Landscape management, parks, ecosystem management, co-management

Notes:
a   Selective logging and small clearcuts to designated areas for large corporations under tenures such as tree farm licences.
Source: Modified from White (1999), 79. Used with permission.

confrontation with a host of new values. A showdown flared between those who wished to clearcut Meares Island and First Nations groups making Aboriginal title claims, along with environmental groups that attempted to stop logging, some by spiking trees. A court injunction in 1985 stopped logging on Meares Island. For the rest of Clayoquot Sound, the battle to preserve environmental and tourist values waged on through a 1991 change in provincial government and a number of government-sponsored round table discussions in an attempt to reach consensus among competing land uses. By 1993, Clayoquot had become an environmental icon of international status, made popular by international figures such as Robert Kennedy Jr. and international rock groups such as Midnight Oil from Australia. Over 800

protesters were arrested in the summer of 1993. The largest mass arrest in Canadian history not only provided publicity but also gave a new signification for the landscape: "This was a place for which ordinary people were prepared to go to jail. It was also giving Canada a bad reputation internationally ('Brazil of the North'), leading to forest product boycotts and international condemnation" (White 1999, 210).

Brian White documents well the various steps involved in the transition from an industrial extractive landscape to one recognized for its many touristic values. By 1999, a new forest company, Iisaak – a joint venture between the Nuu'chah'nulth and MacMillan Bloedel (now Weyerhaeuser) – was created and the major environmental groups, the Western Canada Wilderness Committee and Greenpeace, had agreed to a logging plan for the Clayoquot.

Tofino's economy has diversified considerably from dependence on commercial fishing and logging to major tourist attraction. Tofino has become the tourism gateway to the spectacular setting of Long Beach and Clayoquot Sound, as well as providing whale watching, kayaking, and surfing.

TOURISM IN BRITISH COLUMBIA

How many tourist visits occur each year? From where do tourists originate? To which regions or locations do they go? What attractions and facilities do they use? How much revenue is generated? These questions are of interest from the perspective of geography. Table 14.3 compares 1990, 1996, and 1998 figures for overnight tourists and revenues. Between 1990 and 1996, the number of overnight visitors and the revenue increased in all regions, but figures for visitors from Asia/Oceana and Europe increased significantly more than the others. The 1998 tourism statistics reveal some rather major changes. The number of BC tourists declined considerably, along with their revenues, due largely to an economic recession. The Asia/Oceana region also declined because of a major recession, but figures for other regions were up

marginally, and the short-haul regional tourist market was up sharply. It is a common perception that tourists come from outside the province, but the table clearly demonstrates the importance of British Columbians among tourists in the province. The main tourist market for many communities developing tourist and recreation facilities may be from adjacent communities.

The rapid growth in BC tourism stems from several factors, including the value of the Canadian dollar in relation to the American dollar and other currencies. Canadian tourism is a good buy. British Columbia's growing economic attachment to the Asia-Pacific region also accounted until recently for an increase in international visitors. Asian economies underwent a severe recession in 1997, which affected British Columbia's economy, including tourism. Tourism declined by 9 percent between 1997 and 1998 because of depression in both Asia and British Columbia (Savings and Credit Unions of British Columbia 1999). This reversal in the rapidly growing industry is a stark reminder that tourism is not immune to unpredictable and uncertain global economic conditions: "Many factors affecting Canadian tourism, such as changing economic climate, energy issues, investment capital and heavy taxation burdens on the tourism sector in comparison with other countries,

Table 14.3

Overnight visitors and tourist revenues, 1990, 1996, and 1998

	1990				1996				1998			
Market	Visitors (millions)	%	Revenue ($ billions)	%	Visitors (millions)	%	Revenue ($ billions)	%	Visitors (millions)	%	Revenue ($ billions)	%
British Columbia	15.613	67.1	2.882	53.2	18.158	65.6	3.408	48.9	10.654	48.9	2.380	27.2
Regional[a]	4.163	17.9	0.977	18.0	4.883	17.7	1.261	18.1	6.445	29.6	2.690	30.7
North American long haul[b]	2.696	11.6	1.153	21.3	3.037	11.0	1.372	19.7	3.132	14.4	2.172	24.8
Asia/Oceana	0.484	2.1	0.247	4.6	1.017	3.7	0.598	8.6	0.774	3.6	0.821	9.4
Europe	0.297	1.3	0.151	2.8	0.531	1.9	0.310	4.4	0.623	2.9	0.625	7.1
Other international	0.016	0.1	0.008	0.1	0.033	0.1	0.018	0.3	0.142	0.7	0.072	0.8
Total	23.269	100.0	5.418	100.0	27.659	100.0	6.967	100.0	21.769	100.0	8.759	100.0

Notes: Column figures do not always add to totals, due to rounding.

a Includes short haul regional: Alberta, Saskatchewan, Manitoba, Northwest Territories, Yukon, Washington State, Oregon, Idaho, Montana, and Alaska.

b Includes the rest of Canada, rest of United States, and Mexico.

Sources: British Columbia, Ministry of Small Business, Tourism, and Culture (1994), 6; British Columbia, Ministry of Finance and Corporate Relations (1997), 105-6; (1999), 110-11.

represent challenges which must be dealt with" (Murphy 1983, v).

Keeping track of the number of tourists coming to British Columbia and the revenue they generate is obviously important to the province and to the industry as a whole. Tourists are not distributed equally among regions, though, as each region attempts to attract a share of tourists and their dollars (Table 14.4). Figure 14.4 divides British Columbia into nine tourism regions and the following section briefly describes each in terms of tourism resources and some of the issues affecting the tourism industry.

Tourism Regions

Tourism British Columbia divides the province into nine regions, as can be observed from Figure 14.4. This causes some problems in collecting information because most provincial statistics are gathered in terms of either eight development regions or twenty-eight regional districts. The tourism regions do not co-ordinate well with either of these categories, and delays in acquiring region-by-region statistical data can take years.

The Vancouver Island region fairly consistently increased its tourism revenues between 1989 and 1997 and has placed second to the Vancouver, coast, and mountains region, with 18 to 20 percent of the provincial revenues (see Table 14.4). This region includes the whole of Vancouver Island and the area of the central coast from Mount Waddington to Bella Coola. The Insular Mountains of Vancouver Island and Coast Mountains of the central coast, deep fjords, and mild winter temperatures are the distinguishing features and major strengths for the tourism industry.

The south end of Vancouver Island, and Victoria specifically, attracts the greatest number of tourists. Victoria wears the image and charm of a British past. This capital city also has the distinction of being the most popular city in Canada and rating eighth in a survey of favourite cities in the world (Florence, Italy, ranked first) (British Columbia, Ministry of Small Business, Tourism, and Culture 1991, 9). Its ranking was based on "most pleasant environment." It is interesting to note that Vancouver ranked eighteenth in the same survey.

Recreational opportunities abound on Vancouver Island and the Gulf Islands, with sport fishing, whale watching, kayaking, skiing, bungee jumping, spelunking, camping, and hiking as some of the favourite activities. The popularity of hikes such as the historical Ship Wreck Trail, also known as the West Coast Trail, is clear from the reservation system that the provincial government

Table 14.4

Room revenue by tourism region, 1989-97

Tourism region	1989 ($ millions)	1990 ($ millions)	1991 ($ millions)	1992 ($ millions)	1993 ($ millions)	1994 ($ millions)	1995 ($ millions)	1996 ($ millions)	1997 ($ millions)
Vancouver Island	131.5	145.1	142.1	146.2	154.8	177.6	190.5	194.6	197.8
Vancouver, coast, and mountains	344.4	396.1	390.9	401.6	416.6	473.8	539.2	623.4	663.9
Okanagan/Similkameen	48.4	51.2	53.9	55.1	60.1	65.1	69.1	70.7	n/a
Kootenay country	10.2	11.3	11.5	11.6	12.8	14.8	15.1	15.1	n/a
High country	34.9	39.8	38.3	42.8	49.2	52.8	57.7	61.2	n/a
Cariboo country	11.3	11.7	12.3	12.6	14.3	15.5	17.5	17.5	n/a
North by northwest	34.2	37.2	35.4	39.5	42.7	47.8	52.2	54.3	n/a
Peace River/Alaska Highway	11.2	13.6	13.4	10.8	12.6	19.5	19.2	19.9	n/a
BC Rockies	26.4	26.4	26.8	27.5	31.0	31.8	35.0	36.6	n/a
Total	625.5	732.3	724.8	747.6	794.2	898.7	995.5	1,093.4	1,147.9

Note: Column figures do not always add to totals, due to rounding. Figures for 1997 are for provincial development regions, which overlap but do not correspond precisely to tourism regions.
Source: Internet, BC Stats: Tourism Statistics.

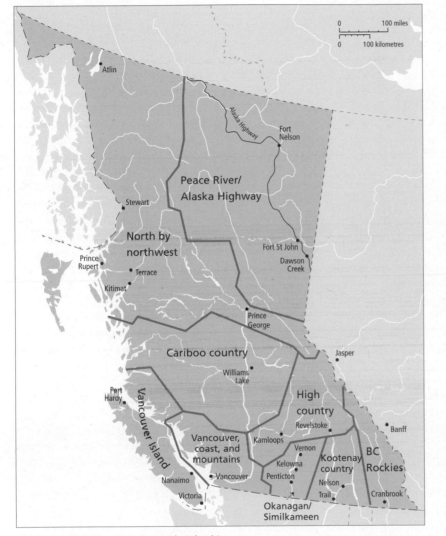

Figure 14.4 Tourism regions in British Columbia

remote, but that is its appeal to many tourists who journey the coast by water craft.

A contentious area that affects Vancouver Island tourism primarily is salmon sport fishing. The sport fishing industry brings in the highest return per fish caught but is allotted the fewest fish. Some areas of coastal British Columbia have had moratoria on catching chinook – the largest and most sought-after species – while the commercial fleet has not had the same curtailment. ARA Consulting Group (1996, S.4) estimates that the 1996 closure on chinook meant losses of 2,175 jobs and $135.7 million in revenues to the sport fishing industry. The Canada-United States agreement of June 1999 will probably bring some better means of managing salmon for tourism and other interests in this resource.

An increasing population of both tourists and residents in the southern half of the island has led to a new Vancouver Island Highway project that will accommodate a greater volume of traffic north to Nanaimo, Campbell River, and beyond. Transportation to Vancouver Island is mainly via the provincial government's ferry system, which connects at Victoria, Nanaimo, Comox, and Port Hardy. The Victoria and Nanaimo crossings are extremely popular and frequently marked by the frustration of one or two sailing waits during the tourist season and holiday weekends.

Table 14.4 reveals the dominance of the Vancouver, coast, and mountains region in the tourism industry. It captures over 50 percent of tourism revenue. Vancouver, by far the largest city in British Columbia, is the focal point of many transportation networks, including highways, railways, and ferries. Its international airport and

had to implement to keep the number of hikers at a manageable level. Pacific Rim National Park is another popular tourist destination site, attracting 700,000 a year to the west coast of the island (British Columbia, Ministry of Tourism and Provincial Secretary 1989, 5-4). The north end of the island, which has its own spectacular attractions from Cape Scott hiking to marine activities, receives considerably fewer tourists than the better connected south end. The central coast region is even more

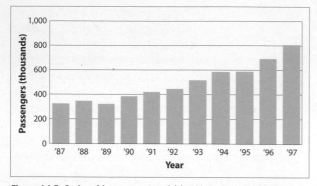

Figure 14.5 Cruise ship passengers visiting Vancouver, 1987-97
Source: Data from British Columbia, Ministry of Finance and Corporate Relations (1998), 101.

cruise ship facilities are some of the most important transportation modes for the industry. Cruise ship tourism has been particularly successful; Vancouver is a main port on the Alaskan cruise circuit. Such success has caused a corresponding problem, in that docking and embarking facilities are now inadequate. Figure 14.5 graphs the number of passengers visiting Vancouver via cruise ships. A 1,000-room hotel, convention centre, and cruise ship berth were potential new developments for the next millennium, but economic and political complications have shelved the proposal. Nevertheless, the demand for cruise ship travel will probably increase as the North American population ages, living longer and having considerable disposable income.

The Sunshine Coast, accessible only by ferry, and the excellent ski resort of Whistler are also part of this region. The municipality of Whistler has a convention centre, golf courses, and hiking and biking trails and is a year-round tourist-based community.

The Okanagan/Similkameen is famous for its many orchards with a great variety of fruit and for its many lakes, which attract water recreationists. More recently, its reputation has been enhanced by the production of high quality wines. Wine tours and wine festivals are a major attraction in this region. Sport fishing, many fine golf courses, and the Silver Star, Big White, and Apex ski resorts have helped this region become a year-round tourist attraction. The Okanagan/Similkameen region has easy access to the large Vancouver population via the Coquihalla Highway and a regional airport at Kelowna. Within the Okanagan, the main highway that

links communities has been improved by a number of bypasses. Even so, the volume of summer tourist traffic still results in delays. Another struggle for this region has been the invasion of Eurasian milfoil into the many lakes, spoiling the beaches for swimming.

Kootenay country gains the least tourism revenue of the nine tourist regions in British Columbia. An examination of its boundaries in Figure 14.4 reveals that the region encompasses only the West Kootenays. The region is mountainous, with relatively narrow valleys and long lakes because of numerous hydroelectric dams. Skiing, fishing, hunting, and hiking are just some of the attractions. There are also hot springs, a Doukhobor village museum, and ghost towns to attract tourists. One of the main drawbacks to Kootenay country is access: the region has poor air service and is not on a major arterial highway (British Columbia, Ministry of Tourism and Provincial Secretary 1989, 5-5).

Kamloops is the regional centre for the high country and is known as the gateway to Alberta. From Kamloops one can head either east to Revelstoke and Banff or Calgary or northeast up the North Thompson River via the Yellowhead Highway to Mount Robson and on to Jasper or Edmonton. Tourism for this region relies on the major transportation systems. The Trans-Canada Highway, Coquihalla Highway, and Yellowhead Highway converge and radiate out from Kamloops. This centre also has a regional airport and both the Canadian Pacific Railway (CPR) and Canadian National Railway (CNR) lines pass through this city. The high country offers many tourist attractions: hiking, heli-skiing, downhill skiing, fishing, boating, swimming, and sightseeing. Some specific sites are the Revelstoke Dam, Craigellachie (where the last spike of the CPR was driven), the world-famous run of sockeye salmon heading for Adams Lake, or Helmcken Falls in Wells Grey Park. The challenge for tourism operators in this region is to make these attractions a destination, since most tourists use the transportation systems to pass through the area.

The Cariboo, or Cariboo country, covers the region where the famous Cariboo gold rush occurred in the 1860s. The historic Cariboo wagon road has been widened into Highway 97 and links the various Mile Houses, which were stagecoach stops. The houses also serve to tell us the distance from Lillooet on the way to the gold fields and communities such as Barkerville. Cariboo country is a large region, most of which occupies the central interior plateau, but its tourist revenues are nearly the lowest of all nine regions. The tourist infrastructure has not been well developed throughout the region.

The Cariboo is cattle country, and tourists are attracted by the western culture of dude ranches, cattle drives, and rodeos. Cross-country skiing, fishing, hunting, and a scenic landscape are also important attractions. The historical site Barkerville brings the gold rush era to life as it engages tourists in gold panning, live theatre, and life in the 1860s. Forestry activities are still very important to the economy of this region, however, and interfere with some tourism. Another drawback for tourism is poor access. As well, the main highway from Vancouver leads tourists through this region and on to Prince George and other destinations.

North by northwest is one of the largest and most diverse tourist regions in the province, as it spans the area from the Rockies, through Prince George, to Prince Rupert and the Queen Charlottes off the coast. Its revenues have increased steadily to maintain approximately 5 percent of the provincial total. Prince George is the largest city in the region and serves as a gateway to much of the central interior and northern British Columbia. The port facilities of Prince Rupert are an important component of tourism, serving ferry traffic to Port Hardy, the Queen Charlottes, and Alaska. (A blockade of Alaskan ferries by BC fishers in August 1997 interrupted much of this trade for the year.) Much of the region north of Highway 16 – the highway between Prince George and Prince Rupert – is remote and herein lies its attraction. Ecotourism, rafting on the Tatshenshini, hiking, fishing, kayaking, and big game hunting are all components of the wilderness experience. As well, various exhibitions of First Nations culture are rapidly becoming a major focus of tourism in this region.

Another very large northern tourism region is the Peace River/Alaska Highway region. Physically and economically it differs from all other regions in the province. The area has within its boundaries a northern portion of the Rockies, the eastern foothills of the Rockies, and the start of the Prairies. The oil and gas industries are important to this economy and so is wheat. The landscape is similar to Alberta. The Peace River has the W.A.C. Bennett Dam, and there are numerous attractions for outdoor activities especially, but the main incentive for tourists is the Alaska Highway, of which Dawson Creek is Mile 0. This highway attracts a great number of tourists – mainly in the summer months – who travel to Yukon and Alaska and enjoy the wilderness or a stop at the popular Liard hot springs.

The final region, the BC Rockies, contains the spectacular mountainous environment of the southern portion of the Rockies and the Rocky Mountain Trench, which runs between the Rockies and the Purcell Mountains. Table 14.4 indicates that revenues from tourism did not increase greatly between 1989 and 1996, but they are nonetheless important as the industry represents an important economic diversion from predominantly mining- and forestry-based communities, many which have been struggling recently.

The "big" tourist parks are mainly on the Alberta side of the Rockies at Banff and Waterton Lakes, but the BC side of the Rockies offers many tourism opportunities. This region has skiing, hot springs, golf, hunting, fishing, and hiking, and many of the communities have heritage sites as well as annual festivals. The geographic location, most often referred to as the East Kootenays, finds many tourists coming from Alberta, and Calgary specifically, or up from the United States.

From this brief overview it can be seen that all areas of the province have touristic values and that tourism has become an important economic component to many local economies. The region dominates in terms of tourist numbers and overall revenues, but for other regions tourism is increasingly being recognized as an important diversification from traditional resource-based economies.

SUMMARY

In many ways, Expo 86 was the catalyst for the most rapidly developing industry in this province. An equally rapid expansion of global tourism has taken place through the 1980s and '90s. British Columbia offers many attractions for tourists: a great variety of saltwater and freshwater landscapes, spectacular alpine vistas, and an array of unique flora and fauna. As more and more people live in urban environments, the wilderness experience of British Columbia becomes increasingly desirable. Convention centres have been developed throughout the province to accommodate the internationalization of business activities that have a clear tourism dimension. All regions and communities in the province are involved in tourism, but not equally.

Tourism is a labour-intensive service industry. For the foreseeable future, tourism employment will increase, although many of the positions will pay only minimum wage. It is a knowledge-based industry in which information and organizational skills are essential commodities. The BC tourism industry is tied to global economic conditions, however, and these are marked by uncertainty. As well, the industry must work with all levels of government because governments provide the infrastructure it needs.

Tourism also relies on the quality of the environment. Since the landscape is part of the product being sold to tourists, it is inevitable that land use conflicts arise, particularly as the industry intensifies. Traditional industries have an impact on forests, fish, and water quality that can directly affect tourism, and they therefore need to be managed. Tourism itself can also have a negative impact on the landscape and on the way of life for residents of British Columbia. It, too, needs to be managed.

REFERENCES

ARA Consulting Group. 1996. *Fishing for Answers: Coastal Communities and the British Columbia Salmon Fishery.* Vancouver: ARA Consulting Group.

British Columbia, Ministry of Small Business, Tourism, and Culture. 1991. *Urban Tourism in British Columbia.* Victoria: The Ministry.

–. 1992. *The Economic Impact of Tourism Industries in British Columbia.* Victoria: The Ministry.

–. 1994. *Tourism Highlights, 1994.* Victoria: The Ministry.

British Columbia, Ministry of Finance and Corporate Relations. 1997. *1997 British Columbia Financial and Economic Review.* Victoria: The Ministry.

–. 1998. *1998 British Columbia Financial and Economic Review.* Victoria: The Ministry.

–. 1999. *1999 British Columbia Financial and Economic Review.* Victoria: The Ministry.

British Columbia, Ministry of Tourism and Provincial Secretary. 1989. *An Approach to Regional Tourism Development.* Victoria: The Ministry.

Canadian Regional Forecasting. 1991. *British Columbia Economic Forecast and Guide.* Richmond, BC: Canadian Regional Forecasting.

Cooper, C., F. Fletcher, D. Gilbert, and S. Wanhill. 1993. *Tourism: Principles and Practice.* Essex: Longman.

Dearden, P. 1983. "Tourism and the Resource Base." In *Tourism in Canada: Selected Issues and Options,* ed. P.E. Murphy, 75-93. Western Geographical Series vol. 21. Victoria: University of Victoria.

Mathieson, A., and G. Wall. 1982. *Tourism: Economic, Physical and Social Impacts.* Essex: Longman.

Murphy, P.E. 1983. "Tourism: Canada's Other Resource Industry." In *Tourism in Canada: Selected Issues and Options,* ed. P.E. Murphy, 3-23. Western Geographical Series vol. 21. Victoria: University of Victoria.

–. 1987. "Tourism." In *British Columbia: Its Resources and People,* ed. C.N. Forward, 401-30. Western Geographical Series vol. 22. Victoria: University of Victoria.

Savings and Credit Unions of British Columbia. 1999. *Economic Analysis of British Columbia* 19 (3): n.p.

White, B. 1999. "Authorizing the Tourism Landscape of Clayoquot Sound." Ph.D. thesis, Simon Fraser University.

INTERNET

BC Stats: Quick Facts – The Economy, "Tourism," www.bcstats.gov.bc.ca/data/QF_econo.HTM

BC Stats: Tourism Statistics, www.bcstats.gov.bc.ca/data/bus_stat/Tourism.htm

British Columbia, Ministry of Small Business, Tourism, and Culture, Home Page, www.tbc.gov.bc.ca/tourism/tourismhome.html

Coho Monitoring Bulletins, www.pac.dfo-mpo.gc.ca/ops/fm/Salmon/Coho/Coho.htm

Single-Resource Communities: Fragile Settlements

15

The arrival in British Columbia of Europeans, Asians, and other non-Natives was largely motivated by the expectation of wealth through resource exploitation. There was also money to be made in building the infrastructure needed to export these commodities and in constructing resource-based communities. Many communities were based on the harvesting or processing of a single resource such as fish, forests, minerals, agricultural products, and much more recently, tourism. These **single-resource communities** took a number of forms and their populations varied, reflecting the economic and social values of the day. Haphazard mining communities such as Emory Creek, Boston Bar, and Barkerville were constructed almost overnight in response to the gold rush of the late 1850s and early 1860s. A silver boom in the Kootenays in the 1890s resulted in similar disorganized, hastily constructed mining communities such as Silverton and Sandon. These communities were made up largely of individual miners and small companies operating with relatively simple technologies.

Company towns were also part of the early settlement landscape of British Columbia. These single-resource communities were built and controlled by large corporations that owned the mine, smelter, mill, or cannery. Coastal pulpmills such as Powell River, Port Mellon, and Woodfibre were all company towns that appeared in the early 1900s. Mines and smelters also required large investments, and many of them became company towns (e.g., Britannia Beach, Cumberland, and Trail). A company town was distinguished from any other single-resource community in that it was built by the company, which "exercised ownership and control over land, housing, recreation, plant, and all assets and facilities of the community including hotels, hospitals, golf courses, churches and movie theaters" (Bradbury 1979, 54). These communities were also not regulated by the Municipal Act.

Single-resource communities, and the province as a whole, were subject to a host of influences over time: changing demands for resource commodities, technological developments, political decisions affecting resource development, and unpredictable global events. Nonetheless, these developments have not had a uniform effect either throughout the province or on the development of single-resource communities. Figure 15.1 illus-

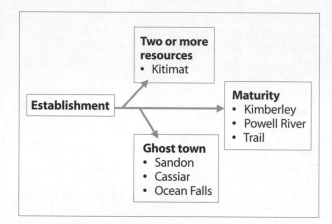

Figure 15.1 Development of single-resource communities

trates the typical pattern of development for these communities. The fate of some was sealed from the start, as the resource on which they were based was non-renewable or not in great supply. Many single-resource communities became ghost towns, while others grew and expanded their economic base until they were no longer classified as single-resource communities. Still others, such as Powell River, Trail, and Kimberley, have remained economically viable as mature, incorporated single-resource communities.

Lucas (1971) found in his study of single-resource communities throughout Canada that there were considerable differences among them based on the resource and on structure of employment: "Agricultural communities, for example, are made up of independent capitalists, and the concerns and patterns of life are quite different from those in a mining community, where miners work as employees of a large bureaucratic organization" (p. 12). He also recognized that geographic isolation was a common factor for single-resource communities, almost guaranteeing that they remained with a single industry (p. 394). In the mountainous landscape of British Columbia, isolation was certainly a factor.

The influence of large corporations and their control of company towns and employees eventually became the target of government policy. A major reassessment of urban and rural policy in British Columbia coincided with the significant expansion of the mining, energy, and forestry resource sectors from the 1950s to the early 1970s. This period is referred to as the megaproject era:

a time of very large resource extraction projects with large and expensive technologies and, most often, foreign corporations in control of the investment. The provincial government, recognizing that its economic well-being came from the harvesting and processing of resources, saw new single-resource communities on the resource frontier as inevitable.

The Instant Towns Act of 1965 was enacted by the provincial government largely in response to the negative conditions of the company town: "In the boom years between 1962 and 1972 a total of eight new industrial towns, or **instant towns,** as they came to be called, were constructed by resource companies in isolated areas of British Columbia" (Bradbury 1978, 117). Bradbury also points out the differences between company and instant towns. Instant towns had "a much greater degree of advance planning than had been the case in earlier single enterprise communities ... [This] reflected an attempt to correct some of the systemic problems of the traditional company town such as high population turnover, unbalanced demographic profiles, settlement impermanence, social instability, isolation and corporate dominance and control; it was also an attempt to be rid of the stigma of the company town label" (p. 118).

The company town reflected the times and conditions in the first half of the century, but times and conditions were quite different by the 1960s. Large multinational corporations were in control and investing in capital-intensive technologies, markets were much more global and uncertain, and governments at all levels played a greater role in everyday life. In many ways these were good times economically – and the instant town reflected these new opportunities. Yet the boom economy was not to last. The destabilization of currencies with the removal of the gold standard, the energy crisis, recessions, and the transition to post-Fordism resulted in the restructuring of resource-based economies (Chapter 7). In this turbulent economic climate, instant towns and all resource-based communities were highly vulnerable, with their basic survival often in doubt.

This chapter compares life in and the conditions of the historical company town to the much more recent instant town. Were the instant towns much better communities in which to live and work? As well, it examines urban policy for the future. What are the options for living conditions and communities for workers as the resource frontier develops?

THE COMPANY TOWN

During the gold rush of 1858, 25,000 to 30,000 people came to British Columbia to find gold (Chapter 4). Numerous communities were rapidly constructed, the largest of which was Barkerville. As fewer and fewer gold discoveries were made, many became ghost towns. These were some of the first single-resource communities in British Columbia. They were hastily built affairs, mainly of frame structure, and most were at high risk of fire. They were not company towns. Even later, with the silver boom of the West Kootenays, newly established communities such as Silverton, Sandon, and Three Forks did not qualify as company towns even though some rather large companies operated the mines adjacent to them. The company town was defined by the fact that the company not only owned and operated the mine or mill but also built the town, owned the housing, and ran the company store.

The company town became very common on the landscape of British Columbia by the early 1900s. The early coal mining communities of Nanaimo and Cumberland were two of the first company towns. Coastal pulpmills soon followed, and later mines such as Britannia Beach and Cassiar became company towns. In the early days, almost all of these single-resource communities existed in remote settings. Few were linked by rail, and coastal communities were serviced only by slow-moving vessels. Even Britannia Beach, today only minutes away from Vancouver on the way to Squamish and Whistler, was hours by boat in the 1920s, and Ocean Falls was days away. Movement was not easy.

Company towns did not have a normal distribution of population. There were few women and children, and no elderly people. Single men dominated the workforce and the community, although the corporations began to realize that married workers were more stable and less likely to move. This was of particular importance as jobs became more skilled and the turnover of labour caused a decrease in productivity. The trade-off for the company was to provide housing for married men and their families and to invest in one-room schools. Nevertheless, education was not a priority.

The work day and work week were much longer than today, with twelve-hour days six days a week and low wages. This left little time for social activities, except on Sundays. The long hours of work were often in hazardous conditions, especially in underground mines, where there was the danger of cave-ins, blowouts, and diseases related to the inhalation of dust particles. Even minor accidents were often fatal because of the lack of proper medical knowledge and facilities and the inability to transport people to the few hospitals the province had (McGillivray 1980, 3).

These rather dismal working and social conditions describe a general way of life that was not confined to company towns. The struggle of the working class to improve its lot was not easy on the resource frontier, and in the company town the struggle for change was at its most difficult.

Distinguishing the company town from other single-resource communities was the control that the corporation had on its labour force. These were "closed" communities, exempt from the Municipal Act, and the company set the rules: "Most company towns existed outside the municipal laws applied to other urban settlements" (Bradbury 1979, 54). The provincial government did try to make them more open: "Attempts had been made, in 1919 and 1948, to penetrate the overall control maintained by the companies over their towns, but little was attempted in terms of changing the ownership of the settlement or altering the ownership of the means of production by large corporate bodies" (p. 54). Unions were seen by corporations as a threat to profits and management rights. They were banned and workers agitating to form unions within company towns were quickly fired and sent packing (Walker 1953, 3). There was no room for organized labour in a company town.

Housing was another source of company control, and it reflected class, status, and ethnicity. Company towns were planned on the principle of residential segregation (Bradbury 1978; Porteous 1987). This form of spatial organization separated workers, with dormitories for single men, small cottages for married labour, larger houses for foremen, and substantial housing (preferably on a hill overlooking the town site) for managers. Chinese, First Nations, and other visible minorities were housed in separate locations, such as across a stream. All rents were deducted from the paycheque.

The company store characterized even greater control as it became the only source of both basics and luxuries. With a monopoly on retail goods, it was not uncommon for the company to charge exorbitant prices, and workers were encouraged to use a system of credit to pay for any purchases. A "bob-tailed" cheque was a pay slip without any funds, as the worker owed more to the company than he received in wages at the end of the pay period. The control over labour was powerful and omnipotent.

The struggle to change the system was not confined to British Columbia. It was much more universal. New political and social philosophies were tempered and tested by world war and depression. And conditions did change. Unions formed in British Columbia and became ever stronger and more militant in their demands, political parties advocated economic and social change, and technologies changed inside and outside the workplace. Better rail and road systems reduced isolation and placed even greater pressure on company towns to become more open. Slowly, the towns transformed as workers commuted from adjacent or nearby communities, breaking the company's control over housing and its monopoly on retail sales. Conditions of work, with improved safety, fewer hours, and improved pay, also became part of the transition as mines and mills reluctantly accepted a unionized workforce. Mergers and takeovers resulted in larger and larger corporate entities, with greater global connections and investments.

By the end of the 1950s and '60s, British Columbia was in the process of an unprecedented expansion of resource development. Large multinational corporate investment was encouraged by the provincial government through tax incentives and provision of roads and energy supplies. As the resource frontier pushed forward into relatively isolated locations, new communities were needed to house workers. Not wanting to carry on the negative company town experience, the provincial government initiated the Instant Towns Act in 1965.

INSTANT TOWNS

Figure 15.2 shows the main instant towns across the province. As stated earlier, initially eight instant towns

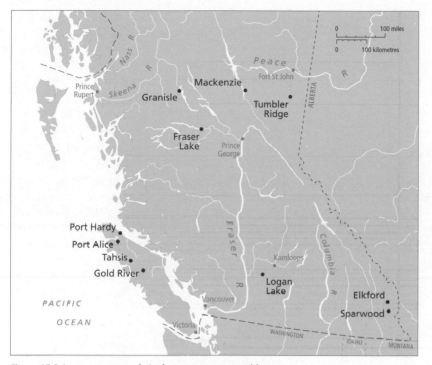

Figure 15.2 Instant towns and single-resource communities
Source: Modified from Porteous (1987), 385.

were built between 1965 and 1971: Gold River (1965), Mackenzie (1966), Sparwood (1966), Fraser Lake (1966), Logan Lake (1970), Tahsis (1970), Elkford (1971), and Granisle (1971). Tumbler Ridge (1981), in the Peace River region, was the only other instant town to be created that still remains. Other instant towns were planned, mainly around mineral extraction, but because of changes to world market prices for metals these communities never materialized. The communities were planned with the involvement of corporations and the provincial government.

Instant towns came under the Municipal Act and were therefore open communities with locally elected councils, unlike company towns. Curvilinear streets, smart three-bedroom houses, apartment complexes for single employees, and mobile home parks characterized their layout and housing options. Many workers owned their own homes. Retailing in the instant town differed greatly from the company town's monopoly via the company store. The instant town included a small shopping centre with independent retail franchises that suited small communities. Most instant towns did not have sufficient population to warrant a hospital, but clinics and dental offices were set up for regular visits by the medical profession. Modern elementary schools were constructed, and when a community became large enough, a high school was added. A recreation centre with a curling rink, an arena, and a gymnasium was a standard facility in the instant town.

A comparison of the old company town with the new instant town reflects differences in time, structure, and attitudes. At the surface, the instant town appears to represent a radical departure from the company town era. Unionization guarantees safer work conditions, the forty-hour work week, and high hourly wages. Recreational opportunities abound, especially for those who enjoy the outdoors. Vastly improved road systems and satellite television receivers connect the most isolated communities. Yet the fundamental existence of the instant town, dependent on the economic viability of a single resource along with heavy corporate investment in the main source of employment, has not changed.

The film *No Life for a Woman*, produced by the National Film Board in 1979, was based mainly on the instant town of Mackenzie. As its title implies, the film depicts a male-dominated town where councils prioritize decisions that facilitate the industry, where there are no women's centres or daycare facilities, and where women have few employment opportunities outside the home. Living in a trailer park, where units are crammed together and offer little privacy, causes anxiety and tension, especially in the winter when heavy snowfall makes it impossible to get out. On the social scale, these communities tend to have high rates of violence and suicide, and of drug and alcohol abuse. Many of these social problems have been recorded by Lucas (1971), Bradbury (1978), McGillivray (1980), and Porteous (1987). Patricia Marchak, in her examination of instant towns related to the forest industry, recognized similar problems: "By their nature, such towns are isolated, with shallow roots" (1983, 303).

The instant town was created by corporate economic interests based on resource extraction. This is critical because the world market price for resource commodities is rarely stable and a corporation may decide to shut its plant down temporarily or permanently when market price drops. The instant town of Kitsault, northeast of Prince Rupert, is a case in point. No sooner had it been planned and the first buildings erected than it became an instant ghost town. The world market price of molybdenum, the impetus for building the town, was in decline. With little prospect of an increase in the short or long run, the corporation pulled out.

The landscape of uneven development results from the multiple influences of wars, recessions, depressions, new technologies that make resources obsolete, and decisions by multinational corporations about resource procurement. The impact is both unpredictable and uneven. Some mines and mills close while others remain in operation. The inevitable vulnerability and uncertainty is the basis of many of the social concerns within dependent communities.

Forestry and mining are capital-intensive industries, and employees sometimes face being replaced by machinery. The Sullivan Mine at Kimberley employed 2,122 people in 1949, for example, but its current workforce is 660 and the mine is scheduled to close in 2001 (Ward 1998, C4). Figures for pulp-and-paper mills and sawmills showing shrinking employment numbers due to capital-intensive technologies are similar (Marchak 1983; Barnes et al. 1992; Rees and Hayter 1996).

The new instant towns, sanctioned by public policy, were "sold" to the public of British Columbia as permanent communities, and workers were encouraged to take out mortgages to purchase housing. Writers such as Bradbury (1979) suggest that the provincial government used the Instant Towns Act to facilitate corporate investment, trapping the worker in the process by shifting the burden of housing from the company to the employee while considerably reducing the employee's

option of leaving. For the worker with a mortgage, anxiety increased as British Columbia experienced restructuring, layoffs, and, in some cases, closure of the resource operation. With the serious economic recession of the 1980s, the federal government recognized the vulnerability of single-resource communities in a report entitled *Canada's Single-Industry Communities: A Proud Determination to Survive* (Canada, Ministry of Supply and Services 1987). The report suggested that planning for single-resource communities had been inadequate: "The absence of long term planning and involvement in the past, and to a lesser extent in the present, has resulted in a rather haphazard development of the country's resource frontier" (p. v). Government policy in the 1960s did not address the fundamental vulnerability of single-resource communities. Rather, by encouraging the construction of instant towns the provincial government gave these communities the illusion that they were permanent.

Randall and Ironside (1996) question the generalizations and stereotypes about single-resource communities and shed new light on their composition. They show that the communities are not homogenous, that they have far more women in the workforce than has been assumed, and that they do not depend totally on a single resource for employment. An obvious strategy for single-resource communities, however, is still to develop a broader economic base, both for the existing resource and for other resources. The town of Chemainus, the "Little Town That Did," is often held up as a prime example of a successful response to the recession of the early 1980s (Rees and Hayter 1996; see also Chapter 7).

Three other examples represent far more of a challenge: the permanent 1999 closure of Gold River pulpmill on the west coast of Vancouver Island, the temporary closure of the Highland Valley copper mine that supports Logan Lake in the south central interior, and the reduced coal production from the mines of Tumbler Ridge in the Peace River region. These three relatively isolated communities will find difficulties in switching to tourism, the retirement industry, or other economic activities to prevent them from becoming ghost towns. Communities such as Kimberley are already on notice that their main employers will soon be gone, and they too are attempting to diversify into tourism, recreation, retirement, and forestry-related employment.

The world market price for metals, forest products, coal, or even new resources is likely to rise, at which point it will be necessary to build new communities on the resource frontier. What lessons have been learned from past experience and what form will new urban policy take? Porteous (1987) has proposed several alternatives to instant towns: communities deliberately structured for commuters to go to; expansion of existing centres; and new large, permanent cities in these isolated regions. Permanence is the key; if it had been applied as a decisive principle, government policy would not have permitted instant towns such as Logan Lake. In that particular example, the existing community of Ashcroft would have been expanded. The federal government has made this transition with its policy of resource development in the Yukon, Northwest Territories, and Nunavut. Single-resource communities in these territories are dwindling, and new mining operations "now tend to rely on flying in the bulk of the labour force on a rotational basis, rather than establish[ing] permanent communities" (Usher 1998, 380). Alternatively, if the frontier has numerous resources to develop where no communities currently exist, then a relatively large, permanent instant city could be developed with a broad range of services based on several resources. Randall and Ironside (1996) suggest a similar strategy for existing single-resource communities, namely, the expansion of the economic base into two or more industries.

As discussed in Chapter 7, British Columbia has evolved a core-periphery relationship in which the periphery relies to a large degree on resource exploitation. The uncertain global economic climate, especially of the 1980s and 1990s, has revealed the difficulties in mines or mills remaining economically viable. The problem has unhinged the economic stability of single-resource communities in particular. One lesson for the provincial government is to review its urban policies as new resources develop in isolated locations. For existing single-resource communities, it is essential to explore ways to diversify the economic base. Provincial policy could establish a reserve fund to assist in relocation, restructuring, and other hardships endured by members of single-resource communities when the mine or mill closes.

SUMMARY

Earlier chapters focused on individual resources such as forests, fish, and minerals, tracing the factors responsible for change and the impact these industries have on the landscape of British Columbia. An examination of single-resource communities shifts the focus from resource development to the human factor of workers and their living and working conditions.

The early single-resource communities, particularly mining towns, were by and large unplanned and hastily built efforts to house miners and provide services. Many became ghost towns. Company towns, with much greater investment in plant and equipment and a complete infrastructure – housing, company store, services, and so on – became popular by the early twentieth century. The provincial government encouraged both the investment and the employment. The first half of the century saw changes in work conditions, unionization, and expectations about working and living conditions. The company town, with its overwhelming control of workers and all who lived within its borders, became an anachronism and an embarrassment to the provincial government.

New single-resource communities were required for the megaproject era that began in the 1960s. Consequently, a new urban model was conceived and promoted by the provincial government: the instant town. These open communities began to appear by 1965. They were different from company towns in all respects except their dependence on a single resource, and in most cases, a single corporation. Changing economic conditions from the 1970s onward – an increasingly global economy, competition, and uncertainty – led to tough economic times and the realization that it was the corporations that got the most benefit from the Instant Towns Act.

There should be caution in creating any single-resource communities in the future; they come with expectations of permanence, and dependence on a resource is anything but permanent. For existing single-resource communities, however, the struggle is to diversify their economic base. Not to do so could add more ghost towns to the map of British Columbia. Of course, the ability to diversify depends on many factors, such as geographic location, accessibility, and the availability of other resources. It is a challenge.

REFERENCES

Barnes, T.J., D.W. Edgington, K.G. Denike, and T.G. McGee. 1992. "Vancouver, the Province, and the Pacific Rim." In *Vancouver and Its Region,* ed. G. Wynn and T. Oke, 171-99. Vancouver: UBC Press.

Bradbury, J.H. 1978. "The Instant Towns of British Columbia: A Settlement Response to the Metropolitan Call on the Productive Base." In *Vancouver: Western Metropolis,* ed. L.J. Evenden, 116-29. Western Geographical Series vol. 16. Victoria: University of Victoria.

–. 1979. "Towards an Alternative Theory of Resource-Based Town Development in Canada." *Economic Geography* 55 (2): 47-66.

Canada, Ministry of Supply and Services. 1987. *Canada's Single-Industry Communities: A Proud Determination to Survive.* Ottawa: The Ministry.

Lucas, R.A. 1971. *Minetown, Milltown, Railtown: Life in Canadian Cities of Single Industry.* Toronto: University of Toronto Press.

McGillivray, B. 1980. "Single Resource Communities in Canada: The Vulnerability of People and Communities, Past and Present." *Canadian Studies Bulletin* (November): 3-6.

Marchak, P. 1983. *Green Gold: The Forest Industry in British Columbia.* Vancouver: University of British Columbia Press.

Porteous, J.D. 1987. "Single Enterprise Communities." In *British Columbia: Its Resources and People,* ed. C.N. Forward, 382-99. Western Geographical Series vol. 22. Victoria: University of Victoria.

Randall, J.E., and R.G. Ironside. 1996. "Communities on the Edge: An Economic Geography of Resource-Dependent Communities in Canada." *Canadian Geographer* 40 (1): 17-35.

Rees, K., and R. Hayter. 1996. "Enterprise Strategies in Wood Manufacturing, Vancouver." *Canadian Geographer* 40 (3): 203-19.

Usher, P. 1998. "The North: One Land, Two Ways of Life." In *Heartland and Hinterland: A Regional Geography of Canada,* 3rd ed., ed. L. McCann and A. Gunn, 357-94. Toronto: Prentice-Hall.

Walker, H.W. 1953. *Single Enterprise Communities in Canada.* Report to Central Mortgage and Housing Corporation by the Institute of Local Government. Kingston: Queen's University.

Ward, D. 1998. "A Town That Wouldn't Die." *Vancouver Sun,* 18 October, C1, C4.

FILMS

Kreps, B., director. 1979. *No Life for a Woman.* Documentary, 26 mins. Serendipity Films and National Film Board of Canada.

Urbanization: A Summary of People and Landscapes in Transition

16

British Columbia entered the new millennium as a highly urbanized society, and Vancouver stands out as a world city – a hub in the global network (Short and Kim 1999, 53) – with a population expected to reach 3 million within the next two decades. No other urban centre in the province will come close in size, but it is increasingly recognized that the spatial order of employment and most of the decisions that affect the province will be made in cities generally. Vancouver, a major urban complex with links to the Pacific-based economies, will show the greatest growth. A hierarchy in which one large urban agglomeration dominates over many smaller communities is not random. The evolution of this urban system is best understood from an historical perspective.

The unique physical characteristics of the province, with its vertical and rugged landscape, has played a significant role by confining settlement and urbanization mainly to river valleys and coastal inlets. The processes resulting in the various patterns of urbanization reflect 200 years of non-Native settlement and development. The antecedents of the current urban system are the fur trade forts erected to secure and control territory for British colonial interests, principally for material gain. Competition for territoriality was initially between the British and the Spanish and, to some degree, the Russians. Later, there was even more serious and sustained competition with the Americans. These struggles shaped and reshaped the political boundaries that eventually became British Columbia. They also refined the basis of urbanization and the important connecting links between centres.

The early fur trade posts represented the first semi-permanent presence of non-Natives and the disintegration of the traditional way of life for First Nations. The forts increased dependence on European goods and were focal points for trade, missionary zeal, and the spread of diseases, although European diseases had reached the west coast of North America long before forts appeared. With the discovery of coal, gold, and silver, and the development of forests, fish, and other resources, the forts gave way to resource- and transportation-based communities. For First Nations, the arrival of Europeans caused an accelerating process of deterritorialization as well as cultural destruction through waves of disease,

alcohol, a reserve system without treaties, and major pressures to assimilate (Table 16.1). First Nations were reterritorialized in a landscape shaped largely by British values of civilization that were well represented in the emerging urban centres, often with adjacent reserves and residential schools. First Nations were set aside on small parcels of land but became part of the urban fabric over time as they sought economic opportunity that was rarely available on the reserve. Asian populations were treated in a similarly discriminatory way, relegated to racially segregated regions of the community.

Staples theory, discussed in Chapter 7, can assist in understanding the growth of communities as resource after resource was discovered throughout British Columbia and exploited for export. Political decisions about the geographic location of important backward linkages, such as the Canadian Pacific Railway (CPR) and other rail lines, were factors in time-space convergence, often giving economic growth to the communities created or connected by the railway. In some cases, growth was significant. Railway companies, and the CPR specifically, acquired enormous economic advantage by taking control of much of the best land and resources in the province. The urban system that evolved also reflected corporate land use decisions.

Urbanization is a dynamic process that concentrates both population and numerous functions, including the sale of goods and services, financial transactions, trade, manufacturing, and government and administrative services. In many ways, cities are the repository of societal values. As these values change, they modify the landscape. The growth, and sometimes decline, of

Table 16.1

Native and non-Native population, 1782-1870

	Native	Non-Native
1782	>200,000	n/a
1835	70,000	n/a
1854	n/a	450
1858-60	47,000	25,000-30,000
1870	25,661	10,586

Sources: Harris (1997), 28, 30; Muckle (1998), 37; Duff (1965), 2; Robin (1972), 14; Meen (1996), 97, 109; Census Canada, 1871.

communities, especially smaller ones, is tied mainly to the viability of a changing economic base. In British Columbia, a fragile economic base turned some communities into ghost towns. Others experienced rapid and often unplanned growth from new industries, transportation developments, and an expanding economic base. Urban centres also grew because of amalgamation, or annexation. The boundaries of many communities have been redrawn, in some cases many times, to incorporate more and more of the surrounding territory.

Population growth and urbanization accelerated after the Second World War. Economic conditions resulted in an unprecedented demand for the forests and minerals of British Columbia. New capital-intensive technology and control by large corporations significantly influenced migration from rural to urban settings. New resource towns were created, major hydroelectric dams were built, pipelines criss-crossed the province, and roads, railways, and airports expanded throughout the megaproject era. Most urban centres, new or old, expanded during this boom period.

The dynamics of post-Fordism – including increased international competition, corporate control and concentration of resources by multinationals, and a host of global economic crises – particularly challenged resource-dependent communities. Global conditions also provided opportunities for banking, finance, and services in the new knowledge- and information-based economy. A spatial pattern emerged that favoured large urban centres. New technologies of transportation and communication also developed at this time, radically changing the traditional form of urban centres and resulting in location decisions that blurred the distinction between urban and rural. Automobile suburbs, shopping centres, and industrial parks were the first developments to cause the relocation of traditionally centralized business functions. The new global communications technologies of fax, e-mail, Internet, e-commerce, and a host of information service providers, all available to the individual household, have only accelerated the move to often less expensive dwellings at ever greater distances from the metropolitan centre.

The overview of British Columbia presented in this chapter, traced mainly through the development of communities and an evolving urban system, is developed within three periods. The first is the British control of the territory through the Hudson's Bay Company until the province joined Confederation in 1871. This period sets one pattern for urbanization. The second, from Confederation to 1951, saw an explosion of urbanization influenced by railway development in combination with resource exploitation and land speculation. The third period spans the post-Second World War boom to the present. These three are tied to census data periods to show clearly the statistical basis of population change and urbanization.

The terms "community" and "urban centre" are used synonymously here, although the census defines urban centres more specifically. Up to 1951, incorporated communities were considered urban, and all other population was rural. An incorporated community has a geographically defined boundary that encompasses fairly dense population and provides a level of local autonomy with an elected mayor and council. Each province has its own way of defining incorporated urban areas, and British Columbia is no exception. The first two incorporated municipalities pre-date Confederation: New Westminster (1860), and Victoria (1862). The Consolidated Municipal Act (1872) encouraged more incorporations. The Municipal Clauses Act (1896) classified incorporated places into cities and district municipalities. As Donald Higgins states, "while city status was presumably intended for the more densely populated settlements and district municipalities for rural areas, such a distinction became lost over time. Some cities had as few as two hundred residents, and some districts close to forty thousand people" (1977, 36). He also points out that the category of "village" was created in 1920 and "town" was added in 1958. By the census definition, there were only two urban centres in British Columbia in 1871, three in 1881, and four by 1891. Nevertheless, many other unincorporated communities both prior to and during this time were very much a part of the evolving urban system. From 1961, the definition of urban also included unincorporated areas if they had a high enough population density per square mile ratio (initially 1,000 per square mile, although this changed again in the 1981 census). Statistics used in this chapter are from the census and reflect the changing definitions of "urban."

SETTING AN URBAN PATTERN ON THE WESTERN FRONTIER

European colonization and global trade between Europe and Asia (China in particular) were well established before the remote region of the northwest coast of America attracted interest. By the late 1700s, seaborne adventurers with attachments to various European countries plied the waters of the Pacific northwest in search of the Northwest Passage, as well as any resources for trade with China. Initially, Russians had discovered and exploited the sea otter trade by setting up fur trade posts in Alaska and down its Panhandle. The Spanish, with a fort at San Francisco, claimed most of the west coast of North America. The British, who arrived after both the Russians and the Spanish, were interested in the valuable sea otter pelts and made territorial claims on the coast. With the Russian hold on Alaskan territory the main struggle was between the British and the Spanish.

The British succeeded in adding what is now called British Columbia to their colonial empire, but the nature of the trade did not require permanent forts on the coast. Permanent forts did appear in British Columbia by the early 1800s, however, though not in connection with the sea otter trade. Rather, they were part of the network of fur trade forts that the North West Company and later the Hudson's Bay Company extended across the country. These forts served many functions, of which the entrepôt trade with First Nations was the most obvious. Year-round fortified communities of a few Europeans represented a military presence, colonization, and the presence of British values and institutions. It was the beginning of reterritorialization. Maps indicating the location of these forts and their transportation networks became a base of knowledge for others, whether missionaries or fortune seekers (see Figures 4.1 and 4.2, pp. 57 and 58).

The British claim to the territory was never secure and conflicted with American claims throughout the early 1800s. With a US presidential election based on the slogan "54° 40' or Fight," the territory was in serious jeopardy of being annexed by the Americans. The Oregon Treaty of 1846 resolved the conflict by extending the forty-ninth parallel from the Rockies to the coast; Vancouver Island, which extends below the forty-ninth, was retained by the British. These geopolitical changes resulted in the loss to the British of what is now Washington State, the lower Columbia River, and the fur trade forts located in this region.

Fort Victoria, constructed at the southern end of Vancouver Island, was much more than a fur trading post. A.H. Siemens calls it the first permanent coastal settlement in British Columbia (1972, 13). Although it was established to secure British interests, the community also had land for farming and forests, and fish and coal were readily accessible for exploitation by the Hudson's Bay Company. Fort Langley, the Hudson's Bay fur trade fort at the western end of the Fraser Valley, turned to farming and the salmon trade as furs were exhausted. While the interior forts still relied on the fur trade, a number of forts appeared on the coast as a result of the increasing demand for coal. Fort Simpson, near present-day Prince Rupert, and Fort Rupert, at the north end of Vancouver Island, were producing coal by the 1840s. By the 1850s, the Hudson's Bay Company had developed the rich seams of coal on the southeastern coast of Vancouver Island and built the new town of Nanaimo.

The discovery of gold, first on the Queen Charlotte Islands, aroused economic and political interest in securing this territory for the British. These initial discoveries were neither large nor sustained. The major one was on the lower reaches of the Fraser River, and by 1858 some 25,000 to 30,000 gold seekers arrived. Many came from the exhausted gold fields of the San Francisco Bay area, while others came from across Canada, Europe, and China. By the early 1860s, further discovery of significant amounts of gold up the Fraser and into the Cariboo was the main catalyst in opening up British Columbia and changing its landscape permanently.

Gold along the Fraser provided the most important single impetus for early rural and urban settlement in the province (Siemens 1972, 11). Frame-structured towns such as Emory Creek, Boston Bar, Richfield, and Barkerville were rapidly erected adjacent to the gold discoveries. Other communities, such as Port Douglas, Yale, Lillooet, and Quesnel (named Quesnelle Mouth then), acted as service centres on the various routes up the Fraser and into the Cariboo. Other, more minor discoveries of gold were made in the 1860s on the Stikine, Peace, and Columbia Rivers. They did not become large enough to warrant infrastructure investments such as the Cariboo Road, nor were communities developed, but by 1863 these finds were instrumental in shaping the political boundaries of the territory: north to the sixtieth parallel and east to the 120th degree of longitude.

The lower Fraser River and Cariboo gold rush left a significant legacy to settlement and urbanization. This was a turning point for British Columbia. Before the gold strike, the area had been under the control of the Hudson's Bay Company, which was interested in extracting its resources with a minimum amount of permanent settlement. The construction of the Cariboo Road between 1862 and 1864 placed a permanent line of connected communities on the map of British Columbia, from Victoria by paddlewheeler to New Westminster and Yale and then by road to Barkerville. The urban system was evolving with Victoria as the anchor.

Victoria was the largest centre and owed its growth to its location as the main entrance and exit to the territory. It became a service centre for travellers as well as the gateway for the import and export of goods. Victoria was also the main political, administrative, and financial centre, especially after 1866 when the two colonies were joined. Growth also occurred because of manufacturing, "including tanneries, sash mills, a foundry, a soap factory, a ship yard and a gas works," and by 1870, "the brewing and distilling industries were concentrated around Victoria" (Meen 1996, 111). Nanaimo emerged as the main supplier of coal during the early 1860s. New Westminster, while losing a bid to become the capital city, was another important centre, providing services for a growing farming settlement in the adjacent Lower Fraser Valley.

British values incorporated a white supremacist attitude that saw all visible minorities as inferior and groups such as the Chinese, who represented a significant percentage of the non-Native population by the mid- to late 1860s, as undesirable. These racist views resulted in Chinatowns being constructed as separate enclaves in the urban landscape of British Columbia.

Women were scarce in the territory, although Sharon Meen informs us that the few women who lived in the Cariboo "seized this unusual opportunity, running hotels, restaurants, saloons and laundries. In 1869, twelve were listed as business women in their own right" (1996, 111). She also relates the arrival of two bride ships from Britain that delivered approximately a hundred women to Victoria in 1862 and 1863 (p. 116). The effort to correct the imbalance between the numbers of men and women was also an attempt to create a more stable, permanent colony and communities where couples would build homes and raise families.

The gold rush communities were structurally different from corporate-controlled communities such as Victoria and Nanaimo. Placer mining communities were largely haphazard, disorganized, frame-structured, and fire-prone settlements, rapidly constructed and just as rapidly abandoned when the gold ran out. After 1865, the flood of people into the region was rapidly reversed. Emory Creek, Richfield, and eventually Barkerville joined the list of resource-depleted ghost towns, establishing an enduring pattern. Even Victoria, whose population was estimated at 6,000 in 1863, had declined to one-half this size by 1870 (Meen 1996, 117).

By the time British Columbia entered Confederation in 1871, the landscape had undergone major changes. The political boundaries of the province had been shaped and reshaped by international challenges and the lure of gold. The economy had been transformed, as had the communities where most economic transactions occurred. The Hudson's Bay Company, the largest corporate monopoly in Canada, developed many resources other than fur, and its forts served to secure colonial control, but they represented little permanent settlement outside of their enclosures. The Fraser River and Cariboo gold rush broke this corporate pattern of control, and a new system of urban centres emerged. Thousands of individuals settled the land, while thousands more came to make their fortune and leave. The resulting communities evolved with an extremely racist British value system, which was incorporated into the institutions of religion, justice, education, and land use.

The decision to join Confederation was an important political direction, particularly because provincial status came with the promise of a railway linking central and eastern Canada to the very distant Pacific. Many Americans had arrived with the gold rush, rekindling concerns of annexation, and the US purchase of Alaska in 1867 only heightened these fears. Confederation ended the anxiety.

THE SHAPING OF BRITISH COLUMBIA WITH RAILWAY AND RESOURCE TOWNS, 1871 TO 1951

Railway construction and the anticipated economic boom did not materialize immediately. Growth was slow from 1871 to 1881, but the influence of CPR construction can be seen in the figures for the 1881 census. The population nearly doubled over the next ten years, and by 1921 had increased tenfold, with over half a million

people. By 1951, over a million people resided in British Columbia (Table 16.2). Another significant change during this time was the shift from a largely rural-based population in 1871 to a more urban-based one by 1951. Rapid population and urban growth was a response to new technologies of time-space convergence and of resource extraction, new market demands, and major political and economic growth promotion schemes. New urban patterns also evolved in response.

The Pacific Scandal of the 1870s delayed the promised railway for nearly ten years, and population growth was therefore modest during these years. Coal mining activity continued, but employment in placer mining diminished and gold mining communities disappeared. The appearance of fish canneries, initially at the mouth of the Fraser River, provided limited seasonal employment. Only sockeye was considered suitable for canning at this time. The forest industry was gaining importance, with substantial mills on Burrard Inlet and at the south end of Vancouver Island. Pockets of fertile agricultural land, particularly in the southwestern corner of the province, attracted some settlers while ranching in the interior sustained others. Victoria was connected to the regular ocean traffic routes to San Francisco and London, but these were long and slow routes. Transportation within British Columbia consisted of river

Table 16.2

Population and urbanization: Canada and British Columbia, 1871-1951

	Canada			British Columbia		
	Population	% rural	% urban	Population	% rural	% urban
1871	3,689,257	80.4	19.6	36,247	84.8	15.2
1881	4,324,810	74.3	25.7	50,387	72.0	18.0
1891	4,833,239	68.2	31.8	98,173	57.5	42.5
1901	5,371,315	62.5	37.5	178,657	49.5	50.5
1911	7,204,838	54.6	45.4	392,480	48.1	51.9
1921	8,787,949	50.5	49.5	524,582	52.8	47.2
1931	10,376,786	46.3	53.7	694,263	56.9	43.1
1941	11,506,655	45.7	54.3	817,861	45.8	54.2
1951	14,009,429	38.4	61.6	1,165,210	47.2	52.8

Sources: Census Canada, 1871, 1881, 1891, 1901, 1911, 1921, 1931, 1941, 1951.

Table 16.3

Native, non-Native, and Asian provincial population, 1871-1951

	Total population	Native		Non-Native		Asian	
		Population	%	Population	%	Population	%
1871	36,247	25,661	70.8	9,038	24.9	1,548[a]	4.3
1881	50,387	26,849	53.3	19,348	38.4	4,195[a]	8.3
1891	98,173	24,543	25.0	64,720	65.9	8,910[a]	9.1
1901	178,657	25,488	14.3	133,687	74.8	19,482[b]	10.9
1911	392,480	20,134	5.1	341,899	87.1	30,447[c]	7.8
1921	524,582	22,377	4.3	462,715	88.2	39,490[c]	7.5
1931	694,263	24,599	3.5	620,320	89.4	49,344[b]	7.1
1941	817,861	24,882	3.0	752,264	92.0	40,715[b]	5.0
1951	1,165,210	28,478	2.4	1,111,693	95.4	25,039[c]	2.2

Notes:

a   Chinese only.

b   Chinese and Japanese origin.

c   Chinese, Japanese and East Indian origin.

Sources: Census Canada, 1871, 1881, 1891, 1901, 1911, 1921, 1931, 1941, 1951.

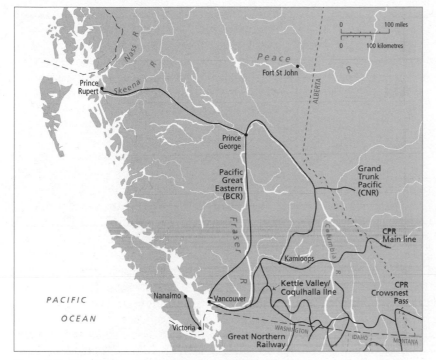

Figure 16.1 Main rail lines in British Columbia to 1952
Source: Modified from Forward (1987), 4; Galois (1990), Plate 21.

for fish, forest, and mineral commodities, all of which were in abundance in the province, expanded greatly. Transportation developments facilitated agricultural production and farming settlements.

Technological progress was not restricted to transportation. All sectors of resource extraction underwent change, and in most cases, increased production. As a result, a great number of people were attracted to British Columbia, and land speculation became a major industry.

Railway companies acquired federal and provincial land grants and cash subsidies for the building of rail lines. Crown grants entitled them to the forest and mineral rights adjacent to the railway. John Belshaw records 116 railway charters between 1883 and 1903, noting that many were for "bogus railways" (1996, 149). By 1914, the number of charters had increased to 212 (Seager 1996, 209). Land and its resources had become the negotiating basis of all levels of government and was a strong motivating force in opening up Vancouver Island and the interior of British Columbia.

navigation, the Cariboo Road, and trails – all slow routes as well.

Building the CPR changed the face of the province in many ways. Over 15,000 Chinese labourers were brought in to build the railway, and many remained in British Columbia because the promise of return passage to China was never fulfilled. First Nations were still in the majority in 1881 but, decimated by disease, they were outnumbered by 1891 and segregated on small reserves (Table 16.3). The majority of immigrants were from eastern and central Canada or Europe. They viewed this newly connected frontier as an economic opportunity.

The CPR initiated rapid growth, but this era also saw the development of international, continental, and regional linkages. The steam engine transformed the economy and the landscape, mitigating isolation and establishing better connections to national and world markets. Steam technology for moving trains and ships reduced the time-distance between locations and therefore the price of moving bulk goods. The export market

The CPR was not a passive bystander once it had completed the line to Vancouver: "Between 1871 (the year of British Columbia's entry into Confederation) and 1913, the province gave out 22 million acres of land to assorted railway companies. Much of this land eventually came into the hands of the CPR, which zealously bought up smaller lines and acquired charters that had never been acted upon" (Chodos 1973, 55). The Crow's Nest Pass Agreement of the mid-1890s not only influenced rail rates (mainly by discriminating against BC transport of commodities such as coal to central Canada) but also secured much of the mineral-rich Kootenays for this corporation (Figure 16.1).

Many individuals and groups of people seeking a better life either developed or speculated in land. The north end of Vancouver Island and the mid-coast (e.g., Cape

Table 16.4

Incorporated communities by region, 1871-1951

	Total	Vancouver Island/ central coast	Lower Mainland	Kootenay	Okanagan	South central interior	North coast/ northwest	North central interior	Peace River/ northeast
1871	2	1	1	–	–	–	–	–	–
1881	3	2	1	–	–	–	–	–	–
1891	4	2	2	–	–	–	–	–	–
1901	22	4	3	10	3	2	–	–	–
1911	26	4	4	10	4	3	1	–	–
1921	34	8	6	9	5	4	1	1	–
1931	48	8	10	12	5	4	3	6	–
1941	52	9	10	12	6	4	3	6	2
1951	73	15	12	18	8	7	3	7	3

Sources: Census Canada, 1871, 1881, 1891, 1901, 1911, 1921, 1931, 1941, 1951.

Scott, Sointula, and Hagensborg) were populated by Scandinavians who had expectations of vastly improved accessibility to markets. Early in the 1900s, the residents of Walhachin, a British settlement on the Thompson River south of Kamloops, had expectations of turning an arid landscape into another Okanagan Valley with fruit farming. Later, settlers in the Peace River region had expectations of farming one of the last agricultural frontiers in North America. Twenty thousand homesteads were claimed between 1928 and 1931 in the area, though it should be noted that this figure includes the Alberta portion of the Peace River region (Strong-Boag 1996, 281).

The pace of development after 1881 spawned new spatial patterns of urbanization in this expanded resource frontier. Table 16.4 shows the total number of incorporated communities for each decade from 1871 to 1951. Few communities were incorporated until 1901, but then there was steady growth until 1941, and twenty-one new communities were added between 1941 and 1951, signalling the beginning of the postwar boom. The table classifies the communities by region to give some sense of the evolution of economic development (see Figure 1.2 for a map of the regions). The regions of Vancouver Island and the Lower Mainland were the first to have incorporated towns, with Victoria and New Westminster, respectively. The development of coal added Nanaimo by 1881, and Vancouver was a significant addition by 1891. The rapid increase in incorporated communities by 1901 shows the development of the

Kootenays, Okanagan, and the railway communities in the south central interior. The north coast/northwest made the list of regions containing incorporated communities in 1911 with the completion of the Grand Trunk Pacific to Prince Rupert. Finally, the geographically isolated Peace River/northeast region was the last to incorporate communities. By this measure, the Kootenay region dominated from 1901 to 1951.

Table 16.5 uses the 1951 census data to rank incorporated communities of 1,000 or more people. Population size tells us a lot more than simply the number of communities in an area and sheds a new light on economic development for the various regions of British Columbia. Comparing the totals at the end of Table 16.5 to those in Table 16.4 reveals that many of the communities incorporated after 1901 were relatively small. Table 16.5 also sections off cities attaining a population of 10,000 or more inhabitants. Three cities reached 10,000 by 1911 – Vancouver, Victoria, and New Westminster – and these were the only three for the next thirty years. By 1951, the number of such centres had doubled, with North Vancouver, Trail, and Penticton added to the list.

Although only three communities in British Columbia were incorporated by 1881, there were many other unincorporated communities at this time, including those connected to the Cariboo Road, farming and ranching communities, mining communities, railway construction camps, and even old fur trade posts. All were small and many were tenuous.

Table 16.5

Incorporated communities over 1,000 by rank, 1871-1951

1871	1881	1891	1901	1911	1921	1931	1941	1951	Community	Region
–	–	13,685	27,010	100,401	163,220	246,593	275,353	344,833	1 Vancouver	Lower Mainland
4,161	5,925	16,841	20,919	31,660	38,727	39,082	44,068	51,331	2 Victoria	Vancouver Island
1,356	1,500	6,641	6,499	13,199	14,495	17,524	21,967	28,639	3 New Westminster	Lower Mainland
–	–	–	–	8,196	7,652	8,510	8,914	15,687	4 North Vancouver	Lower Mainland
–	–	–	1,360	1,460	7,573	3,020	9,392	11,430	5 Trail	Kootenay
–	–	–	–	–	3,979	4,640	5,777	10,548	6 Penticton	Okanagan
–	–	–	–	4,184	6,393	6,350	6,714	8,546	7 Prince Rupert	North coast/northwest
–	–	–	261	1,663	2,520	4,655	5,118	8,517	8 Kelowna	Okanagan
–	–	–	1,594	3,772	4,501	6,167	5,959	8,099	9 Kamloops	South central interior
–	–	–	–	–	1,056	2,356	4,584	7,845	10 Port Alberni	Vancouver Island
–	–	–	802	2,671	3,685	3,927	5,209	7,822	11 Vernon	Okanagan
–	1,645	4,595	6,130	8,306	6,559	6,745	6,635	7,196	12 Nanaimo	Vancouver Island
–	–	–	5,273	4,476	5,230	5,992	5,912	6,772	13 Nelson	Kootenay
–	–	–	–	–	–	–	–	5,933	14 Kimberley	Kootenay
–	–	–	277	1,657	1,767	2,461	3,675	5,663	15 Chilliwack	Lower Mainland
–	–	–	–	–	2,053	2,479	2,027	4,703	16 Prince George	North central interior
–	–	–	6,156	2,826	2,848	2,097	3,657	4,604	17 Rossland	Kootenay
–	–	–	1,196	3,090	2,725	3,067	2,568	3,621	18 Cranbrook	Kootenay
–	–	–	–	–	–	–	518	3,589	19 Dawson Creek	Peace River/northeast
–	–	–	–	–	–	–	–	3,507	20 Westview	Okanagan
–	–	–	–	–	998	702	1,807	3,323	21 Alberni	Vancouver Island
–	–	–	–	–	1,178	1,312	1,539	3,232	22 Port Coquitlam	Lower Mainland
–	–	–	1,600	3,017	2,782	2,736	2,106	2,917	23 Revelstoke	South central interior
–	–	–	–	–	1,178	1,843	2,189	2,784	24 Duncan	Vancouver Island
–	–	–	–	–	–	1,314	1,957	2,668	25 Mission	Lower Mainland
–	–	–	–	–	810	1,219	1,737	2,553	26 Courtney	Vancouver Island
–	–	–	1,640	3,146	2,802	2,732	2,545	2,551	27 Fernie	Kootenay
–	–	–	–	–	627	830	1,786	2,389	28 Salmon Arm	South central interior
–	–	–	–	–	1,030	1,260	1,512	2,246	29 Port Moody	Lower Mainland
–	–	–	746	3,295	1,151	1,443	1,706	2,094	30 Ladysmith	Vancouver Island
–	–	–	–	–	–	–	–	1,986	31 Campbell River	Vancouver Island
–	–	–	–	–	–	–	–	1,979	32 North Kamloops	South central interior
–	–	–	–	–	–	374	515	1,668	33 Hope	Lower Mainland
–	–	–	1,012	1,577	1,469	1,298	1,259	1,646	34 Grand Forks	Kootenay
–	–	–	–	–	–	–	–	1,628	35 Lake Cowichan	Vancouver Island
–	–	–	–	–	–	695	1,153	1,626	36 Creston	Kootenay
–	–	–	–	–	–	446	653	1,587	37 Quesnel	North central interior
–	–	–	–	–	–	–	–	1,350	38 Cranberry Lake	Kootenay
–	–	–	–	–	–	–	–	1,329	39 Castlegar	Kootenay
–	–	–	–	703	1,389	1,296	940	1,251	40 Merritt	South central interior
–	–	–	–	–	–	999	759	1,204	41 Smithers	North central interior
–	–	–	–	–	983	989	977	1,126	42 Armstrong	Okanagan
–	–	–	–	–	–	–	–	1,000	43 Oliver	Okanagan
									Other communities[a]	
–	–	–	732	1,657	2,161	2,371	885	971	Cumberland	Vancouver Island
–	–	–	1,359	778	371	171	363	809	Greenwood	Kootenay
2	3	4	13	19	26	28	29	43	Total[b]	

Notes:
a   Other communities that attained 1,000 population during 1871-1951.
b   Includes only communities of 1,000 or over.
Sources: Farley (1979), 2; Census Canada, 1871, 1881, 1891, 1901, 1911, 1921, 1931, 1941, 1951.

By the 1891 census, the most significant change was the creation of the city of Vancouver, which was incorporated in 1886. According to Robert Chodos, "The city of Vancouver, B.C.'s metropolis and the third largest city in Canada, is entirely a CPR creation, down to its name" (1973, 53). The CPR was responsible for surveying the city into a grid pattern. It also owned much of the land, many of the buildings, and the major infrastructure of a national railway and international port. With these important transportation facilities in place, Vancouver attracted a variety of economic functions: warehousing for import-export activities, sawmills, grain elevators, bulk storage facilities, financial institutions including a stock exchange, and manufacturing. The city became a central location for much of western Canada: "Vancouver's rise to prominence as a symbol of the new industrialism is undeniably related to the expansion of the provincial staple economy and to the intermediary role the city performed within the regional space economy" (McCann 1978, 34). By the 1901 census, the population of Vancouver had surpassed that of Victoria.

The 1901 census shows twenty-two incorporated communities. Four of them were on Vancouver Island: the provincial capital, and the three coal mining communities of Nanaimo, Ladysmith, and Cumberland. Ten were in the Kootenays. This mineral-rich region experienced a mining boom in silver (e.g., Sandon), coal (e.g., Fernie), lead and zinc (e.g., Trail), and copper (e.g., Phoenix and Greenwood). Figure 16.1 shows some of the intricate network of transportation systems, mainly railways, built to serve this region. Three supply and service centres appeared in the Okanagan – Vernon, Kelowna, and Enderby – while Kamloops and Revelstoke, both on the CPR main line, served the south central interior region. The Lower Fraser Valley, with its proximity to Vancouver, attracted a considerable number of settlers.

The urban pattern for 1911 was similar to that of 1901 in that it was dominated by communities in the Kootenays. It should be recognized, however, that the economic volatility of mining resulted in half of these communities losing population. Merritt appeared as one of the centres on the Kettle Valley/Coquihalla rail line, which was under construction in 1911. Vancouver's population exploded – experiencing a nearly 400 percent increase over 1901 – significantly influencing growth in the adjacent areas of New Westminster and North Vancouver. Vancouver Island's four communities continued to grow. Construction of a second national railway, the Grand Trunk Pacific (absorbed into the CNR in 1922), with Prince Rupert as the northern terminus, opened up the north central interior region.

The First World War and enhanced transportation systems gave some impetus to growth for the next decade. Electricity was a new technology, and the increasing demand for it after 1910 resulted in a rudimentary provincial grid of electric lines connecting both communities and industries. Many remote communities and industries, not on the provincial grid, created their own electrical systems. The opening of the Panama Canal in 1914 was a significant factor in time-space convergence internationally and facilitated the movement of bulk goods to and from British Columbia. One of the consequences of new forms of energy and resource competition was a serious reduction in the demand for BC coal. Between 1913 and 1920, both output and employment dropped by 50 percent (Seager 1996, 217). The drop in population in the coal mining communities of Nanaimo, Ladysmith, and Fernie reflects the impact of these dynamics. New technologies of lode mining and smelting developed and initiated smelter communities such as Anyox, in the north coast/northwest region, and Britannia Beach, just north of Vancouver. Neither was incorporated. In the Kootenays, both Phoenix and Sandon dropped off the map of incorporated communities as a result of the decline in their mining fortunes.

The years from 1921 to 1931 saw the postwar expansion of the forest, mining, and fishing industries. Fourteen incorporated communities, most of them small, were added during this time, nine of them along the CNR line and the newly completed (1921) Pacific Great Eastern Railway (PGE), which ran between Squamish and Quesnel. Prince George, with a population of nearly 2,500, was the dominant service centre for the north central interior region. Then the Great Depression began in 1929. These were "bust" times. British Columbia's resource-based economy was hit hard, and many urban centres had slow growth or lost population. Racial tensions during this period affected both the Chinese and Japanese communities as the Exclusion Act of 1923 banned Chinese from entering Canada, and further immigration rules in 1923 and 1928 seriously restricted the Japanese.

The Depression of the 1930s took its toll on employment. These were tough times until the end of the decade and the advent of the Second World War. The coal mining community of Cumberland, for example, lost nearly two-thirds of its population, falling below 1,000. Overall population growth slowed down; only four small communities were incorporated during this period, two of which were in the Peace River/northeast region. Creston, in the Kootenays, was the only addition to communities exceeding the 1,000 population threshold.

The decade from 1941 to 1951 saw a turnaround of the economy due to the demands of the Second World War and the boom following it. Ship building in the Burrard Inlet employed approximately 17,000 workers alone, and sent North Vancouver's population well over the 10,000 mark (Belshaw and Mitchell 1996, 320). Forest, mineral, and fish products were in great demand. Hydroelectric dams were built to tap the tremendous energy potential of the province, and oil and natural gas discoveries prompted growth in the Peace River/northeast region. The Alaska Highway, with Dawson Creek as Mile 0, was constructed by the Americans as an important part of the North American military infrastructure, and it made the whole of northern British Columbia more accessible. The road system throughout the province was expanded and upgraded – newly paved in many areas – as automobile and truck transportation provided competition for rail transportation and diesel fuel provided competition for coal. The addition of twenty-one incorporated communities, increasing the total to forty-three with a population over 1,000 and six over 10,000, reflected provincial growth as a whole. Some population shifts are related to the forced evacuation of Japanese Canadians in 1942 from coastal British Columbia to the interior, mainly in the Kootenays.

Examining incorporated communities from 1871 to 1951 is only part of the urban story for British Columbia in this period. By the 1890s and early 1900s, company towns had appeared on the map. Corporations established mining towns (e.g., Michelle, Cumberland, Britannia Beach), pulpmill towns (e.g., Port Mellon, Woodfibre, Powell River), and cannery communities (e.g., Klemtu, Namu), most of them "closed" communities (see Chapter 15). Some grew and became incorporated, while others became ghost towns.

Table 16.6

Population and urbanization, Canada and British Columbia, 1951-96

	Canada			British Columbia		
	Population	% rural	% urban	Population	% rural	% urban
1951	14,009,429	38.4	61.6	1,165,210	47.2	52.8
1961	18,338,247	30.4	69.6	1,629,082	27.4	72.6
1971	21,568,311	23.9	76.1	2,184,621	24.3	75.7
1981	24,343,181	24.3	75.7	2,744,465	22.0	78.0
1991	27,296,859	23.3	76.6	3,282,061	19.6	80.4
1996	28,846,761	22.1	77.9	3,724,500	18.0	82.0

Sources: Census Canada, 1951, 1961, 1971, 1981, 1991, 1996.

POSTWAR BOOM TO GLOBAL UNCERTAINTY: INSTANT TOWNS TO WORLD CITIES, 1961-96

Urbanization, settlement, resource development, transportation systems, and values – all transformed from the 1950s on. British Columbia experienced prosperous times up until the 1970s, and then a series of unpredictable global economic events, most of them negative, restructured the economy of the province and of the world. The population more than tripled between 1951 and 1996 – one indicator of economic prosperity – but growth was not uniform throughout the province and there was a very distinctive shift to urban areas.

The 1950s to the 1970s are referred to as "the long boom." It was a wave of unprecedented growth in North America, and British Columbia was swept along with it. Higher disposable incomes, increased dependence on automobile use, new homes in the suburbs for the postwar baby boomers, and new consumer goods for all were part of improved living conditions and a move toward greater urbanization. This higher standard of living throughout North America translated into the need for more and more of British Columbia's resources.

Oil, natural gas, and electricity became the most important forms of energy, and regions such as the Peace River/northeast had all three. Oil was discovered just north of Dawson Creek, at Boundary Lake, while natural gas came mainly from the vicinity of Fort St John and Fort Nelson. Employment and growth came not only from resource exploration and development but also from the construction of elaborate pipeline systems throughout the province. The enormous hydroelectric potential of the rivers of British Columbia was recognized by the provincial government in the 1950s and became an important component of their industrial strategy. Construction jobs contributed to the employment and population growth of the Kootenay and Peace River regions in response to the demand for dams and large transmission lines.

The interior had a number of wood processing plants but no pulpmills until 1961. The increased demand for pulp, paper, lumber, and other wood products made the largely untapped forests of the interior economical to exploit by the 1960s. Multinational corporations, most of them foreign, built new pulpmills throughout the province and through consolidation achieved vertical integration of all aspects of the forest industry. The urban growth related to this massive investment was felt in both existing and new towns: Castlegar and Skookumchuck in the Kootenays; Kamloops in the south central interior; and Quesnel, Prince George, and Mackenzie in the north central interior. Prince George alone had three mills. Although the most obvious impact was on the interior, many of the old pulpmills on the coast were either upgraded or phased out. The Ocean Falls mill was shut down and new mills were built at Gold River on Vancouver Island and Kitimat in the north coast/ northwest region.

A major expansion into mining occurred as the world market prices of copper, molybdenum, silver, lead, and zinc increased. By the early 1970s, the price of gold was unpegged and allowed to find its own level; it rose dramatically (see Table 10.2, p. 145). The value of coal made a dramatic rebound as coking or metallurgical coal and thermal coal for producing electricity came into demand, especially in Japan. Large, foreign-owned multinational corporations opened mines on Vancouver Island and throughout the interior, investing in capital-intensive technologies such as enormous earth-moving machinery and open-pit techniques. In some cases, new communities were built (e.g., Logan Lake, Fraser Lake, Sparwood, Elkford), while other mining operations enhanced existing communities.

The initial expansion and upgrading of road systems in the 1950s became "pavement politics" in the 1960s and 1970s. In fact, the whole transportation system was transformed as the provincial government became involved in airport expansions, ferry systems, and the extension of the PGE (renamed the British Columbia Railway, or BCR, in 1971) from Prince George to Fort St James in 1967 and to Fort Nelson by 1971. These developments provided accessibility and assisted in the growth of the expanded urban system.

The provincial and federal governments were also involved in education, health, and developing the social safety net, which not only meant new schools, hospitals, and other government institutions but also a community college system throughout British Columbia.

Most of this rapid growth affected the urban system. Suburbanization went hand in hand with highway and freeway expansion and a growing number of private automobiles, from 153,325 in 1951 to 544,310 by 1971 (Strong-Boag 1996, 288). The move to the rural-urban fringe was prompted by the desire to own single-family dwellings, a prospect that was becoming more difficult as land prices climbed in the growing, established communities. Residential and commercial sprawl moved farther and farther from the urban centres. Most urban functions – retail, wholesale, entertainment, banking, manufacturing – in the form of shopping centres and industrial parks were motivated to find new suburban, commutershed locations. They then became the catalyst for further growth. The process caused enormous problems for high growth regions such as the Lower Mainland, the Victoria-Saanich peninsula on Vancouver Island, and the Okanagan. Agricultural land, never in abundance in British Columbia, was being consumed for urban-industrial uses at a rapid rate. Servicing low density regions with water, sewer, and solid waste facilities presented other challenges. The need for planning was obvious.

The provincial government responded in various ways to all of these developments. The Instant Towns Act of 1965 accommodated construction of new resource towns for the booming resource industry. New instant towns such as Gold River, Mackenzie, Logan Lake, Elkford, and Sparwood were designed as "permanent" incorporated communities and deliberately structured differently from the old company town model. Nevertheless, they were single-resource communities and therefore vulnerable to world market conditions. (See Chapter 15 for a fuller discussion of instant towns.)

Regional districts were also created as part of the new legislation in 1965, creating a federation of incorporated and unincorporated electoral areas, all with political representation and municipal powers under the Municipal Act. This allowed the unincorporated areas of regional districts to become involved in planning, in public utilities for water, sewer, and solid waste, and in many other functions, from parks and recreation to dog control, that had previously been restricted to incorporated urban areas.

By 1972, the provincial government had passed the Agricultural Land Commission Act, which zoned all agricultural land in the province and prohibited much of its conversion to urban uses. Urban amalgamation or annexation was also encouraged by the provincial government, and many urban boundaries expanded (e.g., Kamloops, Prince George, Kelowna, Nanaimo). Vancouver became a Census Metropolitan Area. It was more practical to view the Greater Vancouver Regional District (GVRD) – Vancouver and the rapidly growing adjacent communities of North Vancouver, Burnaby, Richmond, Surrey, and so on – as one large metropolitan area. Vancouver's role in the province was changing as it gained more and more control of resource development throughout the province (Robinson and Hardwick 1968; Denike and Leigh 1972). Victoria, the main urban centre for the Capital Regional District (CRD), had a population of nearly 200,000 by the 1971 census and became the second Census Metropolitan Area in the province.

By the 1970s and '80s, traditional ways of making a living had been radically altered. Internationally, currencies were encouraged to float and find their own levels, and the energy crisis that began in 1973 caused the world to reassess its energy supply and demand. The 1970s were a period of high inflation, increased wages, and speculation in land. By contrast, the first half of the 1980s was a serious reminder of the 1930s Depression, marked with high unemployment and a major reduction in demand for resources. The economy partially recovered in the late 1980s, but the Canada/United States Free Trade Agreement (FTA) of 1989 created another round of uncertainty to lead into the 1990s.

Trade and investment had become much more global by the 1970s and 1980s, as technologies of time-space convergence facilitated the rapid movement of goods, ideas, services, capital, and people. These technologies also aided new global organizational structures that allowed corporations to fragment the production process. Post-Fordism, with its flexible specialization, challenged the traditional, centralized industrial assembly line process of producing goods. (See Chapter 7 for the shift from Fordism to post-Fordism.)

British Columbia felt the impact of these changes. The employment of capital-intensive technologies in mines and mills began reducing the number of unionized workers. Employment in the province fell by 79,000 from 1981 to 1984 with losses of, among others, 3,100 in forestry, 10,400 in mining, 23,600 in construction, and 37,700 in manufacturing (Belshaw and Mitchell 1996, 334). British Columbia suffered more than any other province with the recession of the 1980s. Restructuring, downsizing, and adjusting to the new world order of business meant uncertainty for not only the resource industries but also manufacturing and services.

Governments made substantial investments in an attempt to buoy the economy. One of the new communities to appear in British Columbia during this time was Tumbler Ridge, the centre for northeast coal. The provincial and federal governments spent considerable amounts of money extending the BCR line from Anzac to Tumbler Ridge (the only electric rail line in the province) and double tracking the CNR line to Prince Rupert, where the new coal port of Ridley Island was constructed. (For a fuller explanation of the controversial nature of this investment see Chapter 12.) Other major government investments included the Coquihalla Highway and Expo 86. These two projects had profound impacts on tourism for the province. As the traditional industries were cutting back, the provincial government also became aware of the retirement industry. The Okanagan, the southeastern side of Vancouver Island, and the Lower Mainland became more popular for retirees across Canada and helped these regions grow.

As seen earlier, the population of British Columbia doubled between 1951 and 1971. The economic turbulence of the next twenty years resulted in an increase at half that rate. Nonetheless, the increase in population was almost all found in incorporated communities (Table 16.6).

The 1990s saw the global economic climate of uncertainty continue. First the FTA and then the 1994 North American Free Trade Agreement (NAFTA) affected some industries more than others (e.g., see Chapter 13 on agriculture). Change in the territorial status of Hong Kong from a British colony to a part of China in 1997 increased immigration of Hong Kong Chinese to British Columbia, along with substantial transfers of capital and investment, mainly to the Lower Mainland region. By the end of the 1990s, the "Asian economic flu" affected the strong economic links between British Columbia and the Asia-Pacific countries and resulted in another recession for the province.

The economic polarization between the highly urban and integrated core of the Lower Mainland plus the southern end of Vancouver Island and the rest of the province intensified. As Lewis Robinson remarks, "Georgia Strait links the urban centres rather than separating them; this southwestern region may accurately be described as the heartland of British Columbia" (1998, 342). Characterized by knowledge- and information-based employment, higher value-added manufacturing, and administrative functions, this core has also gained an ever greater share of the incoming population (1998, 344). No longer just a port city with traditionally strong links to the rest of the province and western Canada, by the 1990s Vancouver had become an integrated part of the Asia-Pacific region and a major city (Davis and Hutton 1994; Hutton 1997).

The hinterland of the province, with its reliance on export of resources, struggled through the 1990s. The forest industry had to face a reduction in the amount of wood available and some of the highest production costs in the world for products such as pulp. Nor has mining been spared, as low world market prices for metals such as copper and energy minerals such as coal have reduced production, threatening the closure of communities such as Logan Lake and Tumbler Ridge.

Part of the heartland/hinterland polarization is the perception that the highly urbanized core extends its values and political influence into the resource-oriented remainder of the province. Legislated increases in parks, preservation of wildlife areas, and protection of watersheds is often viewed as driven by "urban environmentalists." Many British Columbians, especially those in resource-based industries, are also concerned about the provincial government's recognition of Aboriginal title by First Nations and the negotiation of the Nisga'a agreement and other modern treaties. (For more details see Chapter 5.)

Table 16.7 shows the number of incorporated communities of 10,000 or more people in 1998 and stands in marked contrast to the list of incorporated communities for 1951, when six had exceeded this number and only forty-three had attained a population of 1,000 or more (Table 16.5). In the fifty years since 1951, the urban system has evolved into twenty-eight regional districts and 152 incorporated communities. In the same time, the population has shifted from 52 to 82 percent urban. The urban hierarchy of two Census Metropolitan Areas (CMAs) and twenty-one Census Agglomerations (CAs, with over 10,000 and under 100,000) for the 1996 census reflects new urban relationships from global to local. More specifically, this pattern reflects the regionalization of British Columbia into a core/periphery distribution of population in which the core is highly urbanized and integrated with diversified employment frequently tied to global economic transactions (e.g., information, banking, and financial services and high-tech industries). The core, or heartland, area includes both CMAs and also five other urban centres listed in Table 16.7: Abbotsford, Nanaimo, Chilliwack, Mission, and Squamish. This relatively small geographic area contains in excess of 60 percent of the population of the province, and is likely to retain an even greater share in the future.

Outside the core is the periphery, or hinterland, of British Columbia, which largely relies on resource extraction and processing. Within this large geographic area three urban centres have emerged to perform important regional service roles: Kelowna for the Okanagan, Kamloops for the south central interior, and Prince

Table 16.7

Municipalities over 10,000 population by rank, 1998

	Municipality	Population	Region
1	Vancouver, CMA	1,972,814	Lower Mainland
2	Victoria, CMA	334,786	Vancouver Island/central coast
3	Abbotsford	113,576	Lower Mainland
4	Kelowna	97,372	Okanagan
5	Kamloops	81,880	South central interior
6	Prince George	80,943	North central interior
7	Nanaimo	75,760	Vancouver Island/central coast
8	Chilliwack	64,597	Lower Mainland
9	Vernon	34,392	Okanagan
10	Mission	32,585	Lower Mainland
11	Penticton	32,582	Okanagan
12	Campbell River	31,459	Vancouver Island/central coast
13	North Cowichan	27,132	Vancouver Island/central coast
14	Cranbrook	19,576	Kootenay
15	Port Alberni	19,533	Vancouver Island/central coast
16	Courtney	19,243	Vancouver Island/central coast
17	Prince Rupert	17,196	North coast/northwest
18	Fort St. John	16,324	Peace River/northeast
19	Salmon Arm	16,222	South central interior
20	Squamish	15,364	Lower Mainland
21	Powell River	13,954	Vancouver Island/central coast
22	Terrace	13,812	North coast/northwest
23	Comox	11,963	Vancouver Island/central coast
24	Dawson Creek	11,817	Peace River/northeast
25	Kitimat	11,712	North coast/northwest
26	Williams Lake	11,296	North central interior
27	Summerland	10,913	Okanagan
28	Parksville	10,269	Vancouver Island/central coast

Sources: Internet, BC Stats: Quick Facts – The People, "Population."

George for the north central interior. The regional centres below 20,000, such as Cranbrook, Port Alberni, Prince Rupert, Fort St John, Salmon Arm, Terrace, Dawson Creek, Kitimat, and Williams Lake, also perform important regional services but they depend much more on primary resources. As well, many of the communities below 10,000 depend on a single resource. These hinterland communities are the most vulnerable of all in the urban hierarchy.

SUMMARY

From the time British Columbia joined Confederation to the present, profound changes have occurred within a relatively short period. In 1871, the population was barely over 36,000, there were only two incorporated communities, and the province was only 15 percent urban. In 1999, British Columbia had over 4 million inhabitants and 152 municipalities and was 82 percent urban. The increase in urbanization and population can be followed regionally, as it has been tied to the discovery and exploitation of resources and development of transportation systems that went hand in hand with corporate and political decisions.

Initially, coastal water routes and interior trails, roads, and river transportation systems linked resource regions. The coming of the CPR was the beginning of the railway era and with it came many new technologies to exploit resources throughout the mountainous landscape of British Columbia. Railways, steam engines, and ships also reduced British Columbia's isolation globally. Urbanization increased dramatically, reflecting the spatial patterns of transportation and resource penetration. From the 1890s to the 1950s, world events such as war and depression also played their role in the economic viability of resource extraction and of many communities.

World events played an even greater role in shaping the geography of British Columbia from the 1960s on, as transportation technologies shrank time and space and business transactions and commerce began to be conducted on an international scale. The way one made a living changed radically, forcing a reorganization of British Columbia into a new regional geography of heartland and hinterland. From an economic and urban perspective, reorganization meant a division into an "old" economy based on primary industry processing and oriented to the export market and a "new" economy based on knowledge and information.

In the future, as the world shrinks and greater pressures are exerted toward freer trade, the new economy will have the advantage. This translates into a disproportionate amount of urban and economic growth, mainly for the southwestern area of the province. The problem will be to manage this growth in order to maintain the quality of life. This does not bode well for the hinterland because many of its communities remain attached to the old economy. Signs of change were evident as primary industry employment shrank from 14 percent in 1987 to 11 percent in 1997 (British Columbia, Ministry of Finance and Corporate Relations 1999, 72). A number of the more isolated regions – the north coast/northwest, the Peace/northeast, and the Kootenays – may actually decline in population. For individual communities in this vast and differentiated geographic hinterland, economic diversification will continue to be the major challenge.

REFERENCES

Belshaw, J.D. 1996. "Provincial Politics 1871-1916." In *The Pacific Province: A History of British Columbia*, ed. H.J.M. Johnston, 134-64. Vancouver: Douglas and McIntyre.

Belshaw, J.D., and D.J. Mitchell. 1996. "The Economy since the Great War." In *The Pacific Province: A History of British Columbia*, ed. H.J.M. Johnston, 313-42. Vancouver: Douglas and McIntyre.

British Columbia, Ministry of Finance and Corporate Relations. 1999. *1999 British Columbia Financial and Economic Review*. Victoria: The Ministry.

Chodos, R. 1973. *The CPR: A Century of Corporate Welfare*. Toronto: James Lorimer.

Davis, H.C., and T.A. Hutton. 1994. "Marketing Vancouver's Services to the Asia Pacific." *Canadian Geographer* 38 (1): 18-27.

Denike, K.G., and R. Leigh. 1972. "Economic Geography 1960-1970." In *British Columbia*, ed. J.L. Robinson, 139-45. Toronto: University of Toronto Press.

Duff, W. 1965. *The Indian History of British Columbia*. Victoria: Royal British Columbia Museum.

Farley, A.L. 1979. *Atlas of British Columbia: People, Environment, and Resource Use*. Vancouver: University of British Columbia Press.

Forward, C.N. 1987. "Evolution of Regional Character." In *British Columbia: Its Resources and People*, ed. C.N. Forward, 1-24. Western Geographical Series vol. 22. Victoria: University of Victoria.

Galois, R. 1990. "British Columbia Resources." In *Historical Atlas of Canada*. Vol. 3, *Addressing the Twentieth Century, 1891-1961*, ed. D. Kerr and D. Holdsworth. Toronto: University of Toronto Press.

Harris, C. 1997. *The Resettlement of British Columbia: Essays on Colonialism and Geographical Change*. Vancouver: UBC Press.

Higgins, D.J.H. 1977. *Urban Canada: Its Government and Politics*. Toronto: Macmillan.

Hutton, T.A. 1997. "The Innissian Core-Periphery Revisited: Vancouver's Changing Relationships with British Columbia's Staple Economy." *BC Studies* 113 (Spring): 69-98.

McCann, L.D. 1978. "Urban Growth in a Staple Economy: The Emergence of Vancouver as a Regional Metropolis, 1886-1914." In *Vancouver: Western Metropolis*, ed. L.J. Evenden, 17-42. Western Geographical Series vol. 16. Victoria: University of Victoria.

Meen, S. 1996. "Colonial Society and Economy." In *The Pacific Province: A History of British Columbia*, ed. H.J.M. Johnston, 97-132. Vancouver: Douglas and McIntyre.

Muckle, R.J. 1998. *The First Nations of British Columbia*. Vancouver: UBC Press.

Robin, M. 1972. *The Rush for the Spoils: The Company Province 1871-1933*. Toronto: McClelland and Stewart.

Robinson, J.L. 1998. "British Columbia: Canada's Pacific Province." In *Heartland and Hinterland: A Regional Geography of Canada*, 3rd ed., ed. L. McCann and A. Gunn, 321-55. Toronto: Prentice-Hall.

Robinson, J.L., and W.G. Hardwick. 1968. "The Canadian Cordillera." In *Canada: A Geographical Interpretation*, ed. J. Warkentin, 438-72. Toronto: Methuen.

Seager, A. 1996. "The Resource Economy 1871-1921." In *The Pacific Province: A History of British Columbia*, ed. H.J.M. Johnston, 205-52. Vancouver: Douglas and McIntyre.

Short, J.R., and Y. Kim. 1999. *Globalization and the City*. New York: Addison Wesley Longman.

Siemens, A.H. 1972. "Settlement." In *Studies in Canadian Geography: British Columbia*, ed. J.L. Robinson, 9-31. Toronto: University of Toronto Press.

Strong-Boag, V. 1996. "Society in the Twentieth Century." In *The Pacific Province: A History of British Columbia*, ed. H.J.M. Johnston, 273-312. Vancouver: Douglas and McIntyre.

INTERNET

BC Stats: Quick Facts – The People, "Population," www.bcstats.gov.bc.ca/data/bcfacts.htm

Glossary

Aboriginal rights. Rights derived from the historical use of land that include the right to use the land and the right to self-government. These rights are constitutionally recognized by the Constitution Act of 1985, which protects them but fails to describe their nature.

Aboriginal title. Ownership or control of the historical territory of various tribes, clans, or bands of First Nations. The geographic boundaries of such territory were not static, and overlaps in territorial use for resource procurement during the year were common among First Nations. Identification of historical boundaries that define Aboriginal title is important in establishing resource rights and compensation in the modern treaty process.

annual allowable cut. The volume of wood allowed by the provincial ministry of forests to be harvested each year from a forest region, or timber supply area. The calculation takes into account the density, species, ages, and growth rates of merchantable timber. It also has to be adjusted over time in relation to changing forest conditions, such as reduction in old-growth volumes, environmental restrictions, removal of forest land for other land uses, fires, and disease.

biomagnification. Process of the food chain whereby low levels of toxic wastes (e.g., parts per billion of mercury) are consumed by one- and two-celled animals and passed up the chain. These tiny animals are consumed by those on the next level, which are in turn consumed. In the process the level of toxic materials magnifies, or accumulates in ever increasing amounts. Because humans are at the top of the food chain and consume food such as fish, contamination levels have in some cases led to serious illnesses and death. These consequences to human health are often referred to as the environmental backlash and raise the serious questions about "safe" levels of toxic wastes allowed into the environment.

clastics. Small fragments of rock created by the weathering process and frequently carried by water and deposited as sediments. The build-up of these sediments results in both pressure and heat, and the process eventually forms sedimentary rock.

common property resource. A resource owned in common by the public and managed by government bodies (e.g., salmon, foreshore, water).

company town. A **single-resource community** created and controlled by a corporation. This was common historically in British Columbia, particularly with mining/smelting communities and coastal pulpmills. The company was the only employer and controlled housing and the company store. Most of the resulting communities were "closed," which means that they were not regulated under the Municipal Act of the province.

comprehensive claims. Claims for compensation made by First Nations for Aboriginal title lands that have not been covered by treaties or were never ceded by First Nations. The federal government recognized the need to negotiate comprehensive claims as a result of the Calder case (1973), although it was not until 1981 that it was willing to exchange undefined Aboriginal land right for concrete rights and benefits. The provincial government of British Columbia did not recognize these claims until 1992. *Contrast* **specific claims.**

convection precipitation. Process by which incoming solar radiation heats the earth and air in proximity to the earth, causing the warm air, which contains moisture from evaporation and transpiration, to rise in the atmosphere, where it cools, condenses, forms clouds, and often results in thunder showers. This is mainly a summer phenomenon in British Columbia, when the solar radiation is its most intense.

cultural region. An area of the populated world defined by having common cultural characteristics. These characteristics can be further subdivided into formal (e.g., common language areas), functional (e.g., timber supply areas), and vernacular/perceptual cultural regions (e.g., sense of place).

deformation. General term to describe processes that produce folding, faulting, and changes to the surface of the earth.

deterritorialization. The process by which a group of people lose their territory and their traditional ways of living. In British Columbia, First Nations have undergone deterritorialization as a result of European colonization and settlement.

earthquake. A tectonic vibration resulting from the clash of two crustal plates. These movements may be a result of volcanic eruption; a transform fault, in which two plates slide past one another; or **subduction,** in which one plate slides under another. The vibrations caused by this movement radiate from the epicentre of the earthquake, and the magnitude of the waves is calculated logarithmically on the Richter scale. A quake reading 3.0 produces little shaking, whereas one reading 8.0 produces major movement, especially at the epicentre or its vicinity.

erosion. The transportation of rock sediments and weathering of rock through the action of water, wind, glaciation, and mass wasting (gravity).

Eurocentrism. The imposition of the value systems of a European culture in order to judge other cultures. The term is derived from *ethnocentrism,* the judgment (often negative) of other cultures in terms of one's own.

fiduciary trust. The legal obligation, under the Indian Act, for the federal government to act in the best interests of First Nations.

flash flooding. Flash flooding is a result of intense and sustained rainfall, which occurs in coastal watersheds, including Vancouver Island and the Queen Charlottes, mainly in the winter months. Flooding is responsible for the greatest costs of destruction in British Columbia.

Fordism. The assembly-line process of manufacturing standardized products, usually at centrally located plants. The term is derived from the process used by Henry Ford in assembling automobiles at the turn of the twentieth century.

frontier mentality. The mentality of people arriving in British Columbia during the Fraser River and Cariboo gold rush period with the intention of getting rich and leaving. It is through this attitude that plenty of damage occurred to the environment, some irreversible.

geomorphology. The study of the processes that change the surface of the earth.

geophysical hazard. The assessment of risk to the earth's forces. These forces can be categorized in relation to tectonic, climatic, and gravitational forces. Geophysical hazards are also referred to as natural hazards. They are assessed in terms of the threat or risk to human property and/or life.

ghost town. A community, usually developed because of a single resource, in which the main employment base has terminated and most or all the residents have left.

hard energy path. A term used by A. Lovins (1977) to describe ever increasing dependence on non-renewable energy sources, a focus on the supply side of energy, future dependence on coal and nuclear energy as oil and gas run out, and enormous costs for energy. In this model, individuals are alienated from energy decisions because large corporations and/or large government bodies control both the energy and the planning for future energy. *Contrast* **soft energy path.**

head tax. A tax on individuals coming into a country. Within the context of British Columbia, the term applies specifically to the levy of a series of head taxes by the federal government in an attempt to reduce Chinese immigration. This became a barrier to the spatial diffusion of the Chinese to British Columbia.

igneous rock. Rock formed from the molten state, either rapidly through exposure to cooler surface environments such as air or water or much more slowly if the material does not reach the surface of the earth. Extrusive igneous rock, such as basalt, has undergone the rapid cooling process and has high density and weight. The slow cooling process results in intrusive igneous, which has much larger grained rock structures with lower density and weight, such as granite.

instant town. A single-resource community created in British Columbia under the Instant Towns Act (1965). Major new investment in the resource frontier in the 1960s and '70s (especially forestry and mining) resulted in the need for new communities. These single-resource communities were created in response to the negative conditions of the company town; they were regulated under the Municipal Act and operated in a similar fashion to other villages and towns in the province.

isostasy. The balance between the weight of continental crusts pushing down into the mantle and the uplifting forces of the mantle itself. Weathering and erosion both wear down and reduce the weight of physiographic features such as the Rockies. This loss of weight, and height, results in an uplifting of the Rockies. The fairly rapid removal of the vast sheets of ice covering much of the northern portion of North America only 10,000 years ago resulted in isostatic rebound, or a similar uplifting of the earth with the removal of this enormous weight.

lode mining. The process of crushing rock and extracting the minerals of value. Most metals occur in "bound" form, found along with other minerals within rock. Lode mining technologies were employed in British Columbia in the late 1800s and early 1900s, mainly to develop silver, gold and copper. Although some placer mining for gold continues today, nearly all metal mining production in the province is through lode mining. The technologies of lode mining are sophisticated, costly, and usually require large corporate financing.

magma. The liquid interior portion of the earth where extreme temperatures melt rock. Magma makes its way to the surface of the earth through plate tectonics and becomes igneous rock.

manifest destiny. A mid-nineteenth-century American belief that the United States was destined to expand through all of western North America and eventually, according to some advocates, to cover the entire continent.

metamorphic rock. Rock formed through the intense pressure and heat from, or chemical infusion of, new molten material intruding into existing rock structures.

Rocks undergoing this process can develop entirely new physical properties and chemical structure.

modern treaties. Contemporary treaties negotiated as compensation – in the form of land, resources and resource management, or money – for the extinguishment of Aboriginal title. These treaties also include options for self-government and greater autonomy for First Nations. Treaty negotiations did not occur until 1992 in British Columbia, at which point agreement between the provincial and federal government resulted in a six-stage process.

Oregon Treaty. An 1846 boundary agreement between the United States and Britain to continue the 49th parallel from the Rockies to the Pacific coast. Vancouver Island, which extends south of the 49th parallel, was allowed to remain part of the British North American territory. British sovereignty over the area, then known as the Oregon Territory, began with this agreement.

orogeny. The process of mountain building caused by tectonic forces that have folded and faulted land masses through compression.

orographic effect. Precipitation resulting from relatively warm, moist air being forced up mountain barriers, where it cools and condenses. It is analogous to a saturated sponge being squeezed.

placer mining. The process of mining stream beds for gold. Since gold can be found in a pure state as dust or nuggets and is one of the heaviest elements, it tends to settle in stream beds. There it can be recovered by some of the simplest and least expensive technologies: a shovel and gold pan. The discovery of gold in British Columbia, and consequent gold rush in the mid-1800s, was the result of placer mining. Many other technologies were also employed, including dams and sluices, hydraulic systems, and dredges, but these are more costly.

plate tectonics. A combination of two older hypotheses, continental drift and sea-floor spreading, plate tectonics theory is essential to the understanding of geomorphology and geology. It asserts that the earth's crust is made up of large and small plates, which move in a manner somewhat analogous to a conveyor belt, through magma being forced to the surface of the earth in some geographic locations and destroyed in others. Regions where molten material comes to the surface and pushes plates apart are known as **rift zones,** and regions where plates are pushed under or over one another are known as subduction zones. Regions where plates simply push past each other in a parallel manner are referred to as **transform faults.** Plate movement results in **earthquakes, volcanic activity,** and mountain building.

post-Fordism. The new technologies and economic conditions that apply to the period after Fordism. From the mid-1960s to the present, multinational corporations have employed techniques of flexible specialization (short-run production through contracting out) to manufacturing goods and services at a global scale. The result has been the bankruptcy, merger, or restructuring of many corporations and uncertainty for all.

private property resource. A resource that can be held, or controlled, by private interests, such as private property.

rain shadow effect. The effect of relatively low precipitation produced in a given region because of mountain barriers. In southwestern British Columbia, westerly winds are forced to rise over the Insular Mountains of Vancouver Island. This **orographic effect** wrings out much of the moisture from the air mass before it passes over the mountains. As it descends on the lee side (east side) of the mountains, the air mass expands and absorbs moisture, leaving the region from Victoria to Vancouver considerably drier than the west side of Vancouver Island.

region. An area of the surface of the earth that can be distinguished through physical and/or human characteristics.

regional geography. A subfield of the discipline of geography in which spatial phenomena are described and studied by dividing the world into areas having common physical and/or human characteristics.

resource. Any naturally occurring substance of value to a society. This definition implies that resources are culturally defined.

reterritorialization. The new set of values, institutions, and "rules" established as one cultural group gains control over the territory of another group. In British Columbia a primarily British value system was imposed on the historical First Nations of the territory.

rift zone. A region of earth where magma reaches the crust's surface and splits crustal plates apart. The ocean floors of the world are where the earth's crust is thinnest and where most rift zones occur, resulting in sea-floor spreading.

rock cycle. The process of rocks constantly being recycled as part of the earth's physical process. Igneous rocks created from the molten state are then subject to weathering and erosion and may become sedimentary rocks; heat, pressure, and chemical action produce metamorphic rock; tectonic processes may cause these rocks to go back to the molten state.

sedimentary rock. A relatively soft rock formed through the bonding, or cementing together, of sedimentary materials (e.g., limestone, shale).

single-resource community. A community in which the main employment is derived from one resource. In British Columbia single-resource communities have developed mainly through the harvest of fish, forests, and minerals, although some agricultural communities have occurred, as have tourist communities such as Whistler.

snow-melt flooding (spring run-off flooding). This type of flooding is associated with the interrelated factors of drainage basin size, snowpack over the winter season, and the spring weather conditions responsible for the rate of snowpack melt. Spring run-off flooding affects communities and built environments adjacent to the many rivers draining the interior of the province.

soft energy path. A term used by A. Lovins (1977) to describe an alternative to the **hard energy path.** The approach involves a focus on the demand side of energy, recognition that energy influences lifestyle, and the necessity for individuals to become responsible for their own energy requirements. Lovins advocates technology changes that both develop renewable energy sources and conserve energy.

spatial diffusion. A concept of movement through time and over space employed to trace the spread (adoption) of ideas or innovations, people, and goods from one geographic location to another. Also known as the spread effect, spatial diffusion identifies the "barriers" (forces that prevent movement) and "carriers" (factors assisting movement) that result in spatial distribution of phenomena.

specific claims. Claims for compensation by individual bands and tribal councils based on an alleged breach of fiduciary duty or responsibility on the part of Canada. These claims are often for reserve lands that have been taken without compensation. Seizure of such land occurred in many ways in British Columbia, from the outright annexation of reserve land to the construction of road, rail, hydroelectric, and pipelines through reserve lands. *Contrast* **comprehensive claims.**

staples theory. A theory of economic development based on the exploitation of five resources: fish, furs, timber, wheat, and minerals. Regional economic growth occurs through the discovery, development, and export of these resources, and some regions undertake resource manufacturing as well. Economic historian Harold Innis suggested this theory in the 1930s to account for the regional development of Canada.

subduction. Plate tectonic activity that occurs where plates collide and one plate overrides the other. The overridden plate – usually the heavier, oceanic plate – bends downward and descends, or subducts, into the mantle. Mountain building occurs as a result of compression along the boundaries where two plates collide. Subduction zones also result in deep oceanic troughs and continental volcanic activity because of the friction of subduction.

tenure. A system of allowing private corporations access to publicly held land and resources. Tenure involves various types of arrangements: licences, leases, and grants. The main issue of tenure in British Columbia is allowing private corporations timber rights to provincially controlled forest land.

terranes. Fragments of oceanic or continental plates. When plates collide these fragments attach themselves to the adjacent continental plate. Much of British Columbia is made up of attached, or accreted, terranes, making the geology of this province very complex.

territory. The boundaries of a geographic area within which political control is exerted.

time-space convergence. Change in transportation technologies that reduces the time required to move or communicate between geographic locations. Also referred to as time-space collapse. Expressions such as "the world is shrinking" recognize that modern satellite communications, airline flights, and expressway systems allow communication and movement on many geographic scales. It is important to recognize that changes to movement are not equally distributed, however, and some geographic locations are therefore more isolated and remote than others.

transform fault. An area of the earth's crust where one crustal plate pushes past another crustal plate in a parallel manner. Both the Pacific Plate and North American Plate are moving north, for example, but the Pacific Plate is moving more rapidly. The two plates are often in a "stuck" position along the transform fault from the Queen Charlottes north to Alaska, and a major earthquake occurs when they move.

volcanic activity. The result of the eruption of molten material, or magma, that has come to the surface of the earth's crust. This activity is frequent in **rift zones,** where magma comes to the surface under the ocean. Volcanoes are also associated with subduction zones, such as the one off the west coast of British Columbia. Many parts of the interior and west to the coast have been active volcanic areas throughout the past 150 million years.

weathering. Breaking down. Two broad divisions, chemical weathering and mechanical weathering, categorize the agents involved in this process. Chemical weathering is the decomposition of minerals through agents such as water, carbon dioxide, and oxygen, which form acids. These agents sometimes combine with organic materials during this process. Mechanical weathering is the breaking down of rocks, mainly through water running into rock cracks or fissures and then freezing, expanding, and breaking the rock into fragments.

Index

Page references for glossary items are marked in **bold**.

Set in Giovanni and Frutiger Condensed by Artegraphica Design Co.

Printed and bound in Canada by Friesens

Copy editor: Camilla Jenkins

Designer: Irma Rodriguez, Artegraphica Design Co.

Cartographer: Eric Leinberger

Proofreader: Darlene Money